UCLA Symposia on Molecular and Cellular Biology, New Series

Series Editor, C. Fred Fox

RECENT TITLES

Volume 84
Technological Advances in Vaccine Development, Laurence Lasky, *Editor*

Volume 85
B Cell Development, Owen N. Witte, Norman R. Klinman, and Maureen C. Howard, *Editors*

Volume 86
Synthetic Peptides: Approaches to Biological Problems, James P. Tam and Emil Thomas Kaiser, *Editors*

Volume 87
Gene Transfer and Gene Therapy, Arthur L. Beaudet, Richard Mulligan, and Inder M. Verma, *Editors*

Volume 88
Molecular Biology of the Eye: Genes, Vision, and Ocular Disease, Joram Piatigorsky, Toshimichi Shinohara, and Peggy S. Zelenka, *Editors*

Volume 89
Liposomes in the Therapy of Infectious Diseases and Cancer, Gabriel Lopez-Berestein and Isaiah J. Fidler, *Editors*

Volume 90
Cell Biology of Virus Entry, Replication, and Pathogenesis, Richard W. Compans, Ari Helenius, and Michael B.A. Oldstone, *Editors*

Volume 91
Bone Marrow Transplantation: Current Controversies, Robert Peter Gale and Richard E. Champlin, *Editors*

Volume 92
The Molecular Basis of Plant Development, Robert Goldberg, *Editor*

Volume 93
Cellular and Molecular Biology of Muscle Development, Laurence H. Kedes and Frank E. Stockdale, *Editors*

Volume 94
Molecular Biology of RNA, Thomas R. Cech, *Editor*

Volume 95
DNA-Protein Interactions in Transcription, Jay D. Gralla, *Editor*

Volume 96
Stress-Induced Proteins, Mary Lou Pardue, James R. Feramisco, and Susan Lindquist, *Editors*

Volume 97
Molecular Biology of Stress, Shlomo Breznitz and Oren Zinder, *Editors*

Volume 98
Metal Ion Homeostasis: Molecular Biology and Chemistry, Dean H. Hamer and Dennis R. Winge, *Editors*

Volume 99
Human Tumor Antigens and Specific Tumor Therapy, Richard S. Metzgar and Malcolm S. Mitchell, *Editors*

Volume 100
Mechanisms of Action and Therapeutic Applications of Biologicals in Cancer and Immune Deficiency Disorders, J. Groopman, C. Evans, and D. Golde, *Editors*

Volume 101
Molecular Biology of Plant-Pathogen Interactions, Brian Staskawicz, Paul Ahlquist, and Olen Yoder, *Editors*

Volume 102
Growth Factors and Their Receptors: Genetic Control and Rational Application, Russell Ross, Anthony W. Burgess, and Tony Hunter, *Editors*

Volume 103
Growth Inhibitory and Cytotoxic Polypeptides, Harold L. Moses, Peter Lengyel, and Charles D. Stiles, *Editors*

Volume 104
Cellular Proteases and Control Mechanisms, Tony Hugli, *Editor*

Volume 105
Molecular and Cellular Mechanisms of Human Hypersensitivity and Autoimmunity, Edward J. Goetzl, *Editor*

Volume 106
Cell Activation and Signal Initiation: Receptor and Phospholipase Control of Inositol Phosphate, PAF, and Eicosanoid, Production, Edward A. Dennis and Tony Hunter, *Editors*

Volume 107
Tissue Engineering, Richard Skalak and C. Fred Fox, *Editors*

Please contact the publisher for information about previous titles in this series.

UCLA Symposia Board

Gene Transfer and Gene Therapy

Gene Transfer and Gene Therapy

Proceedings of an E.I. du Pont de Nemours — UCLA Symposium
Held at Tamarron, Colorado
February 6–12, 1988

Editors

Arthur L. Beaudet
Institute for Molecular Genetics
Baylor College of Medicine
Houston, Texas

Richard Mulligan
Biomedical Research Whitehead Institute
Cambridge, Massachusetts

Inder M. Verma
MBVL
Salk Institute
San Diego, California

Alan R. Liss, Inc. • New York

Address all Inquiries to the Publisher
Alan R. Liss, Inc., 41 East 11th Street, New York, NY 10003

Printed in the United States of America.

The publication of this volume was facilitated by the authors and editors who submitted the text in a form suitable for direct reproduction without subsequent editing or proofreading by the publisher.

Library of Congress Cataloging-in-Publication Data

E.I. du Pont de Nemours-UCLA Symposium (1988 : Tamarron, Colo.)
Gene transfer and gene therapy : proceedings of an E.I. du Pont de Nemours-UCLA Symposium, held at Tamarron, Colorado, February 6-12, 1988 / editors, Arthur L. Beaudet, Richard Mulligan, Inder M. Verma.
p. cm. -- (UCLA symposia on molecular and cellular biology ; new ser., v. 87)
Includes bibliographies and index.
ISBN 0-8451-2686-5
1. Gene therapy--Congresses. 2. Genetic engineering--Congresses. I. Beaudet, Arthur L. II. Mulligan, Richard. III. Verma, Inder M. IV. Title. V. Series.
[DNLM: 1. Genetic Engineering--congresses. 2. Hereditary Diseases--therapy--congresses. 3. Transfection--congresses. W3 U17N new ser. v. 87 / QZ 50 E11g 1988]
RB155.8.E18 1988
616'.042--dc20
DNLM/DLC
for Library of Congress 89-8306
CIP

Contents

Contributors

W. French Anderson, Laboratory of Molecular Hematology, N.H.L.B.I, N.I.H., Bethesda, MD 20892 **[365]**

Donald C. Anderson, Departments of Pediatrics and Cell Biology, Baylor College of Medicine, Houston, TX 77054 **[315]**

A.-C. Andres, Ludwig Institute for Cancer Research, 3010 Bern, Switzerland **[151]**

S.E. Antonarakis, Department of Pediatrics, Genetics Unit, The Johns Hopkins University School of Medicine, Baltimore, MD 21205 **[293]**

M. Antoniou, Laboratory of Gene Structure and Expression, National Institute for Medical Research, London NW7 1AA, United Kingdom **[35]**

D. Armentano, Department of Cell Biology, Howard Hughes Medical Institute, Baylor College of Medicine, Houston, TX 77030 **[355]**

Suzanne Atiee, Department of Cell Biology, Baylor College of Medicine, Houston, TX 77030 **[67]**

M. Bakhanashvili, Department of Biochemistry, Tel Aviv University, Tel Aviv 69978, Israel **[215]**

J. Barber, The Salk Institute, San Diego, CA 92138; present address: Viagene, San Diego, CA 92112 **[129]**

John A. Barranger, Department of Pediatrics, Division of Medical Genetics, Childrens Hospital of Los Angeles, University of Southern California School of Medicine, Los Angeles, CA 90027 **[397]**

W. Bautsch, Cystic Fibrosis Research Group, Zentrum Biochemie, Federal Republic of Germany **[283]**

J.W. Belmont, Institute for Molecular Genetics and Howard Hughes Medical Institute, Baylor College of Medicine, Houston, TX 77030 **[375]**

John M. Belote, Department of Biology, Biological Research Laboratories, Syracuse University, Syracuse, NY 13210 **[1]**

Anton Berns, Division of Molecular Genetics, The Netherlands Cancer Institute and Department of Biochemistry, University of Amsterdam, 1066 CX Amsterdam, The Netherlands **[117]**

Alan Bernstein, Mount Sinai Hospital Research Institute and Department of Medical Genetics, University of Toronto, Toronto, Ontario, Canada **[189]**

R. Michael Blaese, Metabolism Branch, N.C.I., N.I.H., Bethesda, MD 20892 **[365]**

Patrice Blanchet, Unité de Génétique Cellulaire, Institut Pasteur, 75724 Paris Cedex 15, France **[243]**

G. Blom, Laboratory of Gene Structure and Expression, National Institute for Medical Research, London NW7 1AA, United Kingdom **[35]**

Russell T. Boggs, Molecular Biology and Virology Laboratory, The Salk Institute, San Diego, CA 92138 **[1]**

Emiliana Borrelli, Gene Expression Laboratory, Howard Hughes Medical Institute, The Salk Institute, La Jolla, CA 92037 **[163]**

C.D.K. Bottema, Department of Biochemistry and Molecular Biology, Mayo Clinic, Rochester, MN 55905 **[307]**

Florence Botteri, Cancer Biology and Gene Expression Laboratories, The Salk Institute, San Diego, CA 92138 **[179]**

Danielle Boullier, Unité de Génétique Cellulaire, Institut Pasteur, 75724 Paris Cedex 15, France **[243]**

Philippe Brûlet, Unité de Génétique Cellulaire, Institut Pasteur, 75724 Paris Cedex 15, France **[243]**

Martin L. Breitman, Mount Sinai Hospital Research Institute and Department of Medical Genetics, University of Toronto, Toronto, Ontario, Canada **[189]**

J-M. Buerstedde, Department of Biochemistry and Molecular Biology, Mayo Clinic, Rochester, MN 55905 **[307]**

Bruce Bunnell, Department of Microbiology, University of Alabama at Birmingham, Birmingham, AL 35294 **[225]**

D. Canaani, Department of Biochemistry, Tel Aviv University, Tel Aviv 69978, Israel **[215]**

Charles R. Cantor, Department of Genetics and Development, College of Physicians and Surgeons, Columbia University, New York, NY 10032 **[269]**

C.T. Caskey, Institute for Molecular Genetics and Howard Hughes Medical Institute, Baylor College of Medicine, Houston, TX 77030 **[375]**

F. Catala, Laboratory of Gene Structure and Expression, National Institute for Medical Research, London NW7 1AA, United Kingdom **[35]**

S.M-W. Chang, Institute for Molecular Genetics and Howard Hughes Medical Institute, Baylor College of Medicine, Houston, TX 77030; present address: Laboratory of Molecular Genetics, NINCDS, NIH, Bethesda, MD 20814 **[375]**

Shizhong Chen, Cancer Biology and Gene Expression Laboratories, The Salk Institute, San Diego, CA 92138 **[179]**

Jeffrey M. Chinsky, Department of Biochemistry and the Institute for Molecular Genetics, Baylor College of Medicine, Houston, TX 77030 **[255]**

Daniel Chourrout, Laboratory of Fish Genetics, INRA, 78350 - Jouy-en-Josas, France **[9]**

A. Claass, Universitätskinderklinik Kiel, Federal Republic of Germany **[283]**

P. Collis, Laboratory of Gene Structure and Expression, National Institute for Medical Research, London NW7 1AA, United Kingdom **[35]**

Dawn-Marie Coulson, Mount Sinai Hospital Research Institute and Department of Medical Genetics, University of Toronto, Toronto, Ontario, Canada **[189]**

Ross Couwenhoven, Department of Pathology, University of Health Sciences, The Chicago Medical School, North Chicago, IL 60064 **[205]**

Nava Dalyot, Department of Hematology, Hadassah University Hospital, Jerusalem, Israel 91120 **[47]**

G.J. Darlington, Institute for Human Genetics, Baylor College of Medicine, Houston, TX 77030 **[355]**

E. deBoer, Laboratory of Gene Structure and Expression, National Institute for Medical Research, London NW7 1AA, United Kingdom **[35]**

Francesco J. DeMayo, Department of Cell Biology, Baylor College of Medicine, Houston, TX 77030 **[67]**

Zlatko Dembic, Central Research Units, Hoffman-la Roche, Basel, Switzerland **[117]**

P. DeTogni, The Salk Institute, San Diego, CA 92138 **[129]**

Michael Dexter, Paterson Institute for Cancer Research, Christie Hospital, Withington, Manchester, M20 9BX, United Kingdom **[79]**

D. Eli, Department of Biochemistry, Tel Aviv University, Tel Aviv 69978, Israel **[215]**

Glen A. Evans, Cancer Biology and Gene Expression Laboratories, The Salk Institute, San Diego, CA 92138 **[179]**

Ronald M. Evans, Gene Expression Laboratory, Howard Hughes Medical Institute, The Salk Institute, La Jolla, CA 92037 **[163]**

A. Fagan, Department of Neurosciences, School of Medicine, University of California, San Diego, La Jolla, CA 92093 **[409]**

William C. Fanslow, Department of Biochemistry, Rice University, Houston, TX 77251 **[255]**

A. Faras, Department of Microbiology, University of Minnesota, St. Paul, MN 55108; present address: Institute of Human Genetics, University of Minnesota, Minneapolis, MN 55455 **[29]**

Helen Fillmore, Department of Cell Biology and Anatomy, University of Alabama at Birmingham, Birmingham, AL 35294; present address: Department of Anatomy, University of Tennessee, Memphis, TN **[225]**

F.A. Fletcher, Institute for Molecular Genetics and Howard Hughes Medical Institute, Baylor College of Medicine, Houston, TX 77030 **[375]**

F. Flueckiger, Ludwig Institute for Cancer Research, 3010 Bern, Switzerland **[151]**

E. Frömter, Medizinische Hochschule Hannover, D-3000 Hannover 61, Physiologisches Institut, Universität Frankfurt, Federal Republic of Germany **[283]**

T. Friedmann, Department of Pediatrics, School of Medicine, University of California, San Diego, La Jolla, CA 92093 **[409]**

F. Gage, Department of Neurosciences, School of Medicine, University of California, San Diego, La Jolla, CA 92093 **[409]**

Eli Gilboa, Department of Molecular Biology, Memorial Sloan-Kettering Cancer Center, New York, NY 10021 **[365]**

Edward I. Ginns, Molecular Neurogenetics Section, Clinical Neuroscience Branch, DIRP, NIMH, Bethesda, MD 20892 **[345]**

L. Michael Glode, Division of Medical Oncology, University of Colorado Health Sciences Center, Denver, CO 80262 **[189]**

Katherine Gordon, Integrated Genetics, Framingham, MA 01701 **[57]**

Paula Gregory, Department of Cell Biology and Anatomy, University of Alabama at Birmingham, Birmingham, AL 35294 **[225]**

K.O. Greulich, Physikalisch-Chemisches Institut, Universität Heidelberg, Federal Republic of Germany **[283]**

D. Greaves, Laboratory of Gene Structure and Expression, National Institute for Medical Research, London NW7 1AA, United Kingdom **[35]**

B. Groner, Ludwig Institute for Cancer Research, 3010 Bern, Switzerland **[151]**

F. Grosveld, Laboratory of Gene Structure and Expression, National Institute for Medical Research, London NW7 1AA, United Kingdom **[35]**

D. Grothues, Cystic Fibrosis Research Group, Zentrum Biochemie, Federal Republic of Germany **[283]**

K.S. Guise, Department of Animal Science, University of Minnesota, St. Paul, MN 55108 **[29]**

René Guyomard, Laboratory of Fish Genetics, INRA, 78350 - Jouy-en-Josas, France **[9]**

P.B. Hackett, Department of Genetics and Cell Biology, University of Minnesota, St. Paul, MN 55108 **[29]**

O. Hanscombe, Laboratory of Gene Structure and Expression, National Institute for Medical Research, London NW7 1AA, United Kingdom **[35]**

Gail S. Harrison, Division of Medical Oncology, University of Colorado Health Sciences Center, Denver, CO 80262 **[189]**

Lothar Hennighausen, Laboratory of Biochemistry and Metabolism, National Institute of Diabetes, Digestive, and Kidney Diseases, Bethesda, MD 20892 **[57]**

Richard Heyman, Gene Expression Laboratory, Howard Hughes Medical Institute, The Salk Institute, La Jolla, CA 92037 **[163]**

Howard R. Higley, Department of Experimental Biology, Baylor College of Medicine, Houston, TX 77030; present address: Collagen Corporation, Palo Alto, CA 94303 **[255]**

Yorio Hinuma, Institute for Virus Research, Kyoto University, Kyoto 606, Japan **[103]**

Chang Mu Hong, Department of Pediatrics, Division of Medical Genetics, Childrens Hospital of Los Angeles, University of Southern California School of Medicine, Los Angeles, CA 90027 **[397]**

Leroy Hood, Division of Biology, 147-75, California Institute of Technology, Pasadena, CA 91125 **[235]**

Arthur L. Horwich, Department of Human Genetics, Yale University School of Medicine, New Haven, CT 06510 **[325]**

Louis Houdebine, Cell Differentiation Unit, INRA, 78350 - Jouy-en-Josas, France **[9]**

Mary Hsi, Gene Expression Laboratory, Howard Hughes Medical Institute, The Salk Institute, La Jolla, CA 92037 **[163]**

J. Hundrieser, Cystic Fibrosis Research Group, Zentrum Humangenetik, Federal Republic of Germany **[283]**

J. Hurst, Laboratory of Gene Structure and Expression, National Institute for Medical Research, London NW7 1AA, United Kingdom **[35]**

Ursula Just, Heinrich-Pette-Institut an der Universität Hamburg, 2000 Hamburg 20, Federal Republic of Germany **[79]**

Philip Kantoff, Laboratory of Molecular Hematology, N.H.L.B.I., N.I.H., Bethesda, MD 20892 **[365]**

A.R. Kapuscinski, Department of Fisheries and Wildlife, University of Minnesota, St. Paul, MN 55108 **[29]**

Makoto Katsumo, Paterson Institute for Cancer Research, Christie Hospital, Withington, Manchester, M20 9BX, United Kingdom **[79]**

H.H. Kazazian, Jr., Department of Pediatrics, Genetics Unit, The Johns Hopkins University School of Medicine, Baltimore, MD 21205 **[293]**

Rodney E. Kellems, Department of Biochemistry and the Institute for Molecular Genetics, Baylor College of Medicine, Houston, TX 77030 **[255]**

Vincent J. Kidd, Departments of Cell Biology and Anatomy and Microbiology, University of Alabama at Birmingham, Birmingham, AL 35294 **[225]**

Karl Klingler, Heinrich-Pette-Institut an der Universität Hamburg, 2000 Hamburg 20, Federal Republic of Germany; present address: Cancer Research Unit, Walter and Eliza Hall Institute of Medical Research, Victoria, 3050, Australia **[79]**

Norbert Kluge, Heinrich-Pette-Institut an der Universität Hamburg, 2000 Hamburg 20, Federal Republic of Germany **[79]**

Thomas B. Knudsen, Department of Anatomy, Quillen-Dishner College of Medicine, East Tennessee State University, Johnson City, TN 37614-0002 **[255]**

D.D. Koeberl, Department of Biochemistry and Molecular Biology, Mayo Clinic, Rochester, MN 55905 **[307]**

Donald B. Kohn, Metabolism Branch, N.C.I., N.I.H., Bethesda, MD 20892; present address: Department of Pediatrics, Division of Research Immunology and Bone Marrow Transplantation, Childrens Hospital of Los Angeles, University of Southern California, Los Angeles, CA 90027 **[365,397]**

G. Kollias, Laboratory of Gene Structure and Expression, National Institute for Medical Research, London NW7 1AA, United Kingdom **[35]**

Paul Krimpenfort, Division of Molecular Genetics, The Netherlands Cancer Institute, and Department of Biochemistry, University of Amsterdam, 1066 CX Amsterdam, The Netherlands **[117]**

B. Kroll, Medizinische Hochschule Hannover, D-3000 Hannover 61, Physiologisches Institut, Universität Frankfurt, Federal Republic of Germany **[283]**

Gracia Kruppa, Clinical Research Group BRWTI, Max-Planck-Society, University of Göttingen, 3400 Göttingen, Federal Republic of Germany **[89]**

Christine Laker, Heinrich-Pette-Institut an der Universität Hamburg, 2000 Hamburg 20, Federal Republic of Germany **[79]**

Yvan Lallemand, Unité de Génétique Cellulaire, Institut Pasteur, 75724 Paris Cedex 15, France **[243]**

W.W. Lamph, The Salk Institute, San Diego, CA 92138 **[129]**

Carlisle P. Landel, Cancer Biology and Gene Expression Laboratories, The Salk Institute, San Diego, CA 92138 **[179]**

Fred D. Ledley, Department of Cell Biology, Howard Hughes Medical Institute, Baylor College of Medicine, Houston, TX 77030 **[335,355]**

Eric Lee, Laboratory of Molecular Genetics, National Institute of Child Health and Human Development, Bethesda, MD 20892 **[57]**

Kuo-Fen Lee, Department of Cell Biology, Baylor College of Medicine, Houston, TX 77030 **[67]**

Hervé Le Mouellic, Unité de Génétique Cellulaire, Institut Pasteur, 75724 Paris Cedex 15, France **[243]**

Z. Liu, Department of Genetics and Cell Biology, University of Minnesota, St. Paul, MN 55108 **[29]**

G. Maass, Cystic Fibrosis Research Group, Zentrum Biochemie, Federal Republic of Germany **[283]**

G.R. MacGregor, Institute for Molecular Genetics and Howard Hughes Medical Institute, Baylor College of Medicine, Houston, TX 77030 **[375]**

L. MacKenzie-Graham, Department of Cell Biology, Howard Hughes Medical Institute, Baylor College of Medicine, Houston, TX 77030 **[355]**

Anthony Manly, Molecular Biology and Virology Laboratory, The Salk Institute, San Diego, CA 92138 **[1]**

Brian M. Martin, Molecular Neurogenetics Section, Clinical Neuroscience Branch, DIRP, NIMH, Bethesda, MD 20892 **[345]**

Françoise Maxwell, Division of Medical Oncology, University of Colorado Health Sciences Center, Denver, CO 80262 **[189]**

Ian H. Maxwell, Division of Medical Oncology, University of Colorado Health Sciences Center, Denver, CO 80262 **[189]**

Michael McKeown, Molecular Biology and Virology Laboratory, The Salk Institute, San Diego, CA 92138 **[1]**

James V. McMurray, Institute of Neuroscience, University of Oregon, Eugene, OR 97403 **[19]**

V. Mignotte, Laboratory of Gene Structure and Expression, National Institute for Medical Research, London NW7 1AA, United Kingdom **[35]**

K.A. Moore, Institute for Molecular Genetics and Howard Hughes Medical Institute, Baylor College of Medicine, Houston, TX 77030 **[375]**

T. Naiman, Department of Biochemistry, Tel Aviv University, Tel Aviv 69978, Israel **[215]**

Masataka Nakamura, Institute for Virus Research, Kyoto University, Kyoto 606, Japan **[103]**

Kevin Nash, Molecular Biology and Virology Laboratory, The Salk Institute, San Diego, CA 92138 **[1]**

Jan A. Nolta, Department of Pediatrics, Division of Research Immunology and Bone Marrow Transplantation, Childrens Hospital of Los Angeles, University of Southern California, Los Angeles, CA 90027 **[397]**

Risa Ohkawa, Molecular Biology and Virology Laboratory, The Salk Institute, San Diego, CA 92138 **[1]**

Kiyoshi Ohtani, Institute for Virus Research, Kyoto University, Kyoto 606, Japan **[103]**

Ariella Oppenheim, Department of Hematology, Hadassah University Hospital, Jerusalem, Israel 91120 **[47]**

Wolfram Ostertag, Heinrich-Pette-Institut an der Universität Hamburg, 2000 Hamburg 20, Federal Republic of Germany **[79]**

H. Peng, Department of Cell Biology, Baylor College of Medicine, Houston, TX 77030 **[355]**

K. Pfizenmaier, Clinical Research Group BRWTI, Max-Planck-Society, University of Göttingen, 3400 Göttingen, Federal Republic of Germany **[89]**

D.G. Phillips, Department of Pediatrics, Genetics Unit, The Johns Hopkins University School of Medicine, Baltimore, MD 21205 **[293]**

N. Ponelies, Physikalisch-Chemisches Institut, Universität Heidelberg, Federal Republic of Germany **[283]**

Brian Popko, Division of Biology, 147-75, California Institute of Technology, Pasadena, CA 91125 **[235]**

V. Ramamurthy, Department of Biochemistry, Baylor College of Medicine, Houston, TX 77030 **[255]**

Carol Readhead, Division of Biology, 147-75, California Institute of Technology, Pasadena, CA 91125 **[235]**

Hector Rombola, Mount Sinai Hospital Research Institute and Department of Medical Genetics, University of Toronto, Toronto, Ontario, Canada **[189]**

J.M. Rosen, Department of Cell Biology, Baylor College of Medicine, Houston, TX 77030 **[67]**

M. Rosenberg, Department of Pediatrics, School of Medicine, University of California, San Diego, La Jolla, CA 92093 **[409]**

Raul Saavedra, Division of Biology, 147-75, California Institute of Technology, Pasadena, CA 91125 **[235]**

P. Woods Samuels, Department of Pediatrics, Genetics Unit, The Johns Hopkins University School of Medicine, Baltimore, MD 21205 **[293]**

G. Sarkar, Department of Biochemistry and Molecular Biology, Mayo Clinic, Rochester, MN 55905 **[307]**

P. Sassone-Corsi, The Salk Institute, San Diego, CA 92138 **[129]**

C.-A. Schoenenberger, Ludwig Institute for Cancer Research, 3010 Bern, Switzerland **[151]**

Stephen A. Schwartz, Department of Pathology, University of Health Sciences, The Chicago Medical School, North Chicago, IL 60064 **[205]**

A.F. Scott, Department of Pediatrics, Genetics Unit, The Johns Hopkins University School of Medicine, Baltimore, MD 21205 **[293]**

Brent Seaton, Department of Biology and Center for Molecular Genetics, University of California, San Diego, La Jolla, CA 92093 **[417]**

M. Seh, Department of Cell Biology, Baylor College of Medicine, Houston, TX 77030 **[355]**

Barbara Seliger, Clinical Research Group BRWTI, Max-Planck-Society, University of Göttingen, 3400 Göttingen, Federal Republic of Germany **[89]**

R.F. Shen, Department of Cell Biology, Baylor College of Medicine, Houston, TX 77030 **[355]**

S. Shimohama, Department of Neurosciences, School of Medicine, University of California, San Diego, La Jolla, CA 92093 **[409]**

H. David Shine, Division of Biology, 147-75, California Institute of Technology, Pasadena, CA 91125; present address: Departments of Neuropathology and Neuroscience, Harvard Medical School, Boston, MA 02115 **[235]**

Richard L. Sidman, Division of Biology, 147-75, California Institute of Technology, Pasadena, CA 91125; present address: Departments of Neuropathology and Neuroscience, Harvard Medical School, Boston, MA 02115 **[235]**

L-K. Siew, Laboratory of Gene Structure and Expression, National Institute for Medical Research, London NW7 1AA, United Kingdom **[35]**

Cassandra L. Smith, Departments of Microbiology and Psychiatry, College of Physicians and Surgeons, Columbia University, New York, NY 10032 **[269]**

S.S. Sommer, Department of Biochemistry and Molecular Biology, Mayo Clinic, Rochester, MN 55905 **[307]**

Barbara A. Sosnowski, Molecular Biology and Virology Laboratory, The Salk Institute, San Diego, CA 92138 **[1]**

Elaine Spooncer, Paterson Institute for Cancer Research, Christie Hospital, Withington, Manchester, M20 9BX, United Kingdom **[79]**

Michael Steinmetz, Central Research Units, Hoffman-la Roche, Basel, Switzerland **[117]**

Carol Stocking, Heinrich-Pette-Institut an der Universität Hamburg, 2000 Hamburg 20, Federal Republic of Germany **[79]**

E.S. Stoflet, Department of Biochemistry and Molecular Biology, Mayo Clinic, Rochester, MN 55905 **[307]**

Gary W. Stuart, Institute of Neuroscience, University of Oregon, Eugene, OR 97403 **[19]**

Suresh Subramani, Department of Biology and Center for Molecular Genetics, University of California, San Diego, La Jolla, CA 92093 **[417]**

Kazuo Sugamura, Department of Bacteriology, Tohoku University School of Medicine, Sendai 980, Japan **[103]**

Naoki Takahashi, Division of Biology, 147-75, California Institute of Technology, Pasadena, CA 91125; present address: Department of Medicine, Tokyo University, Faculty of Medicine, Tokyo, Japan **[235]**

D. Talbot, Laboratory of Gene Structure and Expression, National Institute for Medical Research, London NW7 1AA, United Kingdom **[35]**

B. Tümmler, Cystic Fibrosis Research Group, Zentrum Biochemie, West Germany, Federal Republic of Germany **[283]**

T. Teitz, Department of Biochemistry, Tel Aviv University, Tel Aviv 69978, Israel **[215]**

John J. Trentin, Department of Experimental Biology, Baylor College of Medicine, Houston, TX 77030 **[255]**

Shoji Tsuji, Molecular Neurogenetics Section, Clinical Neuroscience Branch, DIRP, NIMH, Bethesda, MD 20892; present address: Department of Neurology, Brain Research Institute, Niigata University, Niigata 951, Japan **[345]**

Yasushi Uematsu, Basel Institute for Immunology, CH-4005 Basel, Switzerland **[117]**

Herman van der Putten, Cancer Biology and Gene Expression Laboratories, The Salk Institute, San Diego, CA 92138; present address: Department of Biotechnology, Ciba-Geigy A.G., CH-4002 Basel, Switzerland **[179]**

M.A. van der Valk, The Netherlands Cancer Institute, CX1066, Amsterdam, The Netherlands **[151]**

I.M. Verma, The Salk Institute, San Diego, CA 92138 **[129]**

J. Visvader, The Salk Institute, San Diego, CA 92138 **[129]**

James Vitale, Integrated Genetics, Framingham, MA 01701 **[57]**

R. Vogels, Laboratory of Gene Structure and Expression, National Institute for Medical Research, London NW7 1AA, United Kingdom **[35]**

L. Wall, Laboratory of Gene Structure and Expression, National Institute for Medical Research, London NW7 1AA, United Kingdom **[35]**

Stanley Welch, Department of Pathology University of Health Sciences, The Chicago Medical School, North Chicago, IL 60064 **[205]**

Monte Westerfield, Institute of Neuroscience, University of Oregon, Eugene, OR 97403 **[19]**

Heiner Westphal, Laboratory of Molecular Genetics, National Institute of Child Health and Human Development, Bethesda, MD 20892 **[57]**

J. Wolff, Department of Pediatrics, School of Medicine, University of California San Diego, La Jolla, CA 92093 **[409]**

C. Wong, Department of Pediatrics, Genetics Unit, The Johns Hopkins University School of Medicine, Baltimore, MD 21205 **[293]**

S.L.C. Woo, Department of Cell Biology, Howard Hughes Medical Institute, Baylor College of Medicine, Houston, TX 77030 **[355]**

N. Wrighton, Laboratory of Gene Structure and Expression, National Institute for Medical Research, London NW7 1AA, United Kingdom **[35]**

S.J. Yoon, Department of Animal Science, University of Minnesota, St. Paul, MN 55108 **[29]**

H. Youssoufian, Department of Pediatrics, Genetics Unit, The Johns Hopkins University School of Medicine, Baltimore, MD 21205 **[293]**

James Zwiebel, Laboratory of Molecular Hematology, N.H.L.B.I., N.I.H, Bethesda, MD 20892 **[365]**

Preface

The articles in this volume describe presentations at a meeting entitled **Gene Transfer and Gene Therapy** held at Tamarron, Colorado, February 6–12, 1988. Additional presentations of abstracts can be found in the *Journal of Cellular Biochemistry*, Supplement 12B:158, 1988. The planning for the meeting arose as an outgrowth of the highly successful 1986 UCLA Workshop **Vectors for Gene Transfer in Animals.** The aim of the meeting was to review research in the area of gene expression and gene transfer as it might relate to the goal of conducting somatic gene therapy in humans. The meeting effectively brought together basic researchers and human geneticists working with diseases.

Progress in gene transfer has been made in various species: mouse, *Drosophila,* zebra fish, rat, and other animals. Numerous examples of germ line gene transfer and somatic gene transfer in the mouse were reported. Important new information regarding control of gene expression, including the delineation of distant elements controlling human β-globin expression was presented. There has been a proliferation of knowledge on the cis-regulatory sequences and trans-acting factors which control gene expression. A detailed understanding of these mechanisms will be required for more sophisticated gene transfer and gene therapy efforts.

The meeting provided an up-to-date overview of vectors for gene transfer. Retroviral vector's predominated by a wide margin, although work with other vectors such as vaccinia virus was reported. Retroviral vectors continue to be modified to improve titer and expression. New retroviral packaging cell–lines designed to minimize production of replication–competent virus were reported. Efforts to carry out somatic gene therapy in animals might be viewed as encouraging or discouraging depending on whether one views the cup as half full or half empty. Significant expression in reimplanted bone marrow cells or fibroblasts was reported by a number of groups, but meaningful expression for the remainder of the life of all animals remains elusive.

Extensive time was devoted to discussions on human genetic diseases. Detailed molecular delineation of mutations is available in near innumerable amounts. Important diseases involving liver, bone–marrow derived cells, central nervous system, muscle and other tissues were described. Although diagnosis for single gene disorders in extremely powerful, there is little ability to treat them. The feasibility of somatic gene therapy varies widely from disease to disease. Strategies for cloning disease genes prior to identification of the gene product represent an important new approach to human genetic diseases. A roundtable discussion was devoted to the ethical considerations for human somatic gene therapy.

Exciting progress was reported for homologous recombination in mammalian cells. Homologous recombination would offer major advantages for somatic gene therapy if it could be achieved. In the shorter term, use of homologous recombination in embryonic stem cells might lead to the development of mouse models for numerous human genetic diseases.

It will be of interest to observe whether the promise of somatic gene therapy bears fruit, and if so, what form such treatment might take. Hopefully, progress in the field will someday justify an additional UCLA symposium.

We gratefully acknowledge E.I. du Pont de Nemours & Company for generous sponsorship of this meeting. Additional support was received from: Genentech, Inc. and Pel–Freez Clinical Systems.

Arthur L. Beaudet
Richard Mulligan
Inder M. Verma

Gene Transfer and Gene Therapy, pages 1–8

THE USE OF GERMLINE TRANSFORMATION IN THE STUDY OF SEXUAL DIFFERENTIATION IN DROSOPHILA[1]

Michael McKeown, Russell T. Boggs, Kevin Nash, Risa Ohkawa, Anthony Manly, Barbara A. Sosnowski and John M. Belote[2]

Molecular Biology and Virology Laboratory, The Salk Institute, PO Box 85800, San Diego, Ca 92138

ABSTRACT The *transformer* locus (*tra*) regulates all aspects of female somatic sexual differentiation in Drosophila. A fragment of 2 kilobases is sufficient to confer tra^+ activity in transgenic flies. *tra* is regulated in a sex-specific manner by alternative splicing of its precursor RNAs. This alternative splicing occurs even if the *tra* promoter is replaced by an alternative promoter. The female-specific *tra* RNA contains a single long open reading frame. Expression of this RNA from a heat shock promoter shows that this RNA is capable of carrying out all of the tra^+ functions in flies deleted for the *tra* gene. In addition, expression of this RNA in males leads to female differentiation. This dominant feminizing activity is a result of *tra*-female induced changes in expression of the *dsx* gene. Expression of the non-sex-specific *tra* RNA has no effect in males or in females and does not complement tra^-, indicating that this RNA does not serve as a precursor for the female-specific *tra* RNA.

INTRODUCTION

The excellent genetics, molecular biology and development of *Drosophila melanogaster* have made it a much

---

[1]This work was supported by grants from the NIH to MM and JMB. RTB and BAS were supported by an NIH postdoctoral training grant. MM is a Pew Foundation Scholar in the Biomedical Sciences.
[2]Department of Biology, Syracuse University, Syracuse, New York, 13210.

studied system among experimenters interested in the regulation of gene action during development. These advantages have been greatly augmented by the work of Rubin and Spradling (1,2) in the development of a P element based germ line transformation system.

P element transformation systems are based on two separable components of P element activity, a germ line-specific transposase normally coded for by wild type P elements and transposase sensitive DNA segments normally found as inverted repeats at the ends of wild type P elements. In transformation systems, these two elements are separated into two components, a helper plasmid which is capable of supplying transposase activity but which lacks at least one of the inverted repeat termini and a vector plasmid which contains the inverted repeated termini bounding both a gene of interest and an additional marker gene used to assay the production of transgenic animals (Figure 1).

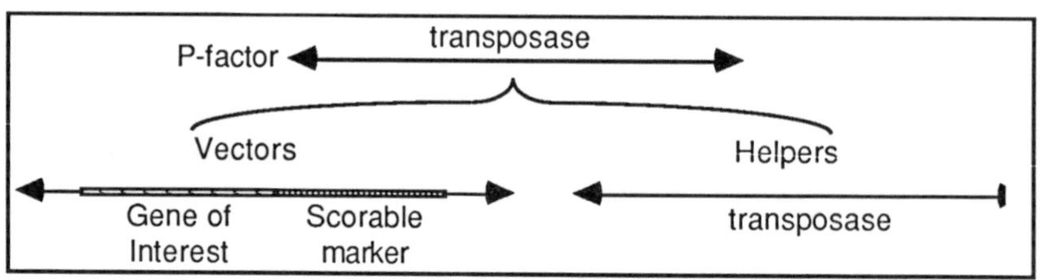

Figure 1. A diagramatic representation of the P factor and its separation into transformation vectors and transposase supplying helpers.

A mixture of these two DNA elements is injected into the posterior pole of precellular embryos, at or near the position at which the precursor cells to the germ line will form (Figure 2). Some of this DNA becomes included in the nuclei of the future germ cells. The action of the germ line-specific transposase causes the transposition of the gene of interest and the associated marker gene out of the vector plasmid and into the chromosomal DNA. Adults which develop after this treatment are referred to as G0 adults. These G0 adults are then mated to appropriate testor strain flies such that the action of the marker gene can be assayed in the progeny. Those G1 flies which express the marker gene found on the vector have arisen from germ cells in which the appropriate parts of the vector plasmid have transposed into the germ line. Note that, unlike transgenic mice, it is not the first generation but the second generation in which transgenic individuals first appear.

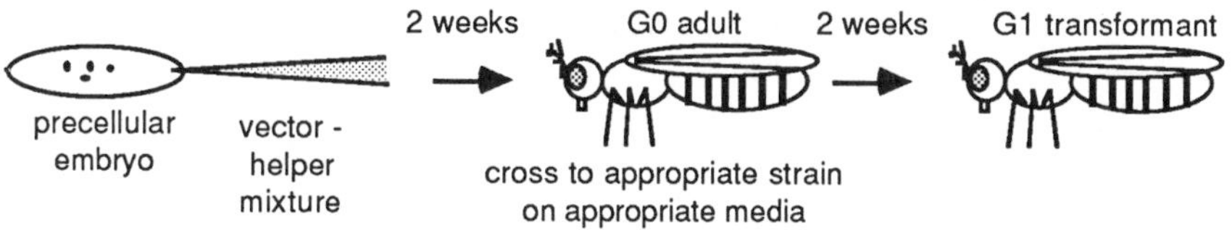

Figure 2. Drosophila transformation.

We have been using germline transformation to study the regulation of somatic sexual differentiation. Genetic results identify many of the regulatory genes which control sexual differentiation and suggest a model of the order in which they interact (for reviews see 3,4). A portion of this model showing the genes of interest for this paper is shown in Figure 3. The primary determinant of sex is the X chromosome to autosome ratio (X:A). This ratio is counted and remembered on a cell by cell basis early in development. In normal females (X:A=2:2), *Sex lethal* (*Sxl*) is turned on, leading to the activation of *transformer* and *transformer-2* (*tra* and *tra-2*). The action of these two genes causes the bifunctional *doublesex* (*dsx*) locus to be expressed in its female mode. In its female mode *dsx* represses male differentiation and allows female differentiation to occur. In males (X:A=1:2) *Sxl* is not active, *tra* and *tra-2* are not turned on and *dsx* remains in its basal, male mode. Female differentiation is repressed and male differentiation occurs.

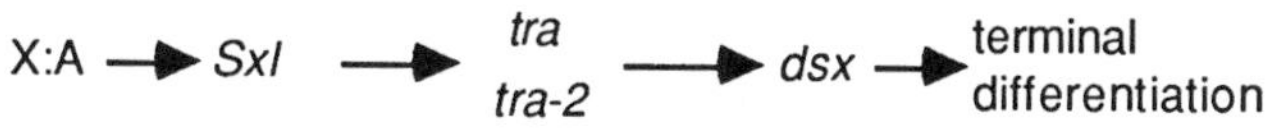

X:A → *Sxl* → *tra* / *tra-2* → *dsx* → terminal differentiation

Figure 3. A model based on genetic data for the way in which those genes of the sex determination regulatory hierarchy which are discussed in this paper interact.

This paper concentrates on *transformer*, a gene in the middle of this regulatory hierarchy. *tra* is required in females but is totally dispensable in males. Loss of function mutations of *tra* lead to male somatic differentiation of chromosomally female individuals.

RESULTS AND DISCUSSION

The *transformer* gene has been cloned (5,6) and shown to be part of a heavily transcribed region. Figure 4 shows a schematic summary of the molecular biology of *tra* (5,7,8). We have used a 2 kilobase (kb) restriction fragment approximately corresponding to the region shown in Figure 4 to phenotypically rescue diplo-X transgenic flies which are otherwise tra^- (5). This shows that we have correctly identified the *tra* gene and that *tra* is quite small (less than 2 kb).

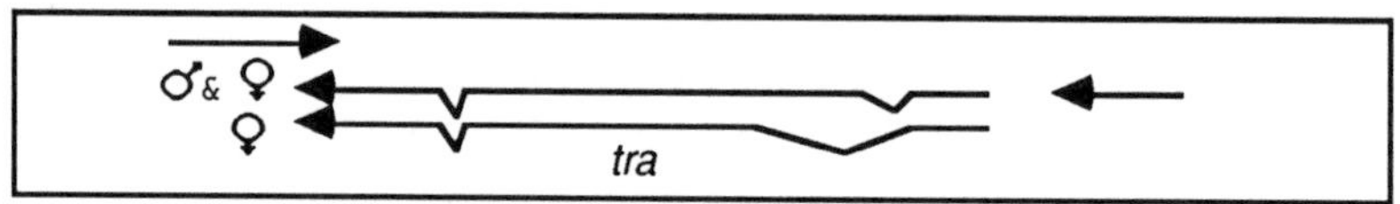

Figure 4. A summary of the molecular data concerning *tra* and its relationship to adjacent genes. The genes are shown in the same orientation as they are found on the left arm of the third chromosome.

A careful analysis of germline and cDNA clones from the *tra* region as well as S1 nuclease and primer extension mapping experiments delineate the likely mechanism for control of *tra* expression and the relationship between *tra* and the two adjacent genes (5,7,8). *tra* gives rise to two different poly(A)-containing RNAs. The smaller of these is 0.9-1.0 kb and is found only in females, consistent with the female-specific function of *tra*. The larger RNA is 175 nucleotides longer and is found in both sexes. These two RNAs are derived from a common set of precursor RNAs by alternative splicing of the first intervening sequence. The non-sex-specific RNA has a relatively short (73 base) first intron while the female-specific RNA uses an alternative splice acceptor which lies 175 nucleotides downstream. The non-sex-specific RNA has no long open reading frame while the female-specific RNA has a single long open reading frame beginning with the first AUG codon. The gene to the 5' side of *tra* ends within 80 bases of the 5' end of the *tra* transcription unit, suggesting the *tra* promoter may be quite small. The 3' end of the *tra* RNAs overlap the 3' end of RNAs from the adjacent gene by about 70 bases.

The short distance between the *tra* transcription unit and the end of the upstream gene suggests either that the *tra* promoter is quite small or that it overlaps the adjacent gene. Our preliminary results using transgenic flies carrying *tra* promoters lacking various parts of their 5' untranscribed regions suggest that the *tra* promoter is in fact quite small and that it has little or no overlap with the upstream gene (Ohkawa and McKeown, unpublished observation).

The anti-sense overlap between *tra* and the gene to its 3' side raises the possibility that these two genes regulate each other by RNA:RNA hybridization. We think that this is not the case (5,7). First of all, a restriction fragment which is truncated in the middle of the downstream gene is still capable of supplying complete tra^+ function. Molecular analysis shows that flies carrying this version of the *tra* gene produce *tra* RNAs of the proper size, at near wild type levels and with the proper ratio between the female-specific and non-sex-specific RNAs. Second, genetic analyses coupled with germ line transformation show that the downstream gene correlates with a lethal mutation (*l(3)73Ah*) that affects both sexes. Third, the RNA from *l(3)73Ah* is present at equal levels in male and female third instar larvae and in male and female pupae. Thus, in these important stages this gene is not sex-specific in its expression. Fourth, this expression of *l(3)73Ah* is female biased in the adult yet the total amount of *tra* expression remains the same in females and in males and the ratio between the two RNAs in females remains the same as in pupae and in larvae, ie., a change in the level of *l(3)73Ah* does not affect *tra* expression.

Given the small size of the *tra* promoter and the similarity of the 5' and 3' ends of the female- and non-sex-specific RNAs, we think that it is quite likely that sex-specific regulation of *tra* occurs solely through alternative processing of the same precursor RNAs in both sexes and that this alternative processing is a property of these RNAs. If this is the case, then sex-specific expression of *tra* should be independent of the particular promoter from which the *tra* precursor RNAs are synthesized. We have been able to test this hypothesis by removing *tra* sequences upstream of -20 and replacing them with a heat shock promoter and its 5' untranslated sequences. This construct has been placed into flies and shown to supply *tra* function to tra^- females. There is no effect in males (7). This is consistent with the hypothesis that *tra* regulation is promoter independent.

One of the advantages of working on Drosophila sex determination is that the system has been characterized genetically. The isolation of *tra* has allowed us to test and extend the genetic conclusions as to the structure of regulatory hierarchy shown in Figure 3 (8,9). One set of results relies upon the fact that the sex-specific RNA patterns from *tra* can be used as RNA phenotypes to examine the effects of upstream mutations on *tra* expression. These experiments verify that *Sxl* regulates *tra* expression and that this regulation involves control of the sex-specific splicing of *tra*. Other experiments of this type show that *tra-2* does not regulate *tra* (ie. it is either downstream of *tra* or on a separate branch of a branched regulatory pathway) and that neither *tra-2* nor *dsx* feed back and alter *tra* expression.

These RNA phenotype experiments with *tra* cannot determine if *tra* regulates *tra-2* in a linear pathway or if they are on different branches of a branched pathway. In addition studies of this kind are dependent upon the existence of mutations in the genes of the regulatory hierarchy, they cannot analyze the affects of mutations in unidentified genes. We have devised an experimental approach that allows us to ask if the pathway of regulated genes is linear at the level of *tra*, or if there are other branches (8). This approach depends upon the fact that we have isolated cDNA copies of the *tra* female RNA (7). We have fused one of these to a heat shock promoter, so it can be expressed in all cell types, and inserted the whole construct into the genome of transgenic flies. XY flies which carry this construct undergo female differentiation rather than male differentiation. This indicates that the genes which are regulated in a sex-specific manner are either upstream of *tra*, as *Sxl* is, or under the control of *tra* female function, as *dsx* is (Figure 5). There are not additional branches of the pathway that are regulated by *Sxl* and which act downstream of the regulation of *tra*. Genetic results using this construct in combination with mutations in other genes of the pathway are consistent with this result. We have also been able to verify, at the molecular level, that expression of the hs-*tra*-female construct in males causes the expression of *dsx* RNAs to switch from its male pattern to its female pattern, consistent with the idea that *tra* acts upstream of *dsx* to control its sex-specific expression.

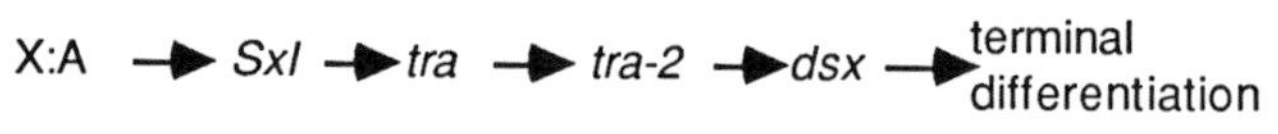

X:A → *Sxl* → *tra* → *tra-2* → *dsx* → terminal differentiation

Figure 5. A model for the interaction of the genes shown in Figure 3 but incorporating the results of both RNA blotting studies and the ectopic expression of the *tra* female RNA. Note that these results extend the genetic model but do not contradict it.

The fact that the longer *tra* RNA is present in both sexes raises the possibility that this RNA performs some function, either in both sexes or specifically in females.

Previous results (5) show that a spontaneous *tra* mutation (10) results from a small deletion which removes *tra* coding sequences. We have recently cloned the chromosomal region surrounding this deletion and shown that all but about 40 nucleotides from the 3' untranslated region of *tra* have been deleted (8). Males homozygous for this deletion are viable and fertile, suggesting that the non-sex-specific RNA has no necessary function in males. We have also expressed the hs-*tra*-female construct in females which carry this *tra* deletion over a larger deletion which eliminates *tra* and a number of nearby genes. These flies have female morphology and behavior (8). This is consistent with the idea that the non-sex-specific RNA has little or no function in either males or females and that the female-specific RNA carries out all *tra* function. As a further test of this idea, we have expressed a non-sex-specific *tra* cDNA from a heat shock promoter in transgenic flies. This construct has no effect on otherwise normal flies of either sex and does not rescue the tra^- phenotype of mutant females (8).

Given that regulation of *tra* is the result of alternative splicing, we want to determine the parts of *tra* that are necessary for this regulation and to infer the mechanism by which *Sxl* exerts its control of *tra*. Among the things we are interested in determining is if *Sxl* acts by repressing splicing at the non-sex-specific site or if it acts by activating splicing at the female site. Our results with the hs-*tra*-female construct suggest that we can use female differentiation of XY flies as a sensitive test of whether female type splicing has occurred in otherwise male individuals. As a first step toward localizing functionally important sequences within *tra*, we have produced a series of small deletions within the first intron. Among the deletions which we are now testing by reinsertion into flies is a small deletion which removes sequences from the branch point site to the splice acceptor site of the non-sex-specific splice. Preliminary results indicate that, for at least some of the lines that have received this construct, XY flies show female differentiation (Sosnowski and McKeown, unpublished). Further study is necessary before we can tell if this sequence represents a site necessary for *Sxl* action on *tra* or if the female type splicing observed represents a general reaction of the splicing machinery to deletion of a splice acceptor site.

REFERENCES

1. Rubin G and Spradling A (1982). Genetic transformation of Drosophila with transposable element vectors. Science 218:348.

2. Spradling A and Rubin G (1982). Transposition of cloned P elements into Drosophila germline chromosomes. Science 218:341.
3. Baker BS and Belote JM (1983). Sex determination and dosage compensation in Drosophila melanogaster. Ann Rev Genet 17:345.
4. Cline TW (1985). Primary events in the determination of sex in Drosophila melanogaster. in Halvorson HO and Monroy A (eds): "Origin and Evolution of Sex." New York:Alan R. Liss, p 301.
5. McKeown M, Belote JM and Baker BS (1987) A molecular analysis of *transformer*, a gene in Drosophila melanogaster that controls female sexual differentiation. Cell 48:489.
6. Butler B, Pirrotta V, Irminger-Finger I and Nöthiger R (1986). The sex determining gene *tra* of Drosophila: Molecular cloning and transformation studies. EMBO J 5:3607.
7. Boggs RT, Gregor P, Idriss S, Belote JM and McKeown M (1987). Regulation of sexual differentiation in D. melanogaster via alternative splicing of RNA from the *transformer* gene. Cell 50:739.
8. McKeown M, Belote JM and Boggs RT (1988). Ectopic expression of the female *transformer* gene product leads to female differentiation of chromosomally male Drosophila. Submitted.
9. Nagoshi RN, McKeown M, Burtis K, Belote JM and Baker BS (1988). The control of alternative splicing at genes regulating sexual differentiation in Drosophila melanogaster. Submitted.
10. Sturtevant AH (1945). A gene in *Drosophila melanogaster* that transforms males into females. Genetics 30:297.

Gene Transfer and Gene Therapy, pages 9–18

PRODUCTION OF STABLE TRANSGENIC FISH BY CYTOPLASMIC INJECTION OF PURIFIED GENES[1]

René Guyomard[2], Daniel Chourrout[2], Louis Houdebine[3]
[2]Laboratory of Fish Genetics, [3]Cell Differentiation Unit
INRA, 78350 - Jouy-en-Josas, France

ABSTRACT Injection of five plasmids bearing mammalian promoters and growth hormone genes into the cytoplasm of fertilized rainbow trout egg has been performed at the first cell stage. Foreign DNA was detected by slot blots and Southern blots in various tissues of the resulting fish. These analyses suggest that injected genes became integrated in the trout genome either as single copies or more often as random concatenates, although part of them persisted apparently extrachromosomally. Foreign genes were transmitted through the germ line to offsprings, in the genome of which they were integrated. Most of the first generation transgenic trouts produced by injection were mosaic. The cytoplasmic injection is very efficient : out of 100 eggs manipulated, 70 provide viable fish, among which 35 are transgenic. No GH could be found in their blood as in muscle and brain cells. Intraperitoneal injections of zinc performed to stimulate the MT promoters were not successful. The promoters used so far may be not appropriate to work in fish cells, but the lack of expression may also be due to other phenomena.

INTRODUCTION

Most reports about gene transfer into animals concern laboratory mice, in which the production of transgenics is routinely achieved by injecting several hundred copies of the genes into the male pronucleus of fertilized eggs. Gigantism of mice expressing growth hormone or GRF genes

---

[1]This work was supported by the Biotechnology Action Program of the Commission of the European Communities.

coupled with poorly tissue-specific promoters is one of the most remarkable physiological modifications due to gene transfer (1, 2, 3). Transgenesis in mouse obtained by microinjection of DNA into pronucleus also proved to be a potent tool for basic studies such as the role of oncogenes or the mechanisms which control the tissue specificity of gene expression (4). The production of transgenic mice by other procedures, such as the infection of embryos with genetically-engineered retroviruses (5) is promising, although often not more than injection of DNA into the pronucleus.

Modifications of economically important traits such as growth rate are of great interest in domestic animals, but positive data in this area are still rare. Our objective, which is the achievement of gene transfer in rainbow trout, encountered several difficulties : the pronuclei cannot be visualized in fertilized eggs, and genetically engineered retroviruses and homologous promoters are not available.

For these reasons, we have tested the microinjection of plasmids bearing mammalian DNA sequences into the cytoplasm of fertilized eggs, a simple method which is poorly efficient in mice (6), but leads to the persistence of foreign DNA in embryos of amphibians and sea urchins (7, 8). In a preliminary study (9), we could obtain a simular result in trout.

However, few reports in amphibians and none in fish or urchin have pointed out the persistence of foreign genes in older individuals and their transmission through the germ line (10). This particular point is shown in the present study, in which it was observed that mammalian GH genes coupled with promoters issued from mammals or from viruses active in mammalian cells were transmitted efficiently to progeny when injected into fish eggs.

MATERIALS AND METHODS

Five plamids have been injected : pH2KhGH (2), pMThGH (11) and pMTrGH (1) which all have led to giant mice ; pSV518 containing SV40 early promoter and hGHcDNA (Lupker et al., unpublished), and pSV-LTRrGH wich contains SV40 early promoter, the LTR of MMTV and the rat growth hormone gene (Crepin, unpublished) which have been tested in vitro in cell culture but never in vivo. All of them were injected with the plasmid sequences.

Injections took place between three and five hours

after in vitro fertilization (before first cleavage division) and were performed as described in our previous work (9). The chorion was first drilled manually over the animal pole with a broken micropipette (50 um diameter). Each egg then received 0.2 ng of foreign DNA in 20 nl (20 to 40 million copies of the genes) using a classical micropipette. Hatching occurs one month later; easy blood collection is possible after six months and 2-5 % of males give sperm after one year.

All expression studies consisted in GH detection in blood using specific radioimmunoassays. Sublethal intraperitoneal injections of ZnS04 were performed to stimulate the MT promoter (two injections at day 0 and day 2 of 5 mg Zn per kilo body weight followed by blood sampling at day 3). Extractions of DNA from blood, muscle and liver were carried out according to the techniques used previously (9), except that they were initiated by an homogenization in guanidium isothiocyanate. Detection of foreign genes was achieved by slot and Southern blots on nylon filters which were prehybridized and hybridized in a buffer containing 20mM phosphate pH7.0, 0.36 M Nacl, 1 % SDS, 0.5 % non fat dry milk and 6 % polyethylenglycol 6000. The probes were corresponding to the entire plasmids (10^8 cpm/ug DNA used at 10^6 cpm/ml).

RESULTS

Persistence of foreign DNA in animals after gene injections

The survival rates from egg manipulation to the age of six months were close to 70 % of that of the non-injected controls, except for one group of eggs which received each 0.5 ng DNA and which gave only 20 % of viable fish. DNA was prepared from blood cells of 119 six to twelve months-old fish. Slot blot analyses permitted the detection of foreign DNA in 62 of them (15/40 for pMTrGH, 18/40 for pH2KhGH, and 29/39 for pSV518). The muscle and liver DNA of the fish resulting from pSV518 injections were also screened by slot blots : presence of the plasmid in the genomic DNA of these organs was observed in a similar proportion of individuals. However, the number of copies per genome greatly varied within most individuals between the three analyzed organs (figure 1). Moreover, some of these signals corresponded to less than an average of 1 plasmid copy per cell. These two facts suggest a mosaic distribution of the foreign DNA.

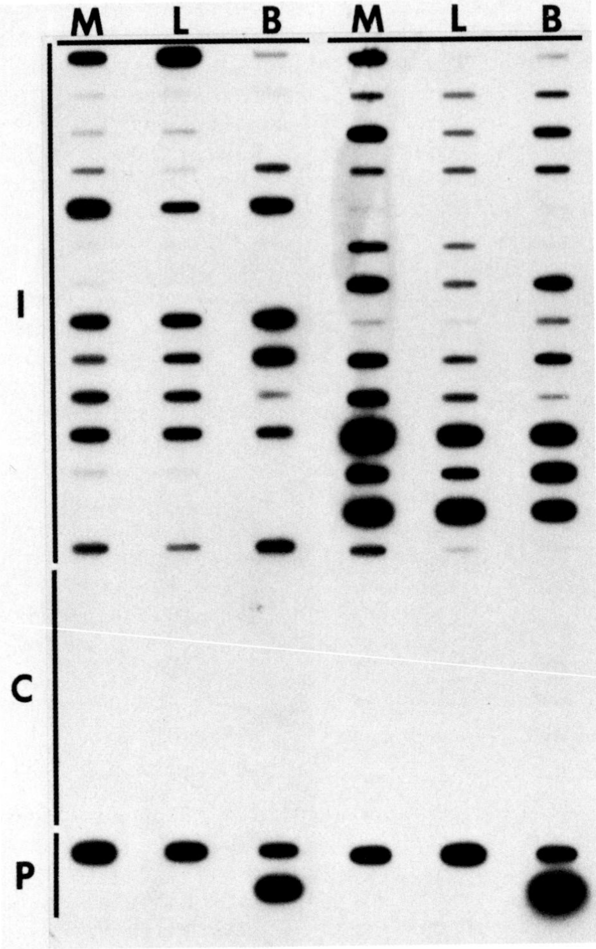

FIGURE 1. Slot blot analysis of DNA extracted from three organs of 42 individuals (M : muscle ; L : liver ; B : blood cells). The I block represents 28 individuals resulting from injections of pSV518. C block corresponds to 14 control individuals which originated from non-manipulated eggs. The average copy number of foreign DNA in the genome of transformed organs can be deduced from the signals of the P block ; P block corresponds to known amounts of plasmid : the signals of the last line correspond to 25 copies per genome ; the six signals of the above line represent 12.5, 5, 2.5, 5, 12.5, 2.5 copies per genome respectively.

Figure 2 represents the Southern blot analysis of DNA extracted from blood and liver of five animals having received pSV518. The genomic DNA were digested by Hind III which cuts the plasmid previously linearized by pVUI in three fragments : their lengths are 1.9 and 0.65 Kb (external fragments A and C), and 1.4 Kb (internal fragment B).

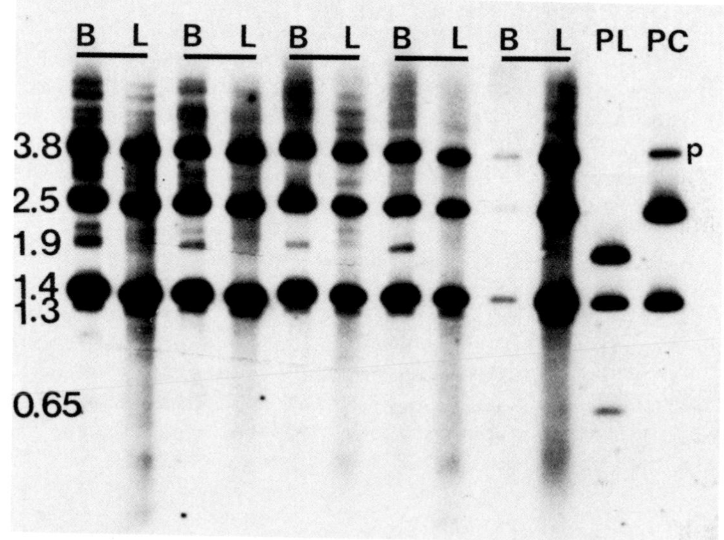

FIGURE 2. Southern blot analysis of five individuals resulting from pSV518 injections : DNA from blood (B) and liver (L) was examined. The PL lane represents the plasmid digested by Hind III after its linearization by PVUI. The PC lane represents the circular plasmid digested by Hind III (the p band corresponds to the entire plasmid, indicating its partial digestion).

The 1.4 kb fragment was found in the ten samples examined. Three bands of 3.8, 2.5 and 1.3 Kb were systematically observed and they correspond to the three possible ligations of two external fragments by their ends (A with A, A with C, and C with C). These bands correspond to an organization of the foreign DNA in random concatenates. Two additional bands (1.9 and 0.65 Kb) were seen in the five blood samples, but not in the liver. This suggests that part of the plasmids can persist extrachromosomally in blood cells. Finally, several bands corresponding to heavier fragments were found in all samples. This might indicate that a large part of the foreign DNA is integrated into multiple sites of the trout genome (either in the same cells, or in different cells of each transgenic animal) : this point is confirmed below.

Lack of GH in animals after DNA injections.

The presence of rGH or hGH was searched in 564 individuals having received the plasmids. The limit of hGH detection in plasma was far higher for hGH than for rGH (5 ng/ml plasma versus 0.5 ng/ml) ; this is due to the presence unexpected of an unidentified hGH-binding factor in trout plasma which competes with the anti-hGH antibody used in the dosage.

In none of the animals, we could find any growth hormone. GH could not be detected either in muscle or brain extracts of animals bearing pSV518. Finally, ZnS04 treatment of animals bearing pMTrGH did not trigger the production of plasmatic rGH. The growth measurements of more than one thousand such animals have not revealed any physiological modification possibly caused by the presence of foreign GH genes (data not shown).

Gene integration and germ line transmission

Seventeen mature males were found one year after the plasmid injections. Foreign DNA could be detected in the sperm of 10 of them (2 to 20 copies per cell). They were individually mated with ordinary females. Total DNA extracts from two-months-old offsprings were examined by slot blots : out of 247 fingerlings issued from 5 males, 42 proved to be transgenic (17 % ; 8 to 29 % depending on the father). Within three progenies, all positive individuals exhibited the same signal in slot blots. In the two others, two groups of fingerlings could be distinguished according to the amount of foreign DNA.

Southern blot analysis of DNA from these offsprings confirmed that the integration of the transgenes into the host genome really occured. Figure 3 shows the patterns observed for 14 fish issued from one male bearing the pSV518 sequence. Thirteen showed the same three bands : one of 1.4 Kb which is the internal fragment (B), and two larger bands which could be the junctions of the plasmid with the

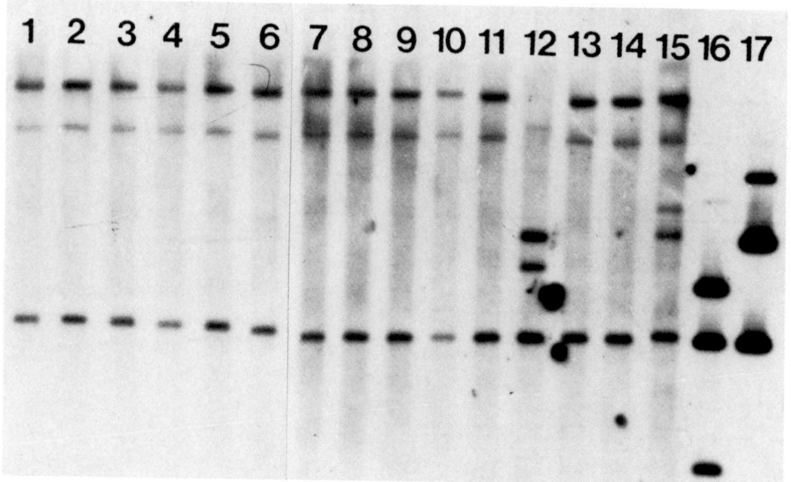

FIGURE 3. Southern blot analysis of DNA extracted from offsprings of one male harbouring pSV518 sequences. Detailed explanations are given in the text. Lane 15 : DNA of father sperm ; lanes 1 to 14 : DNA of 14 offsprings ; lane 16 : pSV518 digested by Hind III after its linearization by pVUI ; lane 17 : circular pSV518 digested by Hind III (partial digestion).

---

trout genome (the ratio of their intensities is similar to that of the external fragments A and C resulting from digestion of the linearized plasmid). Fragments possibly corresponding to concatenates were not observed in this progeny, indicating the presence of single copies. A fourteenth individual exhibited three bands in addition to that of 1.4 Kb : one may correspond to the ligation of fragments A and C of two copies integrated at another site ; the two others might represent their ligation with trout DNA. Finally, the pattern exhibited by DNA extracted from the father's sperm is composed of at least five bands : the three bands of his 13 identical offsprings, one of the two others possibly being the 2.5 Kb band created by concatenation. The fifth band might represent a junction fragment at a third site of integration or a piece of DNA rearranged after injection. This latter band is not observed in any offspring, although its intensity is relatively high in the male sperm. One explanation might be that it corresponds to integrated copies having a lethal effect on the offsprings.

The two putative junction fragments of the fourteenth offspring were not found in the male pattern: this can be explained by their presence in only few sperm cells.

DISCUSSION

The present work is a major advancement of gene transfer in fish. As a matter of fact, it is the first demonstration of a foreign gene integration and a germline transmission. The cytoplasmic injection proved to be remarkably efficient, since it leads from 100 manipulated eggs to 70 survivors, among which 35 are transgenic. Transformed males transmitted the foreign genes to their offsprings. Similar data have been reported in Xenopus (10), although much fewer individuals have been examined in this species.

The high yield of integration in trout genome might be the consequence of the large amount of DNA injected in each egg. Nevertheless, two other phenomena known to occur in Xenopus could have favoured the late persistence of foreign genes : the formation of nucleus-like structures (12) which incorporate the plasmids injected, and their later amplification which proceeds until the gastrula stage (7).

Our data indicate that most of our transgenics are mosaic : a variation of the copy number was found between different organs, and the transformed males provide less than 50 % transgenic offsprings. Nevertheless, the analyses of progenies suggest that this mosaicism remains limited : the males provide on the average 17 % of transformed sperm cells which result from one or several independent events of integration.

The absence of circulating GH in all groups examined does not open yet perspectives for growth stimulation by gene transfer. The high expression levels found in transgenic mice, which result from the utilization of homologous gene regulatory sequences, have not been up to now recorded in lower vertebrates. Expressions have been however detected in Xenopus (7) and in the medaka fish (13), but usually at early developmental stages, when a large amount of extrachromosomal foreign DNA is still present in the embryos : this does not permit reliable predictions of later expression levels, which should depend only upon integrated genes. Moreover, the recent utilization of homologous promoters in Xenopus has not led to high levels of foreign transcripts (14), although the analyses were performed in embryos. This might indicate that other factors than the origin of promoters limit the expression in transgenic lower vertebrates. The fact that GH was not present in muscle and brain extracts (pSV518 batch) indicates that the absence of plasmatic GH is probably not due to a deficient secretion of GH by these organs. On the other hand, it is known that the expression of pMTGH plasmids in mice is enhanced by zinc treatment. (15) This procedure was not efficient in trouts transformed by MTrGH sequences.

Further efforts, including the test of fish promoters, should contribute to solve this problem. Positive data of expression might promote fish as a very interesting model for gene transfer : as a matter of fact, our very efficient and simple method of DNA microinjection, and the high fecundity of these animals allow a very easy production of numerous transgenic fish. Trout virtually also is an attractive biological model to study gene expression during early development.

ACKNOWLEDGEMENTS

We wish to thank C. Puissant and G. Burger for their excellent technical assistance, and Drs D. Morello, R.D. Palmiter and J. Lupker for their generous gifts of plasmids.

REFERENCES

1. Palmiter RD, Brinster RL, Hammer RE, Trumbauer ME, Rosenfeld MG, Birnberg MG, Evans RM (1982). Dramatic growth of mice that develop from eggs microinjected with metallothionein-growth hormone fusion genes. Nature 300: 611.
2. Morello D, Moore G, Salmon AM, Yaniv M, Babinet C (1986). Studies on the expression of an pH2K-human growth hormone fusion gene in giant transgenic mice. EMBO J 5 : 1877.
3. Hammer RE, Brinster RL, Rosenfeld MG, Evans RM, Mayo KE (1985). Expression of human growth hormone releasing factor in transgenic results in increased somatic growth. Nature 315 : 413.
4. Palmiter RD (1987). Germline transformation of mice. Ann Rev Genet 20 : 1.
5. Van der Putten H, Botteri FM, Miller AD, Rosenfeld MG, Fan H, Evans RM, Verma JM (1985). Efficient insertion of

genes into the germ line via retroviral vectors. Proc. Natl Acad Sci USA 82 : 6148.
6. Brinster RL, Chen HY, Trumbauer ME, Yagle MK, Palmiter RD (1985). Factors affecting the efficiency of introducing foreign DNA into mice by microinjecting eggs. Proc Natl Acad Sci USA 82 : 4438.
7. Etkin LD (1982). Analysis of the mechanisms involved in gene regulation and cell differentiation by microinjection of purified genes and somatic cell nuclei into amphibian oocytes and eggs. Differentiation 21 : 149.
8. Flytzanis CN, Mc Mahon AP, Hough-Evans BR, Katula KS, Britten RJ, Davidson EH (1985). Persistence and integration of cloned DNA in postembryonic sea urchins. Dev Biol 108 : 431.
8. Chourrout D, Guyomard R, Houdebine LM (1986). High efficiency gene transfer in rainbow trout (Salmo gairdneri R.) by microinjection into egg cytoplasm. Aquaculture 51 : 143.
10. Etkin LD, Pearman B (1987). Distribution, expression and germ line transmission of exogenous DNA sequences following microinjection into Xenopus laevis eggs. Development 99 : 15.
11. Palmiter RD, Norstedt G, Gelinas RE, Hammer RE, Brinster RL (1983). Metallothionein-human growth hormone fusion genes stimulate growth of mice. Science 222 : 809.
12. Forbes DJ, Kirschner MW, Newport JW (1983). Spontaneous formation of nucleus-like structures around bacteriophage DNA microinjected into Xenopus eggs. Cell 34 : 13.
13. Ozato K, Kondoh H, Inohara H, Iwamatsu T, Wakamatsu Y, Okada TS (1986). Production of transgenic fish : introduction and expression of chicken delta-crystallin gene in medaka embryos. Cell Differ 19 : 237.
14. Wilson C, Cross GS, Woodland HR (1986). Tissue-specific expression of actin genes injected into Xenopus embryos. Cell 47 : 589.
15. Hammer RE, Brinster RL, Palmiter RD (1986). Use of gene transfer to increase animal growth. Cold Spring Harbor Symp Quant Biol 50 : 379.

Gene Transfer and Gene Therapy, pages 19–28

GERM-LINE TRANSFORMATION OF THE ZEBRAFISH[1]

Gary W. Stuart, James V. McMurray, and Monte Westerfield

Institute of Neuroscience, University of Oregon, Eugene, OR 97403

ABSTRACT In an effort to generate stable lines of transgenic fish, early zebrafish embryos were injected with high concentrations of foreign DNA. Foreign sequences could be found in most of the fish three weeks after injection. However, only about 5% of these fish contained foreign sequences in their fins at four months, and only one contained foreign DNA in the germ-line. While only 20% of the offspring derived from this germ-line positive parent inherited the foreign DNA, identified F1 individuals could subsequently pass the DNA on to 50% of their progeny. Genetic screens for mutant phenotypes associated with insertion mutations are also described. These results demonstrate the feasibility of germ-line gene transfer and insertional mutagenesis in fish.

INTRODUCTION

The microinjection of purified DNA into the developing embryo is one method by which the relationship between gene expression and normal development can be explored. Using this technique, purified genes can be expressed in the developing organism as a means of studying the spatial and temporal control of gene

[1]This work was supported by a grant from the Medical Research Foundation of Oregon and NIH grants NS01065, GM22731, and HD22486. G.S. was supported by grants from the American Heart Association, Oregon Affiliate and the Amyotrophic Lateral Sclerosis Association.

expression (1-4) or the role of specific genes in development (5). These experiments can be done using unmodified genes, modified genes with altered activities, or genes which have been designed to be expressed at inappropriate levels or at inappropriate times (5).

In addition, in cases where the injected genes are stably integrated into the genome of the organism, developmentally interesting insertion mutations (resulting from the disruption of normal sequence at the point of insertion) can be isolated and studied. This form of genetic analysis, known as insertional mutagenesis, has been successfully applied in only one vertebrate organism, the mouse (6). However, the requirement for *in utero* development of the mouse embryos makes mutant characterization difficult. In addition, small litter sizes hinder the systematic isolation and analysis of developmental insertion mutations.

Due to the rapid *in vitro* development and the optical clarity of its embryo, the zebrafish, *Brachydanio rerio*, is an excellent model organism for the study of vertebrate embryology and development (7). Furthermore, its fecundity, small size, short generation time, and ease of genetic manipulation make the zebrafish an ideal subject for a molecular-genetic analysis of development (8-10). We have developed procedures to produce stable lines of transgenic zebrafish (11). We wish eventually to use this system not only to analyze the effects of the expression of cloned genes on vertebrate development, but also to generate insertion mutations with interesting developmental consequences.

METHODS

Recently fertilized zebrafish eggs were collected from breeding tanks immediately after the spontaneous spawning period induced by the onset of light. Embryos were manually dechorionated in embryo medium (a modified 10% Hanks containing 1.3 mM $CaCl_2$, 1 mM $MgCl_2$, and 4mM $NaHCO_3$) and placed on a depression slide for injection. The cytoplasm of one or two cell embryos was routinely injected with about 300 pl of a solution containing 2% phenol red and 100 ng/microliter of a SalI digested linear form of the plasmid pSV-hygro (11). This plasmid contained a fusion gene in which the SV-40 early region promoter was joined to bacterial sequences coding for hygromycin phosphotransferase (12). Thus, the injected plasmid had the potential to provide the hygromycin resistance function to the fish. Gynogenotes were produced using heat shock or pressure treatments of eggs fertilized with u.v. inactivated sperm (9). Fish DNA was isolated and prepared for dot blots or Southerns using standard techniques.

RESULTS

Hybridization analysis of injected fish and their progeny

We determined the fate of injected DNA by testing injected fish and their offspring for the presence of pSV-hygro sequences on dot blots (11). Figure 1A shows the results of a dot blot in which DNA samples isolated from individual fish three weeks after the injection were hybridized with radiolabeled pSV-hygro DNA. Almost all injected fish appeared to retain some foreign DNA at this early time. However, comparisons between the signals obtained from these fish and those of a single copy standard indicated that the foreign sequences were present at an average of much less than one copy per cell.

When DNA isolated from the fins of individual adult fish four months after the injection were tested, approximately 5% of the injected fish tested positive for pSV-hygro DNA in their fins (Figure 1B). Again, the average copy number was usually less than one copy per cell. We tested 20 identified fin positive fish for the presence of injected DNA in their germ cells by outcrossing each fish and testing a number of F1 individuals from each cross on dot blots. Only one of the fin positive parents tested in this assay was capable of passing the pSV-hygro DNA on to its progeny.

Figure 1C shows the results of a dot blot assay in which 13 out of 64 F1 progeny (three weeks of age) obtained from the identified germ-line positive fish tested positive for the foreign DNA. The low transmission frequency observed (20%) suggests that the parent was mosaic in that only a fraction of her germ cells contained the foreign DNA. Consistent with this

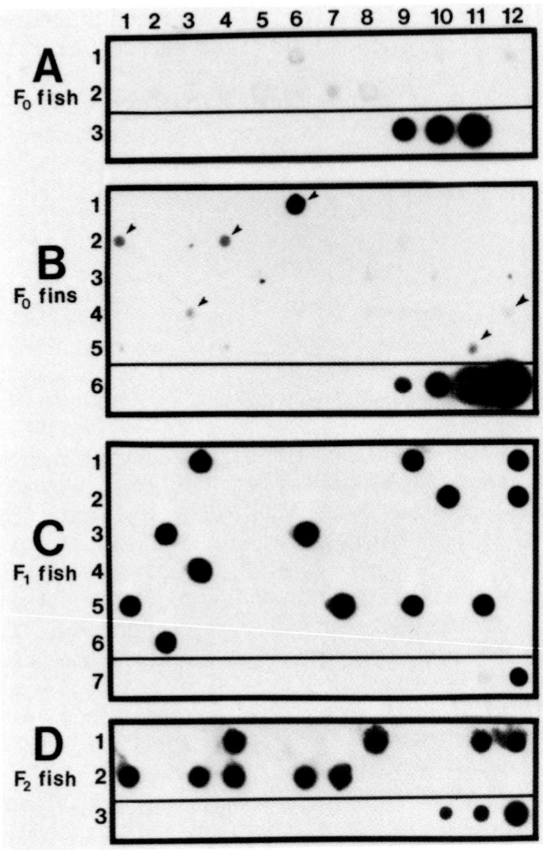

FIGURE 1. Dot blot analysis of injected fish and their progeny. The bottom row of each blot contained control samples; the last few control samples also contained measured amounts of the injected plasmid to serve as standards. A.) Ten out of 16 injected fish at three weeks tested positive for foreign DNA. The standards represent 1, 5, and 50 copies per cell. B.) DNA isolated from the fins of 60 injected fish at four months. Arrows indicate true positives confirmed on a duplicate blot. Standards represent 1, 5, 50, and 500 copies per cell. C.) Thirteen out of 64 F1 fish at three weeks tested positive for foreign DNA. Standards represent 1, 5, and 50 copies per cell. D.) Nine out of 18 F2 progeny at one week tested positive for foreign DNA. Standards represent 5, 25, and 250 copies per cell.

observation is the fact that both whole F0 fish three weeks after injection and fish fins four months after injection contained less than one copy of the injected plasmid per cell. These fish must also have had a mosaic distribution of the foreign DNA. Mosaicism might also explain the poor correlation observed between positive fins and positive germ cells.

Surprisingly, each positive F1 offspring contained approximately 100 copies of the foreign DNA per cell (Figure 1C). Figure 1D shows that nine out of 18 F2 progeny obtained from an identified F1 transgenic fish also inherited about 100 copies per cell of foreign DNA. These results are consistent with the hypothesis that approximately 20% of the germ cells of the founder transgenic fish contained a single genomic insert of a large pSV-hygro multimer. The F1 individuals that inherited this multimer were then fully heterozygous for the insert and could pass it on to about 50% of their progeny. The stable Mendelian inheritence displayed by this line of transgenic fish provides good evidence for the genomic integration of the foreign DNA.

Figure 2 shows the results of a Southern blot in which DNA's from four identified F1 transgenic fish were hybridized with an SV-hygro gene specific probe. SV-hygro gene sequences present in undigested samples were observed to migrate with high molecular weight genomic DNA. Digestion with BglII, which cuts once in pSV-hygro, produced two SV-hygro containing fragments of 5.2 and 8.7 Kb. These two fragments would be expected from BglII digestion of a hypothetical pSV-hygro multimer generated by random ligation of the Sal I digested linear (Figure 3). These fragments corresponded to a monomer plasmid unit (5.2 kb) and an inverted repeat of the SalI to BglII fragment containing the hygromycin gene (8.7 kb). A third BglII fragment (1.7 kb) would also be expected but does not hybridize with the SV-hygro specific probe. Several weak bands possibly representing single copy junction fragments were also observed.

Screening for Mutant Phenotypes with Gynogenetic Progeny

Fertilization in zebrafish can be easily manipulated _in vitro_ with eggs stripped from the female and sperm collected from the male. In addition, eggs fertilized with u.v. inactivated sperm can be made to develop into

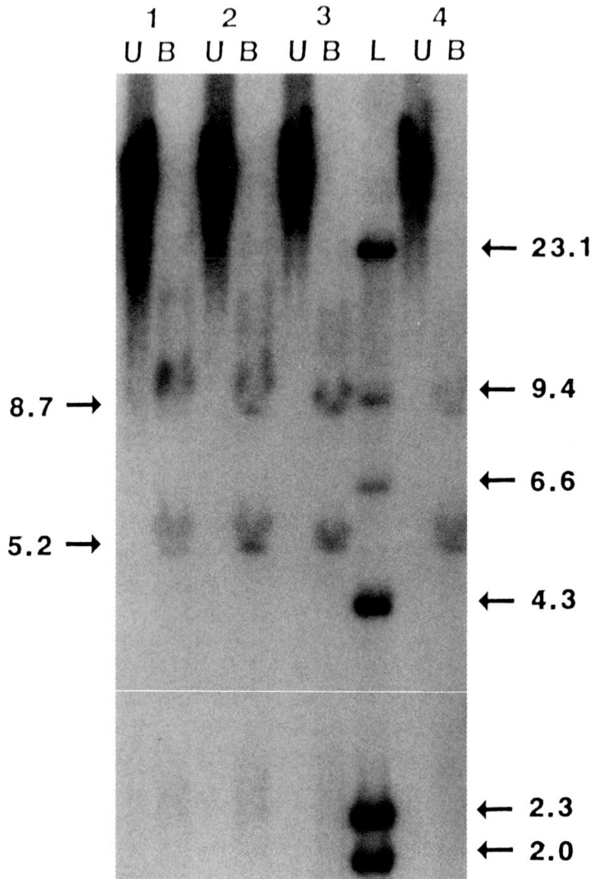

FIGURE 2. Southern analysis of F1 progeny. DNA samples from each of four identified F1 fish were digested with BglII (B) or left undigested (U). Radiolabeled HindIII cut lambda (L) was used as a size standard. A fragment containing a portion of the SV-hygro gene was used as the probe (see Figure 3).

---

fully functional gynogenetic individuals. Heat shock (HS) is used to inhibit the first mitotic division of the embryo and produce diploid fish which are homozygous at all loci. This procedure, when applied over two generations, can be used to generate large numbers of isogenic fish (8,9). Early pressure (EP) is used to

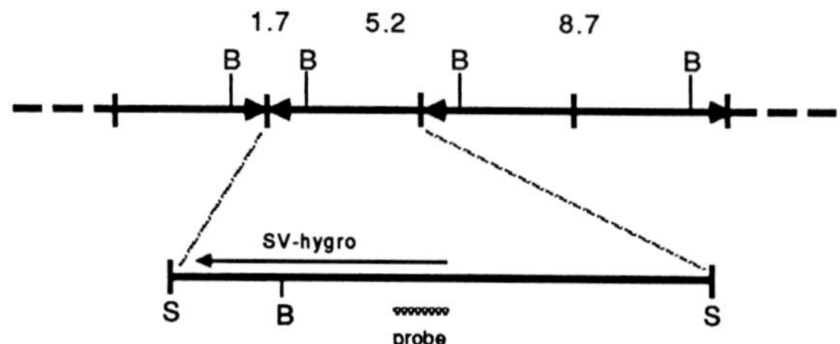

FIGURE 3. Hypothetical multimer of pSV-hygro resulting from random ligation of the SalI digested linear plasmid. SalI (S) restriction sites flank each monomer unit. The predicted size of restriction fragments resulting from digestion with BglII (B) are shown above the diagram. The position of the SV-hygro transcription unit and the fragment used as the probe in Figure 2 are shown in the enlarged monomer unit below.

---

inhibit the second meiotic division of the recently fertilized egg. This procedure produces diploid fish which are homozygous at some loci and heterozygous at others (those most distal to their centromeres). Heterozygosity results from recombination during meiosis I (see Figure 4). The fraction of recombinants observed after EP can be used to estimate gene-centromere distances (13). EP progeny can also be used to screen for recessive visible or recessive lethal mutations hidden in their heterozygous parents. Finally, eggs fertilized with inactivated sperm in the absence of EP or HS develop into haploid embryos which survive for approximately three to six days. If a mutation is capable of producing a visible embryonic phenotype in homozygous diploid individuals, this phenotype should also be observable in hemizygous haploids carrying the mutation.

We examined both haploid and EP progeny generated from the eggs of three identified F1 individuals known to be fully heterozygous for the pSV-hygro insertion. After carefully monitoring these individuals over their first three days of development, no consistent mutant phenotype was observed. Moreover, nine of the individual EP progeny obtained from one of the heterozygous F1 parents were grown to maturity and outcrossed in order to reveal their genotypes with respect to the insertion. Dot blot analysis of the nine sets of progeny obtained revealed

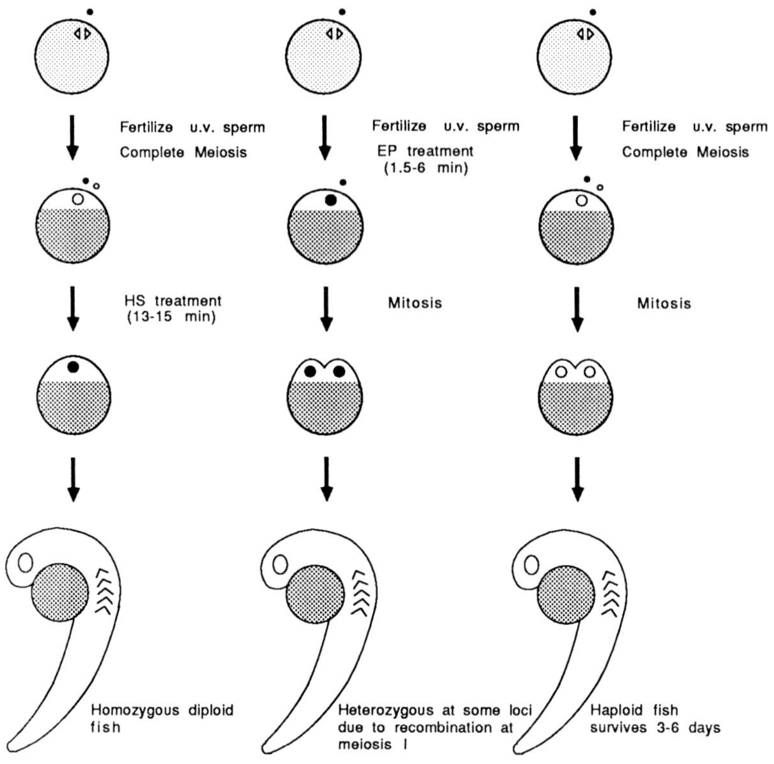

FIGURE 4. Genome manipulation in zebrafish. Filled circles represent diploid nuclei or diploid polar bodies. Open circles and triangles represent haploid nuclei/polar bodies and haploid sets of metaphase chromosomes, respectively. Stipple represents yolk granules. EP = early pressure treatment, HS = heat shock treatment.

---

that one of the EP fish completely lacked the insert, seven were heterozygous for the insert, and one was homozygous for the insert. These results indicate that the pSV-hygro insertion in this line of transgenic fish does not produce an easily observable mutant phenotype.

DISCUSSION

We have succeeded in transferring the SV-hygro gene into a stable line of transgenic fish. However, we have as yet been unable to detect the expression of this gene. Northern blots have failed to detect SV-hygro transcripts, and hygromycin treatments of young fish have failed to select for resistant individuals. Although the reason for this apparent lack of expression is unknown, it is conceivable that the SV-hygro construct functions poorly in fish. Different gene constructs and different modes of selection and screening are currently being evaluated for future use in fish transformation experiments. We have recently discovered a second line of transgenic fish capable of expressing a recombinant CAT gene (manuscript in preparation).

We have shown that the germ-line transformation of zebrafish is possible through the cytoplasmic injection of DNA into early embryos. The utility of this technique should be enhanced by improving the efficiency of germ-line transformation and by developing reporter genes that can be easily detected in transgenic fish without harming the organism. Although our goal is to identify, isolate, and study zebrafish genes involved in development, we also recognize that the establishment of an efficient fish transformation system could also have important implications for applied research in fisheries and food science (14).

ACKNOWLEDGMENTS

We thank Pat Lambert, Thom Montgomery, Rachel Aranoff, Marge Kuhn, and Dorothy Schell for expert technical help. Special thanks to Charline Walker and Molly Rothman for their advice and their help generating and screening gynogenotes. We also thank Charles Kimmel and Tadmiri Venkatesh for commenting on the manuscript.

REFERENCES

1. McMahon AP, Novak AP, Britten RJ, Davidson EH (1984). Inducible expression of a cloned heat shock fusion gene in sea urchin embryos. Proc. Natl. Acad. Sci. USA 81:7490.
2. Wilson C, Cross GS, Woodland HR (1986). Tissue specific expression of actin genes injected into Xenopus embryos. Cell 47:589.

3. Etkin LD (1982). Analysis of the mechanisms involved in gene regulation and cell differentiation by microinjection of purified genes and somatic cell nuclei into amphibian oocytes and eggs. Differentiation 21:149.
4. Kreig PA, Melton DA. (1985). Developmental regulation of a gastrula-specific gene injected into fertilized Xenopus eggs. EMBO J. 4:3463.
5. Herskowitz I (1987). Functional inactivation of genes by dominant negative mutations. Nature 329:219.
6. Palmiter RD, Brinster RL (1986). Germ-line transformation of mice. Ann. Rev. Genet. 20:465.
7. Laale HW (1977). The biology and use of zebrafish, _Brachydanio rerio_, in fisheries research: A literature review. J. Fish Biol. 10:121.
8. Streisinger G (1984). Attainment of minimal biological variability and measurement of genotoxicity: production of homozygous diploid zebra fish. Natl. Cancer Inst. Monogr. 65:53.
9. Streisinger G, Walker C, Dower N, Knauber D, Singer F (1981). Production of clones of homozygous diploid zebra fish (_Brachydanio rerio_). Nature 291:293.
10. Grunwald DJ, Kimmel CB, Westerfield M, Walker C, Streisinger G (1987). A neural degeneration mutation that spares primary neurons in the zebrafish. Developmental Biology (in press).
11. Stuart GW, McMurray JV, Westerfield M (1988). Replication, integration, and stable germ-line transmission of foreign DNA injected into early zebrafish embryos. Development (submitted).
12. Gritz L, Davies J (1983). Plasmid-encoded hygromycin B resistance: the sequence of hygromycin B phosphotransferase gene and its expression in _E. coli_ and _S. cerevisiae_. Gene 25:179.
13. Streisinger G, Singer F, Walker C, Knauber D, Dower N (1986). Segregation analyses and gene-centromere distances in zebrafish. Genetics 112:311.
14. Chourrout D, Guyomard R, Houdebine L-M. (1986). High efficiency gene transfer in rainbow trout (_Salmo gairdineri_ rich.) by microinjection into egg cytoplasm. Aquaculture 51:143.

Gene Transfer and Gene Therapy, pages 29–34

SUCCESSFUL GENE TRANSFER IN FISH[1]

S.J. Yoon*, Z. Liu***, A.R. Kapuscinski**, P.B. Hackett***, A. Faras**** and K.S. Guise*[2]

Departments of Animal Science*, Fisheries and Wildlife**, Genetics and Cell Biology***, and Microbiology****
University of Minnesota, St. Paul, MN 55108

ABSTRACT Successful transfer of the neo gene, conferring resistance to the neomycin analog drug G418, into newly fertilized, dechorionated goldfish eggs was performed by microinjection. Multiple copies of the gene were demonstrably incorporated into the genomic DNA. RNA dot blots indicate specific neo mRNA synthesis.

INTRODUCTION

Novel genes were first introduced into mice in 1979 by Gordon and Ruddle (1). The technology did not receive wide attention until Brinster and Palmiter (2) transfered a rat growth hormone gene linked to a mouse metallothionein promotor, creating a line of mice that grew significantly faster and ultimately larger than control mice. This series of experiments captured the imagination of a wide variety of researchers seeking to improve economic traits in domestic and semi-domestic animals. While gene transfer experimentation is currently progressing in most

---

[1]This work was supported in part by Minnesota Sea Grants R/3 to K.S.G. and RA/6 to P.B.H., by a grant from the Legislative Commission on Minnesota Resources of the State of Minnesota to A.F., K.S.G., A.R.K., and P.B.H., and by the Minnesota Experiment Station (K.S.G.).
[2]To whom correspondence should be addressed.

species of economic importance, including mammalian farm animals and poultry, no group of organisms show more promise for dramatic interaction of transfered genes than fish. For over three decades it has been known that fish are quite responsive to injections of crude or purified growth hormone (3,4). Successful transfer and expression of growth hormone genes in fish are thus expected to produce a similar, dramatic response.

Since 1984, multiple groups worldwide have been pursuing the goal of producing transgenic fish. Laboratories in Japan, England, France and the People's Republic of China have published results of these attempts in medaka, rainbow trout, goldfish and loach (5). Within the U.S., several groups are known to be using microinjection of various plasmid constructs to produce transgenic fish (10). The majority of the groups are pursuing similar goals including the production of transgenic fish by transfer of growth hormone gene constructs using microinjection as the primary transfer technique. Our group reports here the successful transfer of a marker gene, neo, into goldfish via microinjection, as a step toward the goal of transfer of economically important genes.

MATERIALS AND METHODS

Spontaneous ovulation of goldfish was accomplished by the methods of Stacy et al. (7). Eggs were fertilized by mixing with milt in well water in an open petri dish. Ten minutes after fertilization, the eggs were dechorionated by a six-minute incubation in 0.2% trypsin. Dechorionation was stopped by addition of 5% fetal bovine serum in Holtfreter's solution (8). Dechorionated eggs were washed several times in Holtfreter's solution and prepared for microinjection.

Plasmids were linearized with restriction endonuclease KpnI, extracted with phenol/chloroform, ethanol precipitated, and redissolved in TE (9) to a final concentration of 25 ng/µl. Borosilicate glass needles with an inner tip diameter of approximately 2 µm were filled with plasmid solution. Microinjection was performed with a Brinkman MM33 micromanipulator, with injection volume controlled by timing of insertion/withdrawal interval and constant fluid flow. DNA was released into the center of the germinal disc prior to first cleavage. Microinjected eggs were allowed to develop in oxygenated Holtfreter's solution until the blastula stage; and, in gently flowing well water, post-blastula. The primary construct used, pRSVneo, contains the neo gene, whose product, aminoglycoside 3'-phosphotransferase, confers resistance to the neomycin analog G418, under the control of a Rous sarcoma virus (RSV) promoter.

Genomic DNA was isolated from 1-2 month old fish stored at -90°C as per Marmur (10). DNA dot blot analysis was used to detect the neo gene sequences (9). Southern blot analysis was performed by complete digestion of 10 μg of genomic DNA with BamHI followed by electrophoresis on a 0.8% agarose gel (9). Total RNA was isolated as per Maniatis (9). RNA dot blot analysis was used to test for neo gene expression. Serial amounts of RNA were directly dotted onto nitrocellulose. A 2.3 kb BamHI-HindIII fragment including the neo gene was subcloned into pTZ18R under the control of the T7 promoter to produce a transcript complementary to neo mRNA. 1 μg of linear neo DNA was transcribed using the method of Schenborn and Mierendorf (11) to produce cRNA probes used in DNA dot blot, Southern blot, and RNA dot blot analyses.

RESULTS

The survival rate for microinjected fish has ranged from 10% in early experiments to close to 50% currently. The results reported here are from early injection studies, where despite high mortality rates in the injected fish, we were still able to effect transfer and apparent expression of the transferred gene.

Plate 1 shows the results of DNA dot blot analysis of control and 48 pRSVneo-injected goldfish. None of the control fish showed positive hybridization to the neo probe, while two of the pRSVneo-injected fish (fish #30 and #47) showed 1-5 copies of the gene per genome. Southern blot analysis indicated a concatenated integration of the gene copies, but conclusive evidence of integration will await breeding studies. Expression of the transferred neo gene was confirmed in the RNA dot blot. One out of the two putative transgenic fish tested showed expression of neo RNA when probed with a complementary neo RNA probe.

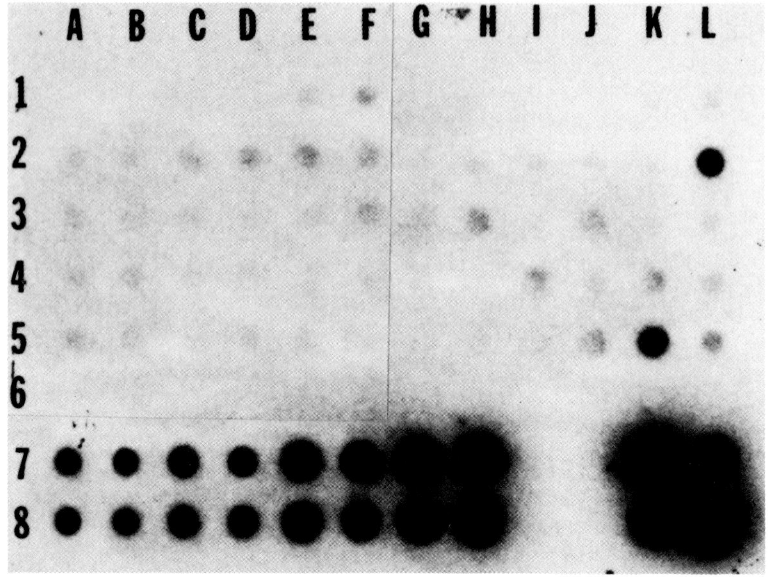

Plate 1. DNA dot blot analysis of 48 goldfish microinjected with pRSVneo. Total genomic DNA probed with neo cRNA. Rows 7 and 8 present dilutions of pRSVneo with (row 8) or without (row 7) added genomic DNA. Dilution series is A-B, 0.5 copies; C-D, 1 copy; E-F, 5 copies; G-H, 25 copies; and K-L, 100 copies per genome. Test fish DNA was spotted on filters producing series 1-5, A-L. Dots, row 1 E, K, and L are of control goldfish DNA. Dots row 2A-2L, 3A-3L, 4A-4L, and 5A-5L are of pRSVneo-injected goldfish DNA. Two fish showed positive hybridization signals; fish #30 and #47 represented by dots 2L and 5L, respectively.

DISCUSSION

The results reported here showed successful transfer of a marker gene neo that encodes resistance to the drug G418. Stable integration of the transferred gene was strongly suggested but could not be conclusively proven at this time. The band pattern upon Southern analysis exhibited by one fish is indicative of either multiple integration sites or possible delayed integration with subsequent production of mosaicism for the neo gene in the transgenic fish. Expression of the marker gene was confirmed through detection of neo mRNA in the transgenic fish.

Our results indicate that microinjection is a viable method for transfer of selectable genes into fish. Microinjection is, however, a labor intensive technique, and promises to remain so. Microinjection works well upon dechorionated eggs where needle placement may be monitored, but is less effective in blind injections, hence our interest in development of a selectable marker for use in either blind injection or in mass transfer techniques. Our group has been exploring the use of mass transfer techniques to overcome the tedium of injecting the substantial numbers of eggs currently necessary to generate the optimal transgenic fish. While mass transfer efficiencies may be orders of magnitude less than those achieved by microinjection, the ability to manipulate tens of thousands of eggs at a time should compensate for the loss in transfer efficiency. A key element necessary to make mass transfer a reality is the development of a selection system to differentiate transgenic from non-transgenic fish. The selectable marker neo meets this criteria, and should prove efficacious in the selection of transgenic fish when used in co-transfer schemes with other genes.

REFERENCES

1. Gordon, J.W., Scangos, G.A., Plotkin, D.J., Barbosa, J.A., and Ruddle, F.H. (1980). Genetic transformation of mouse embryos by microinjection of purified DNA. Proc. Natl. Acad. Sci., U.S.A. 77, 7380-7384.
2. Palmiter, R.D., Brinster, R.L., Hammer, R.E., Trumbauer, M.E, Rosenfeld, M.G., Birnberg, N.C., and Evans, R.M. (1982). Dramatic growth of mice that develop from eggs microinjected with metallothionein-growth hormone fusion genes. Nature 300, 611-615.
3. Pickford, G.E., and Thompson, E.F. (1948). The effects of purified mammalian growth hormone on the killifish Fundulus heteroclitis. J. Exp. Zool. 109, 367.

4. Adelman, I.R. (1977). Effect of bovine growth hormone on growth of carp (Cyprinus carpio) and the influences of temperature and photoperiod. J. Fish Res. Board Can. 34, 509-515.
5. Maclean, N., Penman, D. and Zhu, Z. (1987). Introduction of novel genes into fish. Bio/Technology 5: 257-261.
6. Dunham, R.A., Eash, J., Askins, J., and Townes, T.M. (1987). Transfer of the metallothionen-human growth hormone fusion gene into channel catfish. Trans. Am. Fish. Soc., 116, 87-91.
7. Stacey, N.E., Cook, A.F., and Peter, R.E. (1979). Spontaneous and gonadotrophin-induced ovulation in the goldfish, Carassius auratus L.--Effects of external factors. J. Fish. Biol. 15, 349-361.
8. Grand, C.G., Gordon, M., and Cameron, G. (1941). Neoplasma studies: Cell types in tissue culture of fish melanotic tumors compared with mammalian melanomas. Cancer Res. 1, 660-666.
9. Maniatis, T., Fritsch, E.F., and Sambrook, J. (1982). Molecular Cloning: A Laboratory Manual. Cold Spring Harbor Laboratory, Cold Spring Harbor, New York.
10. Marmur, J. (1961). A procedure for the isolation of deoxyribonucleic acid from microorganisms. J. Mol. Biol. 3, 208-218.
11. Schenborn, E.T., and Mierendorf, R.C., Jr. (1985). A novel transcription property of Sp6 and T7 RNA polymerases: Dependence on template structure. Nucleic Acids Res. 13, 6223-6236.

Gene Transfer and Gene Therapy, pages 35–45

REGULATION OF THE HUMAN β-GLOBIN GENE IN TRANSGENIC MICE AND CULTURED CELLS

F. Grosveld, M. Antoniou, G. Blom, F. Catala, P. Collis, E. deBoer, D. Greaves, O. Hanscombe, J. Hurst, G. Kollias, V. Mignotte, L-K. Siew, D. Talbot, R. Vogels, L. Wall and N. Wrighton.

Laboratory of Gene Structure and Expression, National Institute for Medical Research, The Ridgeway, Mill Hill, London NW7 1AA, U.K.

ABSTRACT We have analyzed the expression of the human β-globin gene in transgenic mice and cultered cells in vitro by DNA mediated gene transfer. The results show that the gene contains erythroid specific regulatory elements inthe 5'- flanking promoter region and two downstream enhancers. Detailed *in vitro* binding experiments on the promoter region and the furthest downstream enhancer indicate that at least one erythroid specific protein (GF1) and several ubiquitous proteins interact with these regions. In addition the entire β-globin domain appears to be controlled by a dominant control region upstream of the ε-globin gene. Inclusion of this region results in copy number dependent full expression of the gene independent of the site of integration.

INTRODUCTION

The human β-like globin genes are a cluster of five active genes in the order 5' ϵ-$^{G}\gamma$-$^{A}\gamma$-δ-β-3' comprising approximately 60kb of DNA on the short arm of chromosome 11. The different genes are expressed in a developmentally and tissue specific manner, i.e. the embryonic ε-gene in the yolk sac, the foetal $^{G}\gamma$- and $^{A}\gamma$-genes, primarily in the foetal liver, and the adult δ- and β-genes primarily in bone marrow (for review, see ref. 1). A large number of mutations have been characterized in this locus, ranging from simple amino acid changes by point mutations to complete deletions of the locus (for review, see ref. 2). This range of mutations has resulted in a large number of haemoglobinopathies which may lead to severe clinical problems and early death (mostly β^0-thalassaemias). The human β-globin locus is different from most species because it has undergone a process of recruitment of a separate set of foetal genes, a phenomenon also observed in other primates and in goat and sheep. This recruitment has probably not taken place at the level of the genes at their trans-acting factors, but at the cellular level by the timing of the production of factors. Thus, when the human adult β- and foetal γ-globin genes are introduced into the germ line of mice, it has been shown that the human β-globin gene is regulated as a mouse foetal/adult β-globin gene, while the γ-globin gene is regulated as a mouse embryonic globin gene (3,4,5,6). This phenomenon has been used to study these genes as a model system of a multigene family undergoing developmental switches. The purpose is to under-stand the regulation of each of these genes at the molecular level, in particular, the β-globin gene because this is clinically the most important gene in this cluster. Eventually, it would be possible to study the regulation of the globin regulators to obtain an insight into the processes of (erythroid) development and differentiation. A number of laboratories have made progress in our understanding of the DNA sequences necessary for the transcription of the β-globin gene, using both DNA mediated gene transfer in tissue culture cells, in particular, murine erythroleukaemia cells (MEL) and in transgenic mice.

MEL cells are Friend virus transformed erythroid cells that are arrested at the proerythroblast stage of differentiation. Cultured MEL cells may be induced to complete erythroid maturation by treatment with a variety of chemicals such as DMSO or HMBA. The maturation resembles normal erythroid differentiation and results in an accumulation of mouse adult globin mRNA in the differentiated cells. This induction of transcription also takes place on a human β-globin gene introduced into MEL cells, either in the form of a complete human chromosome 11 by cell fusion, or as cloned DNA via DNA mediated gene transfer. In a subsequent series of experiments, it was possible to show with the latter method, that the DNA sequences which regulate the human β-globin gene are located both 5' and 3' to the translation initiation site (7,8).

RESULTS AND DISCUSSION

Using MEL cells and β-globin/H-2K major histocompatibility hybrid genes, at least three separate regulatory elements required for the appropriate expression of the human β-globin gene have been identified. Firstly, a globin specific promoter element is present between -100 and -220bp 5' to the initiation site of transcription. Deletion of these sequences, which are homologous to those found in other β-globin genes leads to the loss of induction of transcription, but not the basic level of transcription in differentiating MEL cells (9). DNA binding experiments have shown that at least four proteins interact with this region *in vitro* (Fig. 1); an erythroid specific protein (named GF1) at two positions, -205 and -115bp (deBoer *et al*., submitted), an NF1 containing complex at -215bp (10; deBoer *et al*., submitted), another non erythroid specific protein named a2 binding at -115bp and a CAAT box binding protein (CP1,(11)) at -150bp (deBoer *et al*., submitted). Deletion and insertion of the relevant sequences in the β/H-2K hybrid genes in the absence of the enhancers suggest that at least the -150 CAAT box binding region in combination with either the -200 or the -120 region is required for the induction of transcription of the minimal promoter (TATA box, CAAT box, CAC box) in induced MEL cells (deBoer *et al*., submitted).

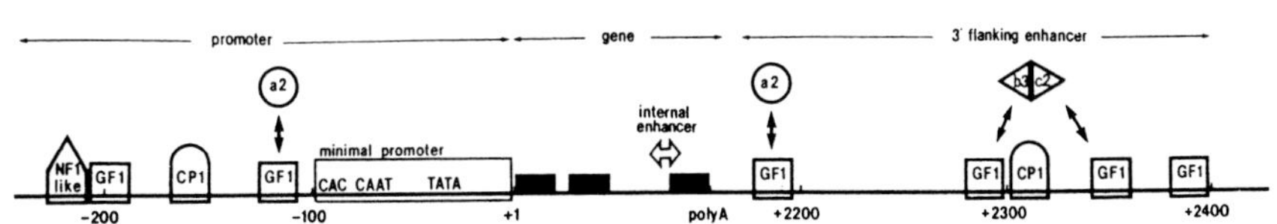

FIGURE 1. Schematic representation of the factors binding to the β-globin promoter and 3' flanking region enhancer.

GF1 represents an erythroid specific factor (deBoer *et al*., submitted, Wall *et al*.,submitted), a2 a factor not described previously (Wall *et al*.,submitted), b3/c2 a ubiquitous factor possibly binding to octamer sequences, CP1 a CAAT box binding factor (11), NF1 (10). CAC, CAAT, TATA represent the respective boxes present in the minimal promoter. The gene is drawn on a different scale than the promoter and enhancer.

Secondly, we and others have identified two downstream regulatory sequences (enhancers), one located in the gene and one approximately 800bp downstream from the gene (9,12,13). Both enhancers act in a developmental and tissue specific manner. The most downstream enhancer contains four erythroid specific GF1 protein factor binding sites and several non erythroid protein binding sites; the same a2 protein as at -115bp in the promoter, a CAAT box binding protein (CP1 as at -150bp in the promoter) and several unidentified proteins (Fig. 1). A similar downstream enhancer sequence has also been identified in the chicken β-globin gene using cultured chicken erythroid cells (14,15). The downstream regulatory sequence has also been shown to be a developmental stage specific enhancer by using transgenic mice (4,6,12,13). All of these results therefore indicate that the developmental specific control regions of the β-globin are located immediately 5', within and 3' to the β-globin gene. They do not, however, indicate how the human β-globin gene cluster is transcriptionally activated during development.

The strongest evidence for the existence of such a control has come from the analysis of a human $\gamma\beta$-thalassaemia. This patient is heterozygous for a large deletion which removes 100kb upstream of the β-globin gene, but leaves the β-globin gene, including all of the control regions described above, intact (16,17). Since the patient is heterozygous and transcribes the normal locus in the same nucleus, it indicates that some control mechanism overrides the functioning of the control sequences immediately surrounding the gene. In the case of the Dutch $\gamma\beta$-thalassaemia, it could indeed be shown that the mutant locus is in an inactive chromatin configuration and behaves like a classical position effect (16). The second piece of evidence is obtained from the position effects observed in transgenic mouse experiments with many genes, including the β-globin gene. In the latter case, the expression of the transgene is highly variable and not correlated to the copy number of the injected gene in the mouse chromosome (3,4,5). The interpretation of this position effect is that the injected DNA does not have sufficient sequence information to be independent from its neighbouring chromosomal localization in the transgenic mouse. Possible candidate sequences for such a control are the regions upstream from the ϵ-globin gene and downstream from the β-globin gene which contain a number of "super" hypersensitive sites (18,19). These sites are more sensitive to DNaseI digestion than the hypersensitive sites found in and around the individual genes when they are expressed. In addition, they are erythroid cell specific and they are present when any one of the globin genes is expressed. We have tested this hypothesis in transgenic mice (19) and MEL cells (Blom *et al.*, submitted).

A mini locus was constructed by combining several parts of the β-globin locus into a single cosmid. Unique linkers were positioned throughout the locus to facilitate future manipulation and mapping. This locus was injected into fertilized eggs of mice as a 40kb SalI fragment. Transgenic 12- and 16-day foetuses were identified by Southern blot analysis as described before (6). Nine transgenic foetuses were further characterized to show that eight were unrearranged and contained the complete 5' and 3' hypersensitive site regions. One foetus had lost the 5' hypersensitive region and behaved like a position effect dependent mouse, i.e. it showed only low levels of human globin gene expression. Seven of the remaining eight mice showed very high levels of human β-globin and showed two unusual aspects not previously observed; firstly, the level of human β-globin mRNA is directly correlated to the copy number of the transgene, secondly, the level of expression of each human gene is as high as that observed for the endogenous mouse gene. Although the eighth mouse had a completely intact injected locus, it nevertheless did not express the human β-globin gene. Subsequent Southern blot analyses showed this mouse to be chimaeric (observed in 10% of the transgenic foetuses).

Transfection of the mini locus in MEL cells also resulted in full position independent expression of the human β-globin gene and shows that the action of the dominant control region (DCR) does not require to undergo the complete developmental pathway to extend its effect. The linearized mini locus also contained a Herpes Simplex Virus thymidine kinase promoter driven neomycin resistance gene (tk-neo) which was used for selection of the transfected MEL cells. Analysis of the tk-neo RNA shows that this is also induced to very high levels of expression (>50 fold) in the presence of the DCR, than in the absence of these sequences (Blom *et al.*, submitted). Interestingly, in one particular clone the β-globin gene (plus enhancers) had been deleted,but the DCR was left intact. This clone shows a reduced expression of the tk-neo gene after induction of the MEL cells. These results show that heterologous genes can be controlled by the globin DCR and suggests that the effect of this control may be mediated through the enhancer(s) of the β-globin gene. Interestingly, when the mini locus was introduced in K562 cells (which express the endogenous ε-globin and γ-globin genes, but not the β-globin gene), the introduced β-globin gene was expressed in a copy number dependent, position independent fashion, albeit at a reduced level per gene when compared to MEL cells (approximately 10%). These results suggest either of two possibilities: 1) the mini locus does not contain a negative regulatory sequence which would normally suppress the expression of the β-globin gene at the foetal stage of expression or 2) the β-globin gene can be expressed at that stage, but its regulatory sequences have a low affinity for regulatory factors when compared to the adult stage. The γ-globin gene, on the other hand, would have a high affinity at this stage and the presence of both genes in the <u>same locus</u> sets up a competitive situation which amplifies the differences in affinities and reduces β-globin gene expression to very low levels at the foetal stage and vice versa. Support for this hypothesis is provided by the fact that such an amplification is observed in bacterial systems and that in several cases of either γ- or β-globin mutations in man (HPFH or β-thalassaemia) an increase in γ-globin expression results in a decrease of β-globin expression and vice versa (see (20) for review).

It is not clear from these data what minimal sequences in the flanking regions are required for the complete and independent expression of the gene. Certainly, the 5' flanking region is required and this region exerts a dominant control on the expression of the β-globin locus (γβ-thalassaemia, (16)). The data from deletion experiments (19; Blom *et al*., submitted) indicate that only the 5' region, but not the 3' region, is required and that the actual active regions are the sequences at the hypersensitive sites.

The control region(s) in the flanking DNA are completely dominant (see above) and the hypersensitive sites are erythroid specific and present at all erythroid developmental stages (18,19). They are even dominant after fusion of a globin expressing cell line (K562) to a non erythroid cell line myeloma (21) when it results in a globin expressing hybrid cell (PUTKO). It is likely that the regions control the accessibility of the β-globin locus to trans-acting factors, perhaps like the border regions flanking the DNaseI sensitive domain originally described for the ovalbumin gene (22). These borders may contain nuclear matrix binding sites which have been poorly defined in mammalian cells, but have been mapped in detail in Drosophila (23). Alternatively, the globin locus flanking regions might contain one or several enhancer-like sequences which can exert their effect over very large distances (>50kb) and would be required to set up efficient transcription complexes, although it should be stressed that they are different from "normal" enhancers because they are completely dominant. An attractive possibility is that such enhancers and matrix binding sites might actually be the same or neighbouring sequences (24). It is likely that similar control regions will be present in many other tightly controlled tissue specific gene systems. Recent experiments carried out with the T-cell specific gene CD2 show that the CD2 locus carries a similar activating sequence (Kioussis *et al.*, unpublished). Low level and ubiquitously expressed genes may, however, be controlled differently.

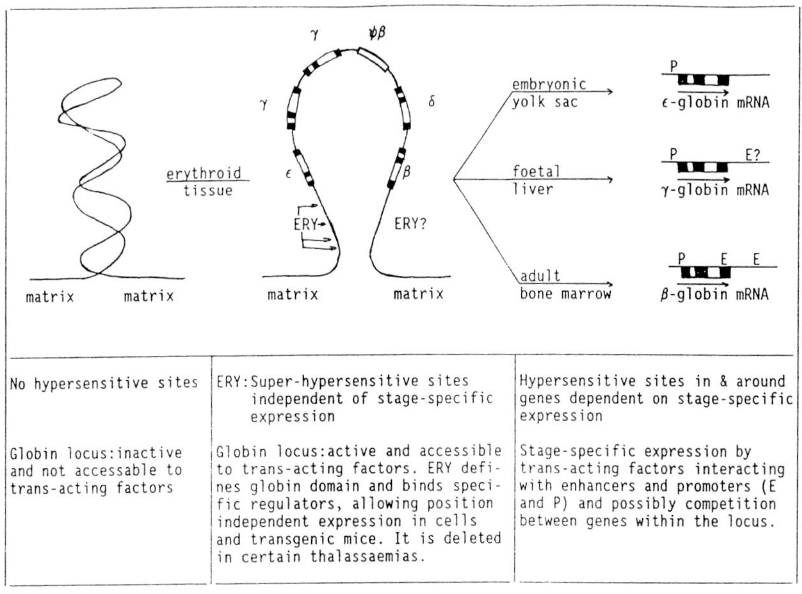
matrix
matrix
erythroid
tissue
γ
ψβ
γ
δ
ε
β
ERY
ERY?
matrix
matrix
embryonic
yolk sac
foetal
liver
adult
bone marrow
P
ε-globin mRNA
P
E?
γ-globin mRNA
P
E
E
β-globin mRNA
No hypersensitive sites
Globin locus:inactive and not accessable to trans-acting factors
ERY:Super-hypersensitive sites independent of stage-specific expression
Globin locus:active and accessible to trans-acting factors. ERY defines globin domain and binds specific regulators, allowing position independent expression in cells and transgenic mice. It is deleted in certain thalassaemias.
Hypersensitive sites in & around genes dependent on stage-specific expression
Stage-specific expression by trans-acting factors interacting with enhancers and promoters (E and P) and possibly competition between genes within the locus.

FIGURE 2. Model of the control of the β-globin gene domain.

Taking all the data together, we propose the following provisional model for the control of globin gene expression during development (Fig. 2). The dominant control region determines the activity of the locus. In inactive and non erythroid tissues there are no hypersensitive sites and the

DNA is not accessible to trans-acting factors. In erythroid cells the dominant control region becomes hypersensitive (possibly not requiring replication, 25) and renders the chromatin of the globin locus accessible to trans-acting factors. The proteins that mediate this action might be matrix bound, such as topoisomerases to change the winding of the locus and/or enhancer binding factors (26,27). This control defines the globin domain and provides position independent expression. Together with the action of the dominant control region trans-acting factors bind to regulatory promoter and enhancer sequences immediately surrounding the globin genes. The latter determine the stage specific expression, i.e. ϵ-gene expression in the embryonic yolk sac, γ-globin gene expression in the foetal liver and β-globin gene expression in the adult. In the case of the β-globin gene, this would possibly involve a negative factor and regulatory sequence to suppress the gene at the embryonic and foetal stages and positive factors acting on the enhancers and the promoters to set up a very active transcription complex. Alternatively, the gene might only be regulated by positive factors and the suppression of the gene is achieved by a lower affinity for factors and amplification of this by competition with the γ-globin genes for a limiting component in the system (perhaps interaction with the locus control region itself) One of the most interesting applications for the properties of the dominant control region(s) is that they might allow completely regulated expression of the β-globin (and possibly other genes) in retroviral vector or transfection systems. Gene expression in such systems to date has been position dependent and inefficient. In particular, retroviral vector systems which are presently the only efficient way to transfer genes into haematopoietic (or other) stem cells, are very sensitive to the integration site of the retroviral vector (28). Inclusion of the globin dominant control regions may solve this problem and allow the efficient transfer of a fully active, single copy human globin gene to haematopoietic stem cells. This would form the basis for somatic gene therapy by gene addition in the case of thalassaemia.

REFERENCES

1. Maniatis, T., Fritsch, E., Lauer, J. and Lawn, R. (1981). Molecular Genetics of Human Hemoglobins. Ann. Rev. Genet. 14, 145-178.

2. Collins, F. and Weissman, S. (1984). The Molecular Genetics of Human Hemoglobin. Prog. Nucl. Acid Res. Mol. Biol. 31, 315-462.

3. Magram, J., Chada, K. and Constantini, F. (1985). Developmental Regulation of a Cloned Adult β-Globin Gene in Transgenic Mice. Nature 315, 338-340.

4. Townes, T. M., Lingrel, J. B., Chen, H. Y., Brinster, R. L. and Palmiter, R. D. (1985). Erythroid-specific Expression of Human β-Globin Genes in Transgenic Mice. EMBO J. 4, 1715-1723.

5. Chada, K., Magram, J. and Constantini, F. (1986). An Embryonic Pattern of Expression of a Human Fetal Globin Gene in Transgenic Mice. Nature 319, 685-689.

6. Kollias, G., Wrighton, N., Hurst, J. and Grosveld, F. (1986). Regulated Expression of Human Aγ-, β- and Hybrid γβ-Globin Genes in Transgenic Mice: Manipulation of the Developmental Expression Patterns. Cell 46, 89-94.

7. Wright, S., Rosenthal, A., Flavell, R. and Grosveld F. (1984). DNA Sequences Required for Regulated Expreesion of β-Globin Genes in MEL Cells. Cell 38, 265-273.

8. Charnay, P., Treisman, R., Mellon, P., Chao, M., Axel, R. and Maniatis, T. (1984). Differences in Human α- and Human β-Globin Gene Expression in Mouse Erythroleukemia Cells: the Role of Intragenic Sequences. Cell 38, 251-263.

9. Antoniou, M., deBoer, E., Habets, G. and Grosveld, F. (1988). The Human β-Globin Gene Contains Multiple Regulatory Regions: Identification of One Promoter and Two Downstream Enhancers. EMBO J. 7, 377-384

10. Jones, K. A., Kadonaga, J. T., Rosenfeld P. J., Kelly, T. J. and Tjian, R. (1987). A Cellular DNA-binding Protein that Activates Eukaryotic Transcription and DNA Replication. Cell 48, 79-89.

11. Chodosh, L. A., Baldwin, A. S., Carthew, R. W. and Sharp, P. A. (1988). Human CCAAT-Binding Proteins Have Heterologous Subunits. Cell 53, 11-24.

12. Behringer, R., Hammer, R., Brinster, R., Palmiter, R. and Townes, T. (1987). Two 3' Sequences direct Erythroid-specific Expression of the Human β-Globin Gene in Transgenic Mice. Proc. Natl. Acad. Sci. USA 84, 7056-7060.

13. Kollias, G., Hurst, J., deBoer, E. and Grosveld, F. (1987). The Human β-Globin Gene Contains a Downstream Developmental Specific Enhancer. Nucl. Acids Res. 15, 5739-5747.

14. Hesse, J. E., Nickol, J. M., Lieber, M. R. and Felsenfeld, G. (1986). Regulated Gene Expression in Transfected Primary Chicken Erythrocytes. Proc. Natl. Acad. Sci. USA 83, 4312-4316.

15. Choi, O-R, and Engel, J. D. (1986). A 3' Enhancer is Required for Temporal and Tissue-Specific Transcriptional Activation of the Chicken Adult β-Globin Gene. Nature 323, 731-734.

16. Kioussis, D., Vanin, E., deLange, T., Flavell, R. A. and Grosveld, F. G. (1983). β-Globin Inactivation by Translocation in γβ-Thalassaemia. Nature 306, 662-666.

17. Taramelli, R., Kioussis, D., Vanin, E., Bartram, K., Groffen, J., Hurst, J. and Grosveld F. (1986). γδβ-Thalassaemia 1 and 2 are the Results of a 100 kbp Deletion in the Human β-Globin Gene cluster. Nucl, Acids Res. 14, 7017-7029.

18. Tuan, D., Solomon, W., Qilang, L. and Irving, M. (1985). The β-Globin Gene Domain in Human Erythroid Cells. Proc. Natl. Acad. Sci. USA, 32, 6384-6388.

19. Grosveld, F., Blom van Assendelft, G., Greaves, D. and Kollias, G. (1987). Position-Independent, High level Expression of the Human β-Globin Gene in Transgenic Mice. Cell 51, 975-985.

20. Poncz, M., Henthorn, P., Stoeckert, C. and Surrey, S. (1988). Globin Gene Expression in Hereditary Persistence of Fetal Hemoglobin and δβ-Thalassaemia. In press.

21. Klein, G., Zeuthen, J., Eriksson, I., Tekasaki, P., Bernoco, M., Rosen, A., Masucci, G., Povey, S. and Ber, R. (1980). Hybridization of a Myeloid Leukaemia Derived Human

Cell Line (K562) with a Burkitt's Lymphoma Line. J. Natl Canc. Inst. 64, 725-738.

22. Lawson, G., Knoll, B., March, C., Woo, S., Tsai, M. and O'Malley, B. (1982). J. Biol. Chem. 257, 1501-1507.

23. Gasser, S. and Laemmli, U. (1986). Cohabitation of Scaffold Binding Regions with Upstream Enhancer Elements of Three Developmentally regulated Genes of D. Melanogaster. Cell 46, 521-530.

24. Cockerhill, P. and Garrard. (1986). Chromosomal Loop Anchorage of the Ig Locus Next to the Enhancer in a Region Containing Topoisomerase II Sites. Cell 44, 273-282.

25. Baron, M. H. and Maniatis, T. (1986). Rapid Reprogramming of Globin Gene Expression in Transient Heterokaryons. Cell 46, 591-602.

26. Angel, P., Imagawa, M., Chin, R., Stein, B., Imbra, R., Rahmsdorf, M., Jonat, A., Herrlich, P. and Karin, M. (1987). Phorbol Ester-Inducible Genes Contain a Common *Cis* Element Recognized by a TPA-Mediated *Trans*-Acting Factor. Cell 49, 729-739.

27. Lee, W., Mitchel, P. and Tjian, R. (1987). Purified Transcription Factor AP-1 Interacts With TPA-Inducible Enhancer Elements. Cell 44, 273-282.

28. Dzierzak, E. A., Papayannopoulou, T. and Mulligan, R. C. (1988). Lineage-Specific Expression of a Human β-Globin Gene in Murine Bone Marrow Transplant Recipients Reconstituted With Retrovirus-Transduced Stem Cells. Nature 331, 35-41.

Gene Transfer and Gene Therapy, pages 47–56

EFFICIENT TRANSFER OF THE COMPLETE HUMAN BETA-GLOBIN GENE INTO HUMAN AND MOUSE HEMOPOEITIC CELLS VIA SV40 PSEUDOVIRIONS[1]

Nava Dalyot and Ariella Oppenheim

Department of Hematology, Hadassah University Hospital
Jerusalem, Israel 91120

ABSTRACT The complete human β-globin gene was cloned into a plasmid vector that carried the SV40 origin of replication. After removing the prokaryotic sequences the plasmid was encapsidated as an SV40 pseudovirion and transmitted into cultured mouse (MEL) and human (K562) hemopoietic cells by viral infection. High level of non-integrated copies of the transmitted β-globin gene was detected in Hirt supernatants of the infected cells after 48 hours.

INTRODUCTION

As the first step in the development of a procedure for gene therapy, we have developed a novel vector for efficient introduction of foreign DNA into human hemopoietic cells (1,2). Gene transmission is based on encapsidation of plasmid DNA as SV40 pseudovirions. The DNA is then transmitted into the target hemopoietic cells via viral infection.

Encapsidation was performed in COS (monkey kidney) cells, which express SV40 T-antigen constitutively (3). The vector, pSO3, was introduced into the COS cells by DEAE-dextran transfection. It carried the SV40 origin of replication (ori), to facilitate replication of the plasmid in the COS cells. The SV40 capsid proteins were supplied in trans by a helper SV40 DNA, co-transfected into the COS cells.

---

[1]Aided by Basic Research Grant No. 1-1073 from the March of Dimes Birth Defects Foundation.

The method was very efficient in introducing the bacterial _cat_ gene, as a model for gene transmission, into the human erythroleukemia cell line K562 and fresh human bone marrow cells (1). Over 40% of the infected K562 cells and 30% of the infected bone marrow cells were observed to contain plasmid DNA 48 hr after the infection. Moreover, later experiments showed that the efficiency of gene transmission by this method can be improved, possibly approaching the theoretical 100%.

In the present communication we describe the introduction of the cloned human β-globin gene into mouse erythroleukemia cells (MEL) and human K562 cells.

RESULTS

Construction of the β-Globin Plasmid.

The first step was to prepare a vector, pSO6, suitable for cloning a variety of genes (Fig. 1). This was accomplished by the removal of the _cat_ gene and expression signals (splicing and polyadenylation) from pSO3_cat_ and the insertion of a multi-site polylinker, which also included a SacI site adjacent to the pBR322 fragment. A second SacI site was inserted on the other side of the pBR322 fragment. The complete human β-globin gene, from BglII to XbaI (4.8 kb), was then inserted into pSO6 at the appropriate cloning sites. The plasmid pSO6β-1 was propagated in _E. coli_. It was used to generate SO6β-1 by excising the pBR322 sequences with SacI and religation at low concentration (Fig. 1). SO6β-1 is 5.2 kb and suitable for encapsidation as an SV40 pseudovirion.

Encapsidation of SO6β-1.

The experimental plan is depicted in Fig. 2. COS cells were co-transfected with SO6β-1 and SV40 DNA, 0.5 μg each per 25 cm^2 culture, by the DEAE-dextran procedure. The transfected cells were incubated for 5 days to allow plasmid replication and encapsidation. The virion mixture was then harvested by repeated freeze-thawing.

To test whether SO6β-1 became encapsidated, an aliquot (200 μl) was used to infect fresh COS cells. Two days later the cells were harvested and plasmid DNA was extracted using the Hirt procedure (4). The Hirt supernatant was treated with RNase, phenol extracted and ethanol

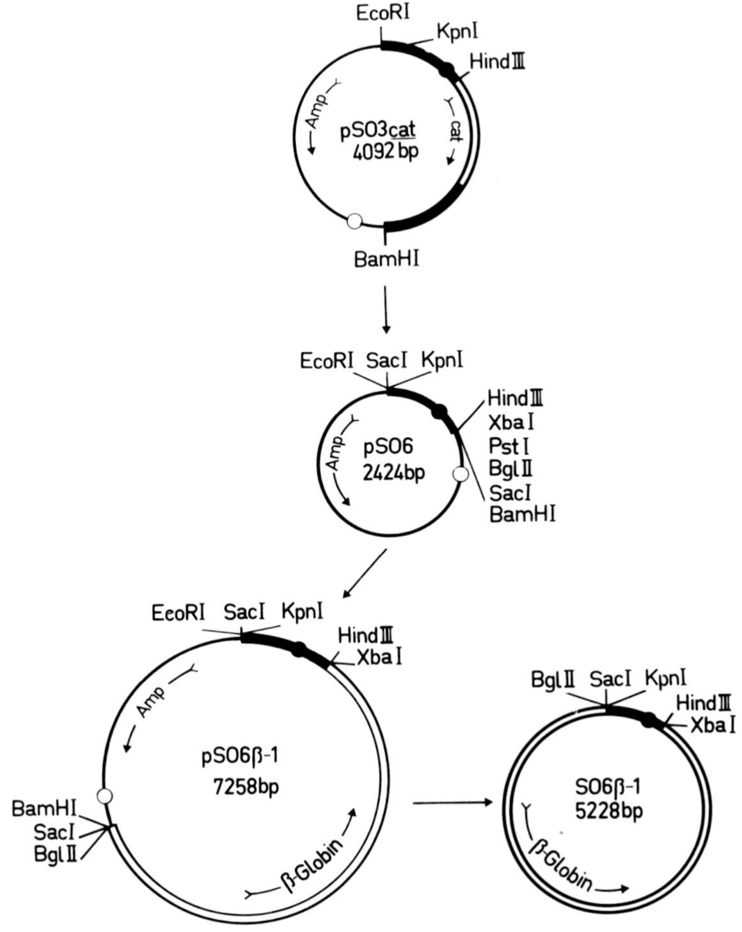

FIGURE 1. Construction of the β-globin plasmid. pSO6 was constructed in 2 steps. First, the cat gene and expression signals were removed by digestion with HindIII and BamHI and the plasmid was religated with a multi-site polylinker. Then, a SacI site was inserted between the EcoRI and KpnI sites, removing a small fragment of SV40 DNA. The β-globin gene (BglII to XbaI) was introduced into pSO6 between the HindIII and XbaI sites to form pSO6β-1. SO6β-1 was generated by digestion of pSO6β-1 with SacI and religation at low concentration.

precipitated. It was then analyzed for S06β-1 DNA by digestion with restriction endonucleases that produce fragments of different size for SV40 and S06β-1.

When the digested DNA was analyzed by gel electrophoresis and ethidium bromide staining, only fragments of the SV40 helper were visible (Fig. 3, top). However, following Southern blotting (5) and hybridization to a labeled S06β-1 probe, S06β-1 specific fragments were detected (Fig. 3, bottom, marked by arrows). Some fragments of the SV40 helper DNA also hybridized to the S06β-1 probe, since it contained the SV40 ori region. From the relative radioactivity of the different fragments we estimated that there was about 20 fold more SV40 DNA than S06β-1 DNA. We concluded that S06β-1 became encapsidated, but at a low efficiency.

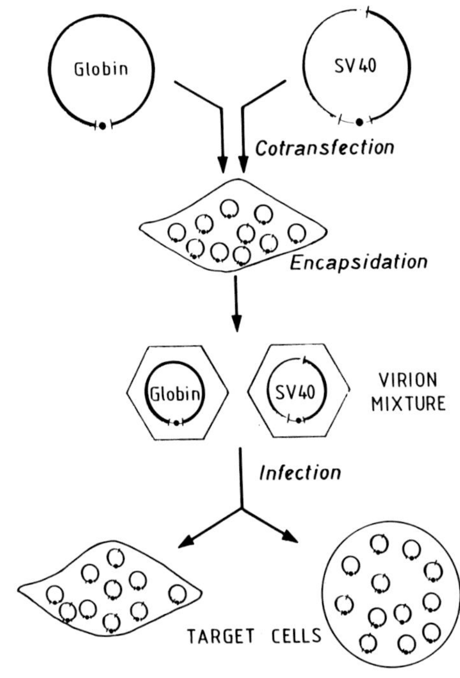

FIGURE 2. Experimental design.

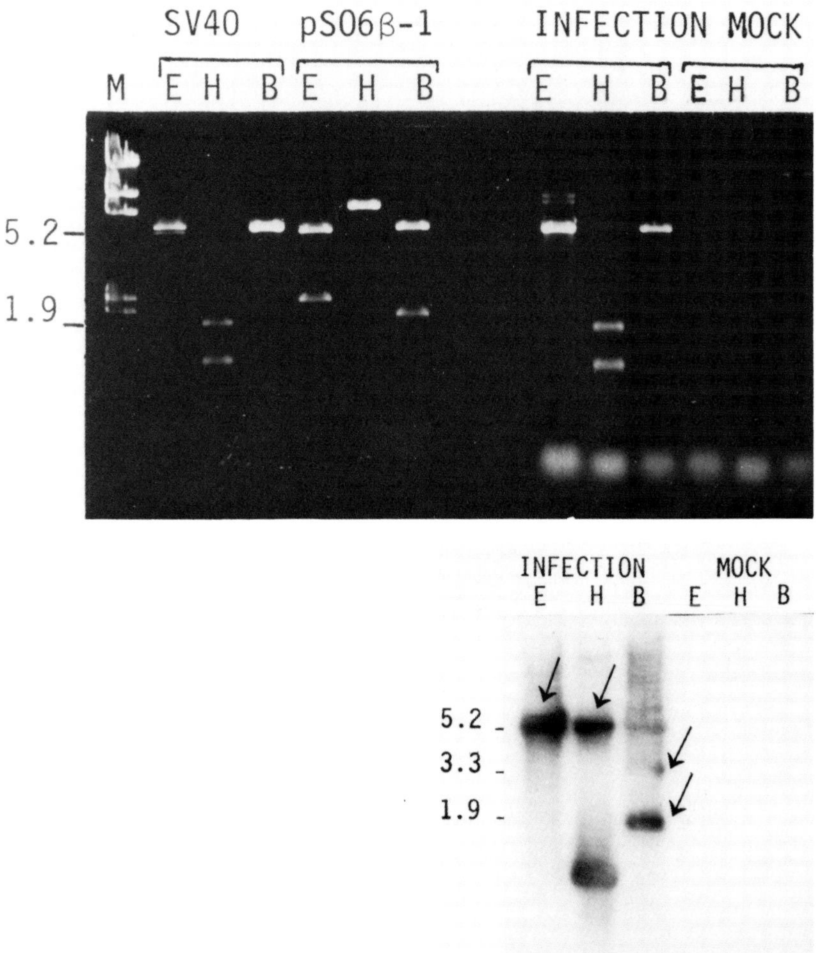

FIGURE 3. Encapsidation of SO6β-1 by transfection of equal amounts of plasmid and helper DNA. The experiment is described in the text. E - EcoRI; H - HindIII; B - BamHI. Top: Ethidium bromide staining. Hirt supernatants from infected and mock infected cells are shown on the right. SV40 and pSO6β-1 DNA digested with the same enzymes are on the left. M - marker DNA. Bottom: Southern blot hybridization of the right part of the same gel. The arrows point to SO6β-1 bands. Note that the fragments generated by digestion of pSO6β-1 are different from those of SO6β-1.

It appeared that SV40 competed with SO6β-1 DNA, probably at the level of DNA replication. We performed a series of experiments, co-transfecting COS cells at different ratios of the two DNA species and followed the efficiency of encapsidation. The conclusion was that using 20 fold excess of SO6β-1 DNA in the transfection is optimal for the encapsidation of SO6β-1. Fig. 4 shows analysis of Hirt supernatants made from cells which were infected with virion mixtures prepared by transfection of 20:1 SO6β-1:SV40 DNA. SO6β-1 DNA in the infected cells was visible with ethidium bromide staining (3.3 kb and 1.9 kb bands), indicating efficient encapsidation.

The Interaction between Viral Infection and Differentiation of MEL and K562 Cells.

Since one of our major interests is to investigate globin gene regulation during erythroid differentiation, it was important to establish whether introduction of DNA by viral infection affected this process. Infection of MEL cells by either SV40 or SO6β-1 + SV40 pseudoviral mixture enhanced growth (Fig. 5). The rate of differentiation remained unchanged (not shown). Differentiation of infected K562 cells was accelerated (table 1), while their growth rate was somewhat decreased (not shown).

Introduction of β-globin plasmid DNA into K562 and MEL cells.

Logarithmic cells were infected with the pseudoviral mixture. After 48 hrs the cultures were analyzed for β-globin DNA by the Hirt procedure and gel electrophoresis as described above. Southern blotting and hybridization to a SO6β-1 probe revealed a significant level of the human β-globin DNA in both K562 and MEL cells. The results for the K562 cells are shown in Fig. 6. The arrows mark the SO6β-1 bands. From the calibration standards (Fig. 6, right), an average of 500-1000 pSO6β-1 molecules per cell was computed. Since the infection was carried out with wild-type SV40 as a helper, this probably reflects gene amplification in the K562 cells. Note that since the analysis was performed on Hirt supernatants, the endogenous gene in K562 did not interfere with the procedure (see noninfected control in Fig. 6.)

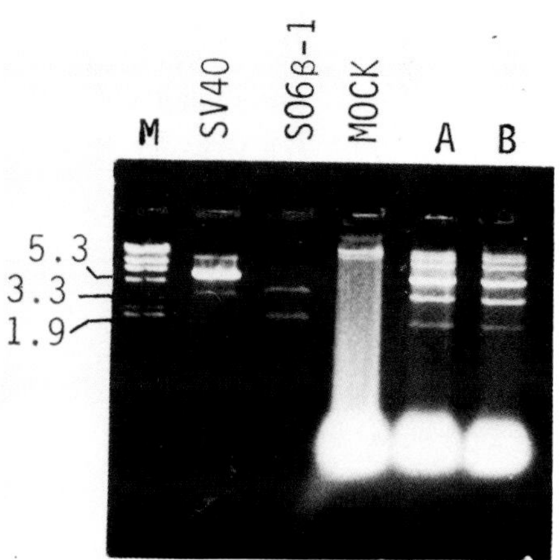

FIGURE 4. Encapsidation of SO6β-1 by transfection of 20 fold more SO6β-1 DNA than SV40 DNA. DNA was digested with BamHI. Mock, A and B are Hirt supernatants of infected cells. The virion mixtures were prepared by transfection of A - 1 μg SO6β-1 + 50 ng SV40 DNA; B - 2 μg SO6β-1 + 100 ng SV40 DNA.

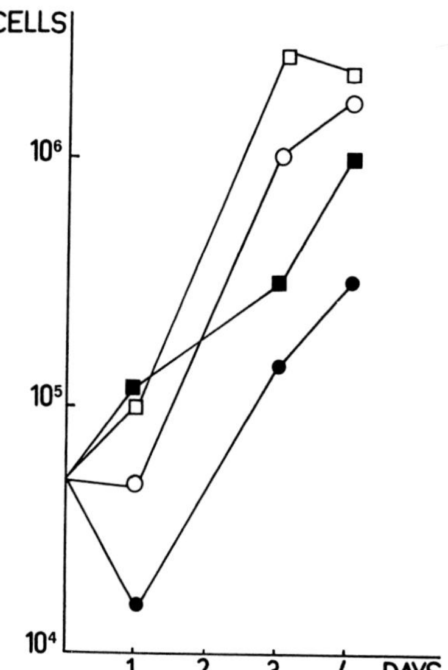

FIGURE 5. The effect of viral infection on growth of MEL cells. Logarithmic cells were infected (for 90 min) with virion mixture. Inducer for differentiation (HMBA, 4 mM) was added 3 hr later.

○ - uninfected, uninduced;
● - uninfected, induced;
□ - infected, uninduced;
■ - infected, induced.

TABLE 1
THE EFFECT OF VIRAL INFECTION ON HEMIN-INDUCED DIFFERENTIATION OF K562 CELLS

	Treatment		% Differentiation		
Culture No.	Infection	Hemin	Day 1	Day 2	Day 3
1	-	-	6%	15%	N.D
2	-	+	1%	62%	90%
3	SV40	-	5%	16%	10%
4	SV40	+	27%	57%	95%
5	S06β-1	-	5%	17%	12%
6	S06β-1	+	60%	93%	80%

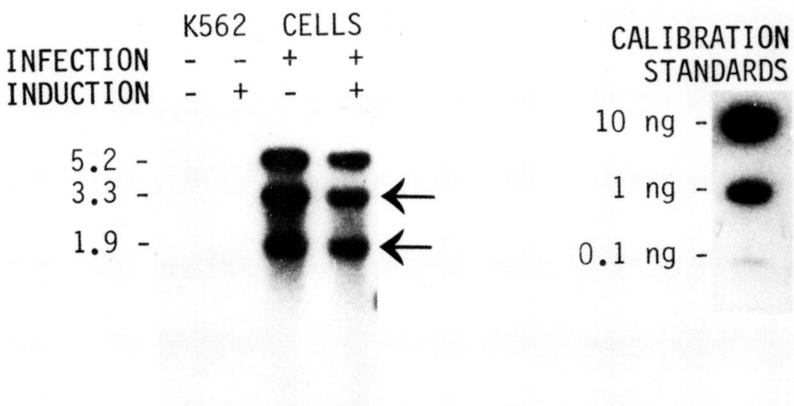

FIGURE 6. DNA from infected K562 cells. Hirt supernatants of 10^6 infected cells were digested with BamHI, and analyzed by Southen blotting. The arrows point to the S06β-1 bands. From the calibration standards (S06β-1 DNA spotted on nitrocellulose and hybridized in parallel) we calculated an average of 500-1000 copies of S06β-1 molecules per infected cell.

DISCUSSION

In the present communication we have shown that the complete β-globin gene can be transmitted efficiently into hemopoietic cells, mouse and human. The construct S06β-1 carries very little non human (SV40) DNA. We think that this approach will be useful for studying the regulation of cloned genes in cells that are not readily amenable to DNA transfection. In another study (2) we have demonstrated that gene expression in the target hemopoietic cells can be investigated with or without gene amplification, depending on whether it is encapsidated with a wild-type SV40 as a helper or with a T-antigen negative helper, SLT3.

The plasmid S06β-1, carrying an SV40 fragment of 370 bp, including the ori and the enhancer, was encapsidated in COS cells as SV40 pseudovirions. In other experiments (Ottolenghi and Oppenheim, unpublished) we observed that plasmids from which the enhancer had been removed, leaving only 204 bp of the SV40 ori region (HindIII to SphI) also became encapsidated. It is thus possible that encapsidation of SV40 does not require any specific DNA signal. If so, many new possibilities for gene transmission via this pathway will be opened up.

It appears from the present studies that optimal conditions for encapsidation may need to be established separately for each new construct. This may depend on the rate of replication of each plasmid, which could be a function of its size as well as of specific DNA sequences. The procedure described above can be used to establish conditions for encapsidation. In addition, we are developing an assay for titration of infectious pseudoviral units.

The next steps in the use of SV40 pseudovirions for the development of gene therapy are to establish conditions for integration and stabilization of the transmitted DNA, to develop a helper free system, and to construct a β-globin plasmid that will be correctly regulated after insertion into the hemopoietic cells. An attractive alternative is the curing of genetic diseases by gene replacement. The feasibility of using SV40 pseudovirions for gene targeting is currently being investigated.

ACKNOWLEDGMENTS

We wish to thank Amos Oppenheim for many helpful discussions, Mrs. Aviva Peleg for competent assistance in some of the experiments, Dr. H. Giladi and Mrs. S. Koby for help in plasmid construction. Part of this work was performed in the Davide and Irene Sala Laboratory of Molecular Genetics.

REFERENCES

1. Oppenheim A, Peleg A, Fibach E, Rachmilewitz EA (1986). Efficient introduction of plasmid DNA into human hemopoietic cells by encapsidation in simian virus 40 pseudovirions. Proc. Natl. Acad. Sci. USA 83:6925.
2. Oppenheim A, Peleg A, Rachmilewitz EA. (1988). Efficient introduction and transient expression of exogenous genes in human hemopoietic cells. Ann. New York Acad. Sci. In press.
3. Gluzman Y (1981). SV40-transformed simian cells support the replication of early SV40 mutants. Cell 23:175.
4. Hirt B (1967). Selective extraction of polyoma DNA from infected mouse cell cultures. J. Mol. Biol. 26:365.
5. Southern EM (1975). Detection of specific sequences among DNA fragments separated by gel electrophoresis. J. Mol. Biol. 98:503.

Gene Transfer and Gene Therapy, pages 57–66

EXPRESSION OF HUMAN TPA IN THE MILK OF TRANSGENIC ANIMALS

Katherine Gordon, James Vitale, Eric Lee*, Heiner Westphal*, and Lothar Hennighausen#,

Integrated Genetics, 31 New York Avenue, Framingham, MA 01701, #Laboratory of Biochemistry and Metabolism, National Institute of Diabetes, Digestive and Kidney Diseases, Bethesda, MD 20892, *Laboratory of Molecular Genetics, National Institute of Child Health and Human Development, National Institutes of Health, Bethesda, MD 20892.

ABSTRACT We have produced a series of transgenic mice which contain a mammary-specific expression cassette in order to target production of tissue plasminogen activator to the lactating mammary gland. The fusion construction contained the promoter from the murine whey acid protein gene and the secretion signal and coding segment of the TPA gene. Biologically active human TPA was secreted into milk in all transgenic lineages which expressed product, though there was much variability in expression levels among the different lineages. The WAP-tPA expression vector appears to be expressed and regulated properly, providing a model system for the production of active proteins in the mammary gland.

INTRODUCTION

It has been demonstrated in numerous model systems that genes injected into mouse embryos may be incorporated into the germ line and be expressed in patterns that mimic those of their endogenous counterparts (1,2). The pattern of spatial and temporal expression of foreign genes in transgenic animals can be controlled by prior manipulation of the signals regulating gene expression. We introduced into mice a construct designed to express a foreign protein in the lactating mammary epithelium in which 5' sequences from the whey acid protein gene were fused with a cDNA coding for tissue plasminogen activator. We demonstrate that such an approach is a feasible means to direct expression of foreign proteins into secreted milk (3). The transgenic mouse system is also useful to define and to characterize control elements governing milk protein gene expression. Appropriately regulated mammary cell lines are not available to evaluate hormonal regulation and tissue-specific elements. The transgenic mouse model also provides data about gene regulation in context of an intact animal.

We chose to utilize WAP upstream DNA as the promoter in our model expression vector. WAP is the most abundant whey protein in mouse milk (4). Its gene is expressed in the lactating mammary gland and is inducible by steroid and peptide hormones. Putative regulatory protein binding sites within the WAP promoter have been described (5). By demonstrating secretion of a foreign protein (tissue plasminogen activator) into milk, we show that 2.6 kb of upstream sequences from the WAP gene are sufficient to target gene expression to the lactating mammary gland in transgenic mice.

RESULTS

A mammary expression vector was constructed in which 5' sequences from the whey acid protein gene were fused with a cDNA coding for tissue plasminogen activator (Figure 1). The TPA gene utilized here was a cDNA clone from a human uterus cDNA library. The TPA DNA sequence was determined previously and the protein expressed in C127 cells using bovine papilloma vectors (6). The secretion signal sequence in this construct derives from the native TPA gene; the analogous signal-encoding region from the WAP gene was removed in the construction.

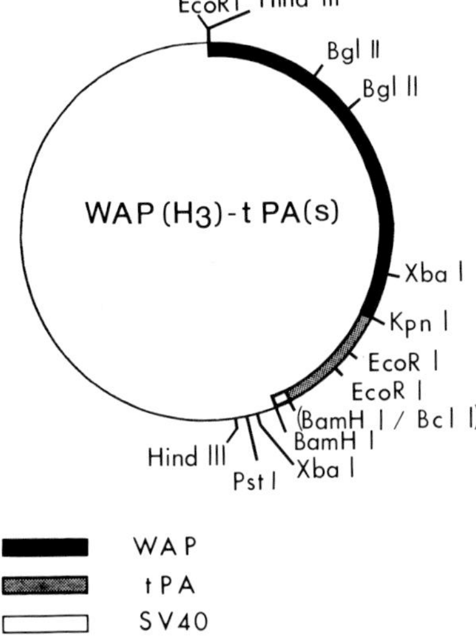

FIGURE 1. Mammary expression vector. The expression construct is a triparite fusion consisting of 2.6 kb of upstreamp stream DNA from the WAP gene through the endogenous CAP site, TPA cDNA beginning in the untranslated 5' region, and the polyadenylation/termination signals from SV40.

This plasmid was injected into one-cell pronuclear mouse embryos as a purified Hind-3/BamH1 fragment containing no procaryotic sequences. Seven mice were identified as being transgenic by diagnostic Southern blot hybridization with a human cDNA TPA probe. Under conditions of high stringency, this probe does not hybridize with the endogenous mouse TPA gene. The blot patterns of three positive mice, #wtl-26, wtl-25 and wtl-7 are shown in Figure 2. By comparison to the hybridization intensity obtained with positive controls, the number of copies of the injected fragment present in the genomes of these transgenic mice was estimated to be between 1 and 50. Digestion with SacI (lanes b-d) yielded a diagnostic band of 1.75 kb that spans the WAP and TPA junction and hybridizes to a tPA probe. The intact plasmid digested with SacI was a positive control (lane a). Exogenous DNA injected into embryos tends to form concatomers even when introduced as a fragment with non-cohesive ends. The 2.3 kb band seen in lanes b-d corresponds to the 3' end of the TPA gene (which does not contain SacI sites), apparently ligated to the 5' end of the WAP promoter, and through to the first Sac I site in the WAP DNA. The presence and size of this fragment is diagnostic of head-to-tail concatomers.

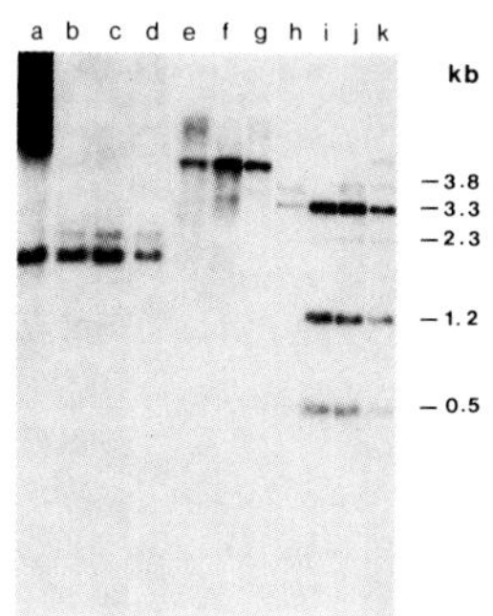

FIGURE 2. Southern blot tails DNA. Lanes a and h show 500 pg of WAP-tPA DNA digested with Sac I or EcoRI, respectively. Lanes b, c, and d contain 5 μg of DNA from mouse wtl-26, wtl-25 and wtl-7, respectively, digested with Sac I. Lanes e-g are from mouse tails of wtl-26, wtl-25 and wtl-7, respectively,digested with Kpn 1, and lanes i-k are these DNAs cut with Eco RI. Lanes a-d were electrophoused on a separate gel than the rest of the lanes. Negative control DNAs did not show any hybridization to this probe under these conditions.

The EcoRI digest (control, lane H; experimental, lane i-k) showed the expected 472 bp band internal to the TPA gene. In addition, a 3.3 kb band can be seen that represents the 5' region of the TPA gene and extends through the WAP gene to 5' boundary EcoRI site. Thus, despite the fact that the WAP EcoRI site was near the end of the injected fragment, it appeared to be intact in the genomic DNA of this transgenic animal. The 1.2 kb band represents the 3'-most region of the TPA gene, which must have ligated head-to-tail to the 5' end of the WAP gene, leaving the TPA gene bounded on its 3' end by an EcoRI site. Interestingly, the weak 2.3 kb band indicates that some of the copies of the fragments formed concatomers in a head-to-head configuration. KpnI digestion (lanes e-g) produced a single band of 4.9 kb, as expected. It is impossible to determine from this Southern blot whether all copies of the concatomer integrated at a single or at multiple sites.

Milk was obtained from female transgenic mice after parturition. Samples were assayed by ELISA.

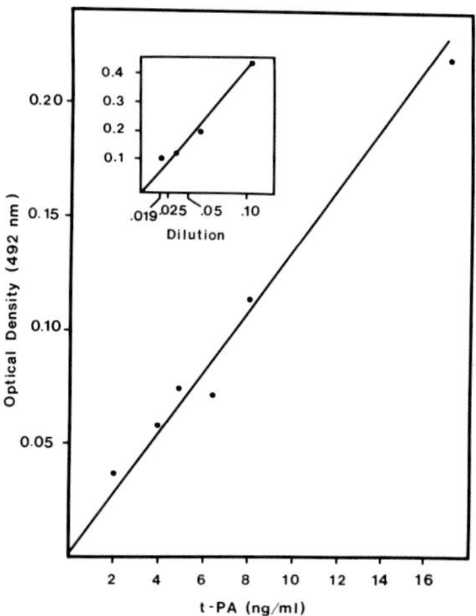

FIGURE 3. Quantitation by ELISA of recombinant t-PA secreted into milk of mouse wtl-26. The standard curve was performed in negative mouse milk diluted to a final concentration of 10% with PBS, to which was added TPA supplied with the kit as indicated. The inset shows milk from mouse #wtl-26 in dilutions as indicated. The dilution of .1 refers to a final concentration of 10% milk. In each dilution of transgenic milk, samples were supplemented with negative mouse milk in order to keep the final concentration

at 10%. All points of the control curve and the experimental (inset) curve have the background value (the value determined for negative mouse milk) subtracted. Assays were performed with the IMUBIND ELISA kit produced by American Diagnostica Inc.

There was no direct correlation between the copy number of WAP-TPA genes and the level of TPA expression (Table 1). In fact, line #27 produced the highest amount of human TPA in the milk, but contained the smallest number of WAP-TPA gene copies. The amount of TPA found in milk of the line #27 was approximately 50 μg per ml milk as judged by an ELISA assay.

Lineage	Sex of founder	DNA copy #	Expression level
wtl-7	male	30	400 ng/ml
wtl-11	female	10	<20
wtl-15	male	10	did not transmit
wtl-25	female	50	80
wtl-26	female	30	350
wtl-27	male	1	50,000
wtl-28	male	<1	did not transmit

TABLE 1. DNA copy number and TPA expression level in milk. Expression levels in milk were determined by ELISA as described in the legend to Figure 3.

Milk from wtl-26 and other mice (data not shown) was shown to contain biologically active tissue plasminogen activator at levels which far exceed in background as determined by the fibrin clot lysis assay (Figure 4). This assay measures the ability of TPA to digest fibrinogen matrices laid down in a background of agarose, thrombin and plasminogen within the wells of a plate. A small hole is bored through the agarose mixture upon hardening and 25 microliters of the samples are loaded into each hole. As TPA diffuses into the agarose, clearing of the fibrinogen is evident visually and the amount of clearing is directly proportional to the amount of active TPA. These assays are extremely sensitive and reproducible.

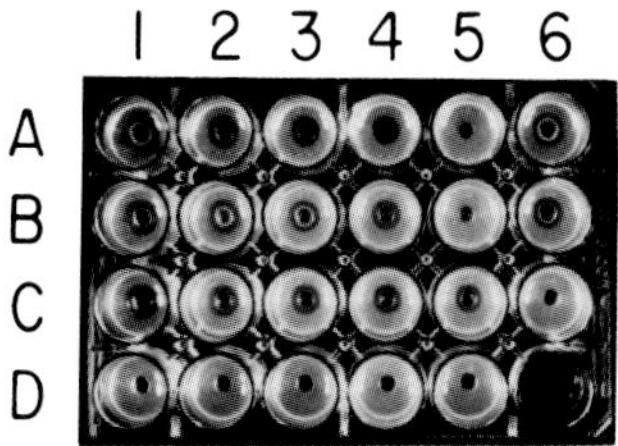

FIGURE 4. Clot lysis bioassay of milk from transgenic mouse #wtl-26. Following identification of mouse #wtl-26 as a positive transgenic, the mouse was mated to a wild type male. Seventeen days after the first litter was born, milk was removed from the lactating female following stimulation with oxytocin. Milk was diluted in PBS by 50% and stored frozen. Milk was diluted further in PBS as indicated below just prior to assay and added to the wells of a fibrin clot lysis plate. The positive controls were generated by addition of recombinant TPA to media composed of either 10% negative mouse milk (row A), 10% negative cow milk (row B), or PBS (row C). Concentrations of TPA in the milk dilution curves, from columns 1 through 5 were: 40,20,10,5, and 0 ng/ml. The concentrations in the PBS dilution curve from column 1 through 6 were: 40,20,10,5,2.5 and 0 ng/ml. In row A, column 6 was the milk from mouse #wtl-26 at a final concentration of 10% and in row B, column 6 is the milk at a concentration of 5%.

This photograph was taken after approximately 8 hours of assay incubation time. The negative mouse milk used for these controls was pooled from outbred CD-1 mice in different stages of lactation. In other negative controls (not shown) milk was used from inbred females of the same strain used for microinjection and was obtained at the same stage of lactation as the positive sample. The specificity of recombinant TPA secreted into mouse milk was shown to be plasminogen in other experiments (not shown) in which plasminogen was omitted from the agarose matrix in similar fibrin clot lysis assay. In this figure it can be seen that milk form wt1-26 cleared the fibrin clot to a significant extent. By comparison with lysis catalyzed by known amounts of added TPA, the concentration was calculated to be about 200 ng/ml. In parallel assays, milk obtained from wt1-25 and wt2-102 was shown to contain 200 ng/ml and 400 ng/ml of TPA (data not shown). When plates were incubated longer than 24 hours, minor clearing was seen in control wells containing milk from untransfected mice, but this was always significantly less than clearing seen from milk of any of the transgenic lineages. The origin of the residual fibrinolytic activity in nontransgenic mouse milk is not known. However, the presence of low levels of plasminogen activator (PA) in the lactating mammary gland of rodents raises the possibility that some fibrinolytic protein is present naturally in milk.

In all the lineages which we have analyzed, with the exception of one mosaic, expression levels and patterns were heritable in subsequent generations. Moreover, sibling female transgenics expressed TPA to equivalent extents. The TPA secreted in milk by these animals is induced at a constant level over the course of lactation. Figure 5 shows this for two sibling animals, 2-59 and 2-62.

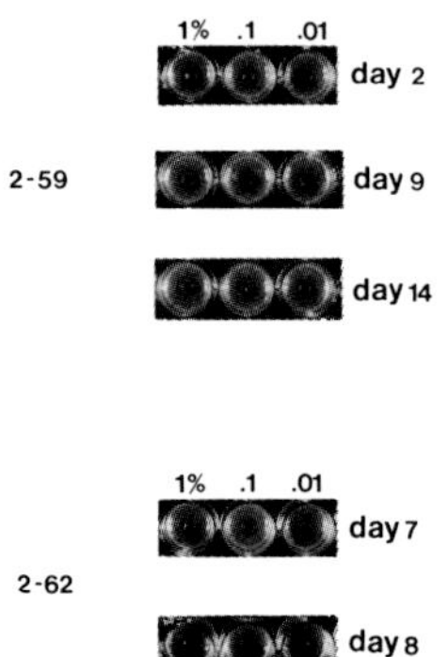

FIGURE 5. Expression of TPA in the milk of transgenic animals at different times of lactation. Samples were diluted such that the final concentrations were 1%, .1% or .01%, as indicated, and applied to a fibrin clot lysis plate as described earlier.

DISCUSSION

We demonstrate that a foreign protein, human tissue plasminogen activator, can be secreted into milk of transgenic mice under the control of a mammary-specific promoter. Moreover, expression appears to be targetted to the mammary gland (data not shown). These observations suggest that key regulatory elements targeting expression to the lactating mammary gland are located within 2600 bp of WAP gene upstream sequence.

Variation in TPA expression levels among transgenic mice containing WAP-TPA is considerable, suggesting that the chromosomal integration site may play a key role in establishing levels of expression from this construction. Analysis of additional animals may identify those which produce more TPA. In addition, intragenic and/or noncoding 5' and 3' sequences from the WAP gene, missing from the construction introduced into mice in these experiments, may play important roles in RNA stability. Considerable work remains to be done to configure the TPA expression vector for maximal expression.

A detailed knowledge of tissue and hormonal regulation of milk protein gene expression provides the basis for a biotechnology aimed at producing large amounts of complex proteins. The advantages of producing foreign proteins in this manner include the fact that milk is well characterized biochemically and that many of the genes encoding key milk proteins have been cloned. In addition, many milk-specific genes are expressed in the lactating mammary gland at high levels under hormonal control and in a tissue-specific manner. Thus, with expression vectors similar to the one described here, it should be possible to target precisely foreign gene expression to the lactating mammary epithelium. Production of foreign proteins in the milk of transgenic dairy animals appears to be an alternative and cost effective means for isolating human pharmaceuticals and other proteins in large quantities. Although many technical hurdles remain, the data presented here demonstrate that transgenic animals may become an attractive alternative for future production of genetically engineered, biologically active proteins.

REFERENCES

1. Gordon K, Ruddle FH (1986). Gene transfer into mouse embryos. In Gwatkin RBL, (ed): "Developmental Biology," New York, p 1.
2. Palmiter RD, Brinster RL (1986). Germ-line transformation of mice. Ann Rev Genet 20:465.
3. Gordon K, Lee E, Vitale JA, Smith AE, Westphal H, Hennighausen L (1987). Production of human tissue plasminogen activator in transgenic mouse milk. Biotechnology 5:1183
4. Hennighausen LG, Sippel AE (1982). Characterization and cloning of the mRNAs specific for the lactating mouse mammary gland. Eur JBC 125:131.
5. Lubon H, Hennighausen L (1987). Nuclear proteins from lactating mammary glands bind to the promoter of a milk protein gene. Nuc Acids Res 15:2103.
6. Reddy VB, Garramone AJ, Sasak H, Wei CM, Watkins P, Galli J, Hsuing N (1987). Expression of human uterine tissue- type plasminogen activator in mouse cells using BPV vectors. DNA 6:461.

Gene Transfer and Gene Therapy, pages 67–78

REGULATION OF RAT β-CASEIN GENE EXPRESSION IN TRANSGENIC MICE[1]

Kuo-Fen Lee, Francesco J. DeMayo, Suzanne Atiee and J. M. Rosen

Department of Cell Biology, Baylor College of Medicine
One Baylor Plaza, Houston, TX 77030

ABSTRACT The rat β-casein gene is a member of a small gene family, whose expression is developmentally and hormonally regulated. In order to understand the mechanisms governing rat β-casein gene expression, initially, lines of transgenic mice bearing a 14 kb genomic clone containing the entire β-casein gene along with 3.5 kb of 5' and 3.0 kb of 3' flanking DNA were established. The transgene was initiated at authentic transcriptional start sites and expressed in a tissue- and stage-specific fashion. However two intragenic MspI/HpaII sites of the transgene whose demethylation has been previously shown to be correlated with expression, were hypermethylated. In order to further define the minimal DNA sequences required for tissue- and stage-specific expression, transgenic mice carrying two fusion genes containing either 2.3 kb or 0.5 kb, respectively, of 5' flanking DNA along with noncoding exon I and 0.5 kb of intron A linked to the bacterial chloramphenicol acetyltransferase (CAT) gene were established. These studies revealed that a minimum of 0.5 kb of 5' flanking DNA along with exon I and 0.5 kb of intron A are capable of targeting CAT gene expression to the mammary gland.

---

[1]This work was supported by USDA grant 86-CRCR-1-2250 and NIH grant CA16303.

INTRODUCTION

Caseins are the predominant milk proteins encoded by a small gene family. Casein gene expression exhibits both tissue- and stage-specificity and is regulated by a variety of factors including peptide and steroid hormones, cell-cell and cell-substratum interactions (1,2). Rat β-casein mRNA increases 250-fold from virgin to day 8 of lactation at which time it is the predominant milk protein mRNA comprising approximately 20 % of the poly(A)$^+$ mRNA (3). Thus, β-casein should provide an excellent model to elucidate the molecular mechanisms by which milk protein gene expression is regulated. While functional definition of cis-acting DNA sequences important for either hormonal or mammary epithelial cell-specific expresion of the rat β-casein gene by DNA-mediated gene transfer into cell cultures has not yet been successful (4,5), the ability to introduce cloned or manipulated genes into mice via microinjection into fertilized eggs provide an alternative approach to studying the regulation of casein gene expression (6). Following germ line integration, the microinjected DNA may be assembled into the appropriate chromatin structure during development. A number of cellular genes have been introduced into mice, and most of them resembling their mouse counterparts have been expressed and regulated correctly (6). It has been demonstrated that cis-acting DNA sequences responsible for site- and stage-specific gene expression can reside in the 5' flanking region, introns, 3' flanking region or a combination of the above. Furthermore,in some cases, expression of the transgenes is correlated with the site-specific demethylation of the transgenes (7,8,9). Previously demethylation of two intragenic MspI/HpaII sites of the rat β-casein gene has been shown to be correlated with its expression (see Fig. 1 and Ref. 10). In order to test these possibilities, initially, a clone containing the entire rat β-casein gene with 3.5 kb of 5' and 3.0 kb of 3' flanking DNA sequences was used to generate lines of transgenic mice. The transgene was expressed in a tissue- and stage-specific manner (11).

To further define the minimal DNA sequences required for tissue- and stage-specific expression, two rat β-casein CAT fusion genes containing either 2.3 kb or 0.5 kb of 5' flanking DNA along with noncoding exon I and 0.5 kb of intron A of the rat β-casein gene were constructed and microinjected into mice. Interestingly both constructs contain a putative mammary consensus sequence between -110 and -140 upstream of the site of transcription initiation previously identified by DNA sequence comparison of a number of milk protein genes

(12,13). This region may be involved in a tissue-specific interaction with mammary nuclear proteins (14). In contrast to the previous inability to show mammary epithelial specific expression of the rat β-casein CAT fusion genes in cell transfection experiments, CAT activity was detected predominantly in the mammary gland of transgenic mice carrying either construct.

RESULTS

Tissue- and Stage-specific Expression of the Rat β-Casein Gene in Transgenic Mice

A 14 kb genomic clone containing the entire 7.5 kb of rat β-casein and 3.5 kb of 5' and 3.0 kb of 3' flanking DNA sequences was isolated and characterized from a Charon 35 rat genomic library as illustrated in Figure 1. This 14 kb fragment was excised in its entirety from the vector with SmaI and microinjected to generate lines of transgenic mice. Eight independent lines of transgenic mice have been established (11).

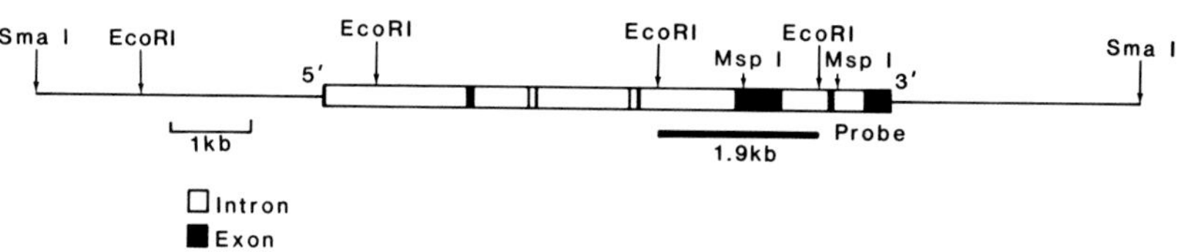

Figure 1. Structure of the rat β-casein gene used to generate transgenic mice as adapted from Lee et al. (11). The 1.9 kb of EcoRI fragment was used to determine the methylation status of the transgene (Fig. 4).

Both mouse and rat β-casein mRNA share more than 80% sequence homology, are of the same length and have a high A and T content. Therefore, in order to determine if the 14 kb clone contains sufficient information to direct tissue-specific expression of the rat β-casein gene, a specific RNase protection assay was developed to distinguish the mouse and the rat β-casein mRNA as illustrated in Figure 2C (11). A 650 NT probe was used which generated a 450 NT protected fragment of the 3' end of rat β-casein mRNA (Fig. 2A and B, lane 2). In contrast, smaller fragments were

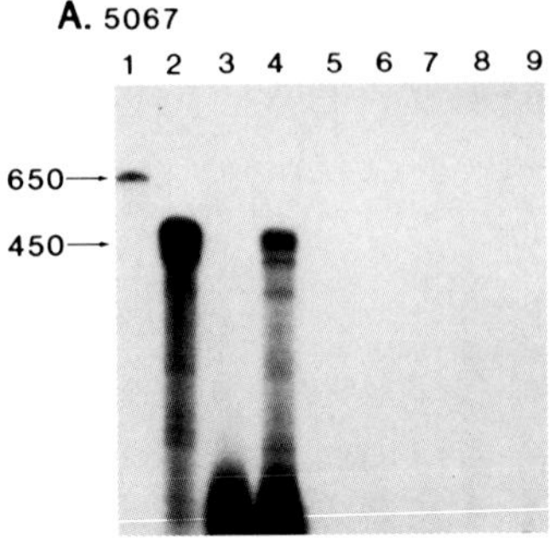

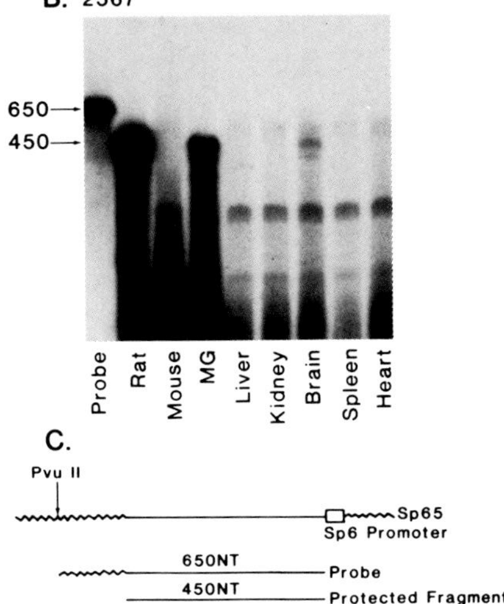

Figure 2. Tissue-specific expression of the rat β-casein gene in transgenic mice as adapted from Lee *et al.* (11). In both panel A and B, lane 1, probe; lane 2, RNA isolated from lactating mammary gland of rats; lane 3, RNA isolated from lactating mammary gland of normal mice; lane 4, RNA isolated from mammary gland (MG) of lactating transgenic mice; lane 5-9, RNA from various tissues as indicated. Note that the exposure times were 12 h and 60 h for A and B, respectively. Mouse 5067 and 2567 are female F1 of founder mouse 1290 and 1287, respectively. Panel C, the RNase protection assay as described (11).

---

protected with mouse β-casein mRNA (Fig. 2A and B, lane 3). Total RNA was isolated from various tissues from F1 females of two lines of transgenic mice, 1290 and 1287, at 7 days of lactation, and subjected to the RNase protection assay. The transgene was expressed predominantly in the mammary gland (Fig. 2A and B, lane 4). The 5-fold longer autoradiographic exposure shown in Figure 2B indicates that the transgene is also expressed in the brain (lane 6) but at a much lower level. Such expression was not observed in mouse line 1290 . However, even in the highest expressing line, the level of transgene expression is only 1% of that of the endogenous mouse β-casein gene (11).

In order to determine if the 14 kb clone also contains sufficient information to exhibit developmental regulation, RNA was isolated from the mammary gland of the offspring of mouse at various stages of development from virgin, mid-pregnancy, late pregnancy to 7 days of lactation and subjected to the same RNase protection assay (Fig. 3, lane D-G, respectively). The level of transgene expression is increased 10-fold from mid-pregnancy to lactation. These data demonstrate that the 14 kb clone contains sufficient information to direct tissue- and stage-specific expression of the rat β-casein gene in the mammary gland of transgenic mice.

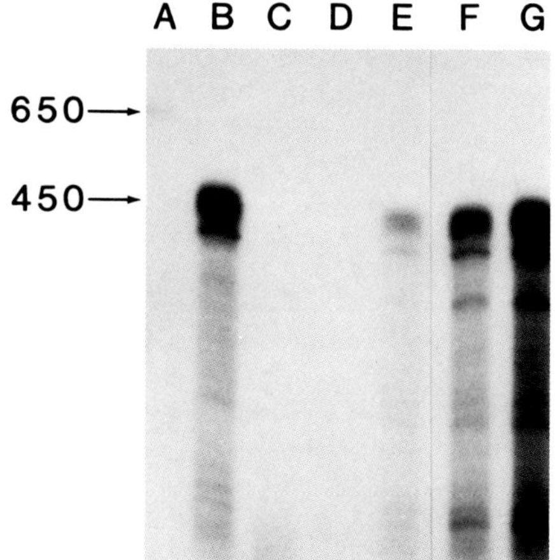

Figure 3. Developmental expression of the rat β-casein gene in transgenic mice. Lane A, probe; lane B-F, RNA isolated from mammary gland of lactating rats (B), lactating normal mice, and transgenic mice at various stages of development from virgin (D), mid-pregnancy (E), late-pregnancy (F) to 7 days of lactation (G), respectively. The data are taken from Lee _et al._ (11).

Methylation Pattern of the Rat β-Casein Gene in Transgenic Mice

It has been demonstrated that the rat β-, γ-, and κ-casein genes are hypomethylated in lactating mammary gland while hypermethylated in virgin mammary gland or liver (non-expressing tissue) (10,15). In particular, demethylation of two internal MspI/HpaII sites (Fig. 1) of the rat β-casein gene was found to be correlated with its level of expression (10). Therefore, lines of mice bearing the 14 kb transgene containing these two sites provide a useful model to determine the relationship between site-specific demethylation and expression of the rat β-casein gene. Restriction enzymes MspI and HpaII are isoschizomers and HpaII is sensitive to methylation of cytidine residue of recognition sequences (CCGG). A 1.9 kb EcoRI fragment was used to probe the methylation status of these two sites (Fig. 1). The presence of the 1.1 kb band in Hpa II digestions indicates demethylation of these two sites otherwise high molecular weight bands are observed (see below). DNA was isolated from both liver (L) and mammary gland (MG) of both expressing and non-expressing mice at 7 days of lactation (Fig. 4, mice 1965 and 5067, respectively), digested with either MspI (odd lanes) or HpaII (even lanes), and subjected to Southern analysis using a 1.9 kb EcoRI fragment as the probe. Following MspI digestion a 1.1 kb band was detected in all samples. Among the higher molecular weight bands, the 10 kb band observed in all samples is the endogenous mouse β-casein gene which cross-reacts with the 1.9 kb probe. The 8.5 kb fragment may result from recognition and cleavage of the regenerated SmaI site by MspI. The regneration of SmaI sites suggests that the transgene integrated into mouse chromosomal DNA in a head-to-tail fashion. The 13 kb and 15 kb bands may represent the junction fragments of the transgene and mouse DNA. When digested with HpaII, no 1.1 kb fragment is detected in any of the DNA samples from transgenic mice. Instead higher molecular weight bands (mostly larger than 23 kb) were observed. These results indicate that regardless of the level of transgene expression, these two sites were highly methylated.

Expression of the Rat β-Casein CAT Fusion Genes in Transgenic Mice

In order to define further the minimal DNA sequences required for tissue- and stage-specific expression, rat β-

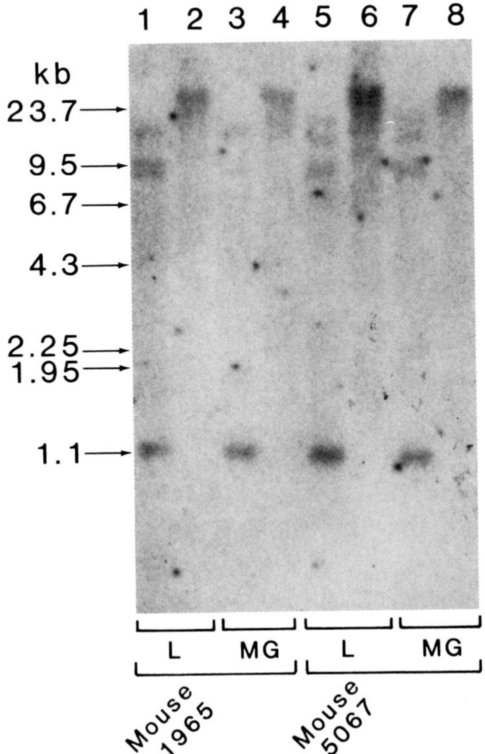

Figure 4. Methylation status of the transgene in transgenic mice. A Southern analysis of DNA samples digested with either MspI (odd lanes) or HpaII (even lanes) using 1.9 kb EcoRI fragment as probe (Fig. 1). DNA was isolated from both liver (L) and mammary gland (MG) of non-expressing (1965) and expressing (5067) mice.

---

casein CAT fusion genes were constructed essentially as described (4). The fusion genes contain either 2.3 kb or 0.5 kb of 5' of intron A of the rat β-casein gene as designated -2300/+490 and -524/+490, respectively. It has been demonstrated in previous transfection experiments that casein-CAT constructs containing the first noncoding exon exhibited greater CAT activity than constructs containing 5' flanking DNA alone and this effect was somewhat cell-specific (4). Twenty and eight lines of transgenic mice, respectively, containing either construct were established. For female founder mice, mammary gland biopsies were perfomed during lactation to facilitate the screening of transgene expression. For male founder mice, positive female F1 were obtained by breeding and subjected to the same biospy procedure. The fourth gland of transgenic mice was surgically removed at 7-10 days of lactation under anesthesia. Mammary tissue-extracts were prepared and the CAT enzymatic assays were performed as illustrated in Figure 5. In Fig. 5A, CAT activity was detected in the lactating gland of three out of five mice containing -2300/+490 CAT transgene (lane B-C, F). In Fig. 5B, CAT activity was detected in the lactating gland of two of three mice containing -524/+490 CAT transgene (lane B and D). These results indicate that a minimum of 0.5 kb of 5' flanking DNA along with noncoding exon I and 0.5 kb of intron A are sufficient for targeting CAT gene expression to the mammary gland. Additional studies have revealed that CAT activity is predominantly detected in the mammary gland of lactating mice (Lee, Atiee and Rosen, in preparation).

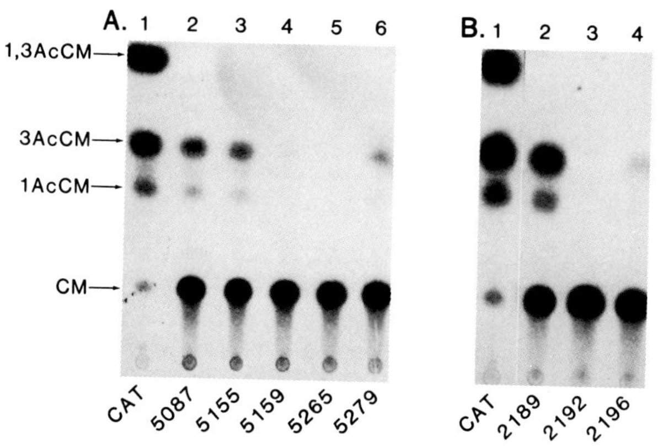

Figure 5. Expression of the rat β-casein CAT fusion genes in lactating mammary gland. Tissue extracts were prepared from biopsied lactating mammary gland from F_0 mice containing either -2300/+490 (panel A) or -524/+490 (panel B) CAT fusion gene and subjected to CAT assays as described (20). Radiolabeled chloramphenicol (CM) and its acetylated derivatives (1AcCM, 3AcCM and 1,3AcCM) are chromatographically separated. A CAT enzyme was used as a positive control (lane 1 in both panel A and B).

DISCUSSION

In contrast to the previous inability to show mammary-specific and hormone-regulated expression of the rat β-casein gene when transfected into cell cultures, transgenic mice carrying the 14 kb construct express the transgene in a tissue- and stage-specific manner. This may possibly be attributed to the requirement of the transgene to be processed through the germ line to acquire the proper chromatin structure throughout development which can be modulated both positively and negatively by trans-acting factors in response to physiological signals. The level of expression of the rat β-casein gene is, however, only 0.01-1% of that of the endogenous gene. These results are consistent with data obtained from other transgenes where the level of transgene expression is usually position-dependent and suggest that other regulatory sequences further upstream/downstream of the rat β-casein gene required for a high level of expression are not present in the 14 kb construct. The need for additional sequences may also be reflected from the methylation results (see below). Recently Grosveld et al. (16) have demonstrated that sequences located 50 kb 5' and 20 kb 3' of the human β-globin gene characterized by the presence of DNase I- hypersensitive sites are able to elicit high-level and position-independent expression of the β-globin gene in the erythroid cells of transgenic mice. It is conceivable that the chromatin structure appropriate for casein gene expression has been achieved only partially with the 14 kb contruct. The casein genes, like the globin genes, are part of a gene family containing as many as five members spread over at least 100 kb (13). Analogous to the β-globin gene, dominant regulatory sequences, therefore, may be quite far from the β-casein gene, which is thought to be in the middle of the casein gene cluster. In order to test this possibility a cosmid library is needed for identifying far upstream and downstream sequences by screening and chromosome walking. Once these sequences are isolated and mapped, DNase I hypersensitive sites, if any are present, can be located. A β-casein gene "minilocus" can then be constructed and microinjected into mice to test for efficient position-independent expression.

The methylation status of the β-casein transgene was also examined since it was previously found that demethylation of two internal MspI/HapII sites was correlated with the expression of the rat β-casein gene (10). In some cases, expression of other transgenes has also been correlated with their methylation status (7,8,9). Regardless of the level of expression, these two sites within the β-casein gene were highly methylated in transgenic mice, suggesting that demethylation at these two sites is not obligatory for the expression of the rat β-casein gene. Alternatively, this observation may partially account for or reflect the low level of transgene expression as compared to the endogenous gene. It has been suggested and partially demonstrated in transgenic mice that the information necessary for methylation/demethylation is an intrinsic property of the injected gene (17,18), perhaps in conjunction with a specific set of trans-acting factors. Therefore, the most likely explanation for the failure of demethylation of these two sites is the absence of these additional necessary sequences, subjecting the transgene to position effects and a reduced level of expression. It is not possible to determine definitely the role of methylation in the expression of the rat β-casein gene until lines of high-expressing mice are generated to investigate if there is then a direct correlation with the demethylation of these two sites.

In order to further define the minimal sequences required for tissue- and stage-specific expression, rat β-casein-CAT fusion genes also were constructed and used to generate lines of transgenic mice. We have shown that a minimum of 0.5 kb of 5' flanking DNA along with noncoding exon I and 0.5 kb of intron A are capable of directing expression of the CAT gene to the mammary gland. It has been demonstrated that casein gene expression is regulated both at the transcriptional and posttranscriptional levels in response to steroid and peptide hormones (19). By contrasting results obtained from mammary explant cultures derived from mice containing the whole casein gene with those containing the casein-CAT fusion genes, it should now be possible to define the boundary of cis-acting DNA sequences reponsible for regulation by individual hormones.

ACKNOWLEGEMENTS

We thank Sally Lee and Monica Phelps for their help in preparation of this manuscript.

REFERENCES

1. Lee EY-H, Lee W-H., Kaetzel CS, Parry G, Bissell MJ (1985). Interaction of mouse mammary epithelial cells with collagen substrata: regulation of casein gene expression and secretion. Proc Natl Acad Sci 82:1419.
2. Wiens D, Park CS, Stockdale I (1987). Milk protein expression and ductal morphogenesis in the mammary gland in vitro: hormone-dependent and -independent phases of adipocyte-mammary epithelial cell interaction. Develop Biol 120:245.
3. Hobbs AA, Richards DA, Kessler DJ, Rosen JM (1982). Complex hormonal regulation of rat casein gene expression. J Biol Chem 257:3598.
4. Bisbee CA, Rosen JM (1986). DNA sequence elements regulating casein gene expression. In Granner DK, Rosenfeld G and Chang S (eds.), Transcriptional Control Mechanisms, UCLA Symposium on Molecular and Cellular Biology, Vol. 52. Alan R. Liss, New York, pp. 312.
5. David-Inouye Y, Couch CH, Rosen JM (1986). The isolation and transfection of the entire rat β-casein gene. Annal N Y Acad Sci 478:274.
6. Palmiter RD, Brinster RL (1986). Germ-line transformation of mice. Ann Rev Genet 20:465.
7. Palmiter RD, Chen HY, Brinster RL (1982). Differential regulation of metallothionein-thymidine kinase fusion genes in transgenic mice and their offspring. Cell 29, 701.
8. Swain JL, Stewart TA, Leder P (1987). Parental legacy determines methylation and expression of an autosomal transgene: a molecular mechanism for parental imprinting. Cell 50:719.
9. Hadchouel M, Farza H, Simon D, Tiollais P, Pourcel C (1987). Maternal inhibition of hepatitis B surface antigen gene expression in transgenic mice correlates with de novo methylation. Nature 329:454.
10. Johnson ML, Levy J, Supowit SC, Yu-Lee L-Y, Rosen JM (1983). Tissue-and cell-specific casein gene expression. J Biol Chem 258:10805.
11. Lee KF, DeMayo FJ, Atiee SH, Rosen JM (1988). Tissue-specific expression of the rat β-casein gene in transgenic mice. Nucl Acids Res 16:1027.
12. Hall L, Emery DC, Davies MS, Parker D, Craig RK (1987). Organization and sequence of the human α-lactalbumin gene. Biochem J 242:735.

13. Yu-Lee L-Y, Richter-Mann L., Couch CH, Stewart AF, Mackinlay AG, Rosen JM (1986). Evolution of the casein multigene family: conseved sequences in the 5' flanking and exon regions. Nucl Acids Res 14:1883.
14. Lubon H, Henninghausen L (1987). Nuclear proteins from lactating mammary glands bind to the promoter of a milk protein gene. Nucl Acids Res 15:2103.
15. Thompson MD, Nakhasi HL (1985). Methylation and expression of rat K-casein gene in normal and neoplastic rat mammary gland. Cancer Res 45:1291.
16. Grosveld F, van Assendelft GB, Greaves D, Kollias G (1987). Position-independent, high level expression of the human β-globin gene in transgenic mice. Cell 51:975.
17. Kolsto AB, Kollias G, Giguere V, Isobe KI, Prydz H, Grosveld F (1986). The maintenance of methylation-free islands in transgenic mice. Nucl Acids Res 14:9667.
18. Jahner D, Jaenisch R (1985). Chromosomal position and specific demethylation in enhancer sequences of germ line-transmitted retroviral genomes during mouse development. Mol Cell Biol 5:2712.
19. Guyette WA, Matusik RJ, Rosen JM (1979). Prolactin-mediated transcriptional and post-transcriptional control of casein gene expression. Cell 17:1013.
20. Gorman CM, Moffatt L, Howard BH (1982). Recombinant genomes which express chloramphenicol acetyltransferase in mammalian cells. Mol Cell Biol 2:1044.

Gene Transfer and Gene Therapy, pages 79–88

Retroviral mediated transfer of growth factor genes into hemopoietic cells: A model for leukemogenesis[5].

Karl Klingler[1,3], Christine Laker[1,4], Carol Stocking[1], Ursula Just[1], Norbert Kluge[1], Wolfram Ostertag[1], Elaine Spooncer[2], Makoto Katsumo[2], and Michael Dexter

[1] Heinrich-Pette-Institut an der Universität Hamburg, 2000 Hamburg 20, FRG. [2] Paterson Institute for Cancer Research, Christie Hospital, Withington, Manchester, M20 9BX, UK.

ABSTRACT

Retroviral mediated gene transfer into hemopoietic cells has long been hampered by the low efficiency using Mo-MuLV based vectors. Vectors based on the myeloproliferative sarcoma virus (MPSV) transfer neomycin resistance to fibroblasts and to hemopoietic precursor cell lines with equal efficiency and with a 10^{-4} lower efficiency to hemopoietic stem cells. Transfer of hemopoietic growth factor genes into factor dependent hemopoietic stem and precursor cell lines leads to autocrine stimulated cells . Subsequently, most of these cells reach autonomy, becoming independent from their own growth factor. Hemopoietic stem cell lines (FDC-Pmix) infected with a vector carrying the IL-3 (Multi-CSF) gene are still capable of undergoing normal differentiation and transplantation of these cells into syngeneic mice results in a marked increase in spleen weight and circulating leukocytes, similar to chromic myeloid leukemia. Tumourigenicity of factor-dependent (IL-3) myeloid

---

[3] Present address Dr. Karl Klingler, Walter and Eliza Hall Institute of Medical Research, Victoria, 3050, Australia.[4] C. Laker is a fellow of the Boehringer Ingelheim Fonds, Stuttgart.[5] This work was supported by the Cancer Research Campaign UK and the Deutsche Forschungsgemeinschaft (Os 31/12-3)

precursor cells (FDC-P2) infected with a vector carrying the IL-3 gene increased strikingly when cells shifted from the autocrine to the autonomous state. Nevertheless cell density dependent clonability of tumour cells in vitro demonstrated that autonomy was not a prerequisite for tumour formation.

INTRODUCTION

Hemopoietic growth factors play a central role in the hemopoietic system, since they are required for the survival and proliferation as well as for the differentiation and functional activation of hemopoietic cells (reviewed: 1). Although these proteins are essential for the normal hemopoietic system, activation of growth factor genes in cells which are dependent on these factors seems to be important for their malignant transformation (review: 2). This is underlined by the fact that several tumour cell lines are known which produce growth factors (3-5). In order to elucidate the role of growth factor genes in the process of malignant transformation, we decided to introduce normal hemopoietic growth factor genes into factor dependent hemopoietic cells.

In general, retroviral vectors are an efficient tool for transferring genes into cells, but their use in the hemopoietic system has been hampered by the low transfer efficiency of Mo-MuLV-based vectors for these cells (6). The use of the myeloproliferative sarcoma virus (MPSV) which has a wider host range than any of the other Mo-MuLV derived vectors (7) as a basis for the construction of retroviral vectors has led to two vector types, M3-neo^R and M5-neo^R, which allow equal transfer efficiency into fibroblasts and hemopoietic precursor cells (8).

We had reported previously that by using these vectors the introduction of the granulocyte-macrophage colony stimulating factor (GM-CSF) gene into the GM-CSF dependent mouse cell line, FDC-P1, leads initially to growth factor producing (autocrine) cells which are independent of externally supplied GM-CSF. By a yet uncharacterised second step most of these cells subsequently became independent of GM-CSF (autonomous) (8).

Here we report the effects of further vectors based on the M3-neo^R and M5-neo^R constructs, carrying several growth factor genes of mouse and human origin.

MATERIAL AND METHODS

Vector construction, virus producing cell clones and viral infection

All recombinant viruses were constructed using standard techniques (9). Transfection of recombinant DNA into the packaging defective helper cell lines Ψ2 and PA-317 (10,11) was performed as described previously (8). Infection of hemopoietic cell lines and determination of transfer efficiency were done according to previously described procedures (8). The Z-GMV producing Ψ2 clone was kindly provided by T. Gonda.

Cell lines

Factor dependent myeloid precursor cell lines FDC-P1 and FDC-P2 (12) were grown in Iscove's modified Dulbecco's medium (IMDM) supplemented with 10% fetal calf serum (FCS) and 5% WEHI-3BD$^-$ conditioned medium containing IL-3. The hemopoietic stem cell line, FDC-Pmix, was maintained in IMDM supplemented with 20% horse serum and 5% WEHI-3BD$^-$ conditioned medium. The human hemopoietic precursor cell line HEL (13), was grown in IMDM containing 10% FCS. The human adenosarcoma cell line EFO and the virus producing cell lines Ψ2, PA-317, and RAT 1 were maintained in minimum essential medium (MEM), containing 10% FCS.

GM-CSF activity

Conditioned media were produced by incubation of factor producing cells (10^6 cells/ml) medium was then concentrated 10-fold by Amicon filtration (exclusion M_r >10,000), dialysed and tested for GM-CSF activity on 10^4 FDC-P1 cells by [^{3}H] Thymidine incorporation. GM-CSF activity was defined as 50 units per ml, the amount of mouse lung conditioned medium (MLCM) giving 50% of maximum stimulation of 10^4 FDC-P1 cells.

Colony formation

FDC-Pmix cells (10^3) either uninfected, infected with M3-neo^R or infected with M3-MuV were plated in soft agar in culture conditions which allow the proliferation of different types of hemopoietic colonies. Individual colonies were isolated after 10 days, cytospin preparations were made and cells were stained with benzidine plus May-Grünwald Giemsa. At least 30 colonies were examined for each type.

Tumorigenicity assays

FDC-Pmix cells (5-10 x 10^6) were suspended in Fischers medium, cells were injected intravenously into syngeneic mice which had received 10 Gy prior to transplantation.

Infected FDC-P2 cells (10^5-10^7) were washed and

injected subcutaneously into 3 month old Balb/c nu/nu mice.

Clonability of tumour cells

Tumour cells were washed and placed in agar or methylcellulose (3 x 10^1-3 x 10^5 per dish) with or without WEHI-3BD$^-$ conditioned medium as a source of IL-3. Triplicate cultures were scored for colony formation after 9 days of incubation.

RESULTS

MPSV-based constructs

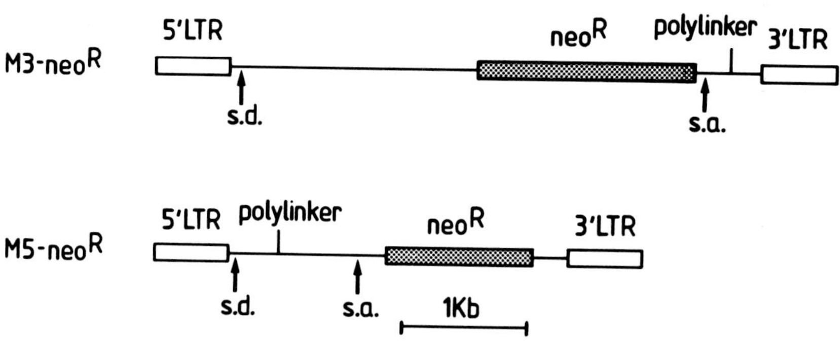

Figure 1

Moloney murine leukemia virus (Mo-MuLV) based constructs allow gene transfer into hemopoietic cells only at a very low efficiency (6). We therefore decided to use the myeloproliferative sarcoma virus (MPSV) as a basis for our vectors. MPSV has a wider host range than any of the other Mo-MuLV derived viruses. This wider host range is at least in part due to alterations within the U3-region of the LTR (14)

The M3-neo^R construct was produced by deleting the mos-oncogene and pol-sequences and insertion of the Tn5 drug resistance gene (neomycin resistance) which allows selection of infected cells. Polylinker sequences were inserted near the 3′-LTR (Fig 1). In these constructs, neo-resistance is expressed from a full length genomic mRNA, whereas genes which are inserted into the polylinker should be expressed from a spliced mRNA.

The M5-neo^R construct was obtained by additional

deletion of gag sequences, the insertion of polylinker sequences near the 5′-LTR and the insertion of the Tn5 gene 3′ to the splice acceptor (Fig 1). Expression of genes which are inserted into the polylinker occurs from a full length mRNA whereas neomycin-resistance should be expressed from a subgenomic mRNA.

The following growth factor cDNA′s were inserted into both vector types: the mouse cDNA′s for GM-CSF (with and without the poly (A) sequence), IL-3 and IL-4 as well as the human cDNA′s coding for M-CSF, GM-CSF and IL-3 (all without the poly (A) sequence).

Growth factor production of cells infected with different MPSV vectors

Infection of fibroblasts with supernatant from helper cell lines carrying either the M3 or the M5 construct containing the mouse GM-CSF gene leads to G418 resistant, growth factor producing cells. Growth factor production of infected fibroblasts was dependend upon the type of construct used for the infection. Cells infected with M5-based constructs produced up to five times more growth-factor compared to mouse lung cells, whereas cells infected with M3 based constructs produced 100-fold less growth factor than M5 infected cells.

Efficiency of G418-resistance transfer of MPSV based vectors into different cell types

To test the transfer efficiency of MPSV based vectors, different cell types of mouse and human origin were infected, cells were then selected in G418 concentrations necessary to kill uninfected cells: (fibroblasts were selected at 400 µg/ml, FD-cells at 1 mg/ml and human cells were selected at 1,6 mg/ml). Table 1 shows typical transfer efficiencies of G418 resistance by MPSV based vectors into different cell types compared to NIH3T3 fibroblasts. Transfer efficiency to the hemopoietic precursor cell line FDC-P1 was always comparable to mouse fibroblasts. Infection of other hemopoietic precursor cell lines (FDC-P2, HEL) was less efficient and transfer efficiency into hemopoietic stem cell lines was 10^{-4} compared to fibroblasts. These data show that there is no species specific difference in the efficiency of MPSV mediated gene transfer.

Table 1

Transfer efficiency of MPSV based bectors

Cell line	Cell type	Transfer efficiency compared to NIH3T3*
NIH3T3	mouse, fibroblast	1.0
RAT 1	rat, fibroblast	0.3
FDC-P1	mouse, myeloid precursor	0.6
FDC-P2	mouse, myeloid precursor	$3.7x10^{-2}$
FDC-Pmix	mouse, hemopoietic stem cell	$1.4x10^{-4}$
EFO	human, adenosarcoma	0.:2
HEL	human, myeloid precursor	$1.0x10^{-3}$

* Transfer efficiency was determined as the number of genetecin (G418) resistance transferring units (GTU) per ml of conditioned medium.

Tumourigenicity of M3-MuV infected hemopoietic stem cells

Infection of continuously growing but non-leukemic multipotent hemopoietic cell lines (FDC-Pmix) (15), with the M3-vector carrying the mouse IL-3 gene (M3-MuV) gene gave rise to G418 resistant cells which grew in the absence of IL-3. These cells showed a density dependent growth pattern which indicated that these cells were dependent on their own secreted growth factor. Conditioned medium from these cells contained small amounts (less than 50 units/ml) of IL-3 (data not shown). Colony formation by infected FDC-Pmix cells was similar to that of uninfected FDC-Pmix cells with the exception of a reduced number of mixed/erythroid colonies.

Transplantation of FDC-Pmix cells carrying the M3-MuV vector into irradiated syngeneic mice caused a 5 to 10 fold increase in the spleen weight and a dramatic increase in circulating leucocytes in conjunction with a reduced hematocrit. Morphological analyses of spleen and blood cells from infected animals showed a pattern similar to chronic myeloid leukemia (CML): spleen (blood) 16 (2)% blasts, 21 (10)% promyelocytes/myelocytes, 55 (75)% metamyelocytes and polymorphonuclear granulocytes, 5 (1)% nucleated red cells, 2 (12)% other cells.

Karyotypic analysis and plating of cells from the leukemic spleen *in vitro* in the presence of G418 and in the absence of IL-3 confirmed that 80%-100% of the proliferating cells were of donor origin. We therefore draw the conclusion that FDC-Pmix cells infected with M3-MuV produce CML.

Tumourigenicity of FDC-P2 cells infected with M3-GMV

Similar to previously reported observations regarding the infection of the GM-CSF dependent myeloid precursor cell line FDC-P1 with the M3 vector carrying the gene for GM-CSF (8), the IL-3 dependent myeloid precursor cell line FDC-P2 infected with M3-MuV gave rise to G418 resistant cells which grew in the absence of externally added IL-3. Immediately after infection these cells were dependent upon their own secreted IL-3 and their proliferation could be blocked by using anti-IL-3 antibodies. Within a few weeks most of these cells shifted to autonomy showing unaffected proliferation in the presence of anti-IL-3 antibodies (data not shown).

Subcutaneous injection of M3-MuV infected FDC-P2 cells either autocrine or already autonomous, into nude mice, gave rise to tumours. Mice injected with autonomous cells showed tumour formation after 4 weeks with an average tumour weight of 0.5g., 6 weeks after injection average tumour weight was more than 2.5g. In contrast to these results, mice infected with autocrine cells showed tumours with an average weight of 0.1g 4 and 6 weeks after injection. These data suggest that transformation and autonomous growth are coupled.

Clonability of tumour cells

Tumor formation 4 to 6 weeks after injection of autocrine FDC-P2 cells might arise directly from these cells or may be caused by a subset of autonomous cells which emerge during the latent period in the mice.

In order to elucidate whether autonomous growth is a strict prerequisite of transformation, tumour cells were recovered and replated. Most tumour cells showed density independent growth in the absence of exogenously added IL-3, however some tumor cells were still factor dependent (Fig 2). This suggests that although tumour formation is much more efficient if autonomous cells are used, autonomous growth is not a prerequisite for the tumourigenicity of FDC-P2 cells infected with M3-MuV.

DISCUSSION

The use of retroviral vectors for introducing genes into hemopoietic cells has long been hampered by the low efficiency of available vectors. Here we report that MPSV based vectors allow gene transfer into some hemopoietic

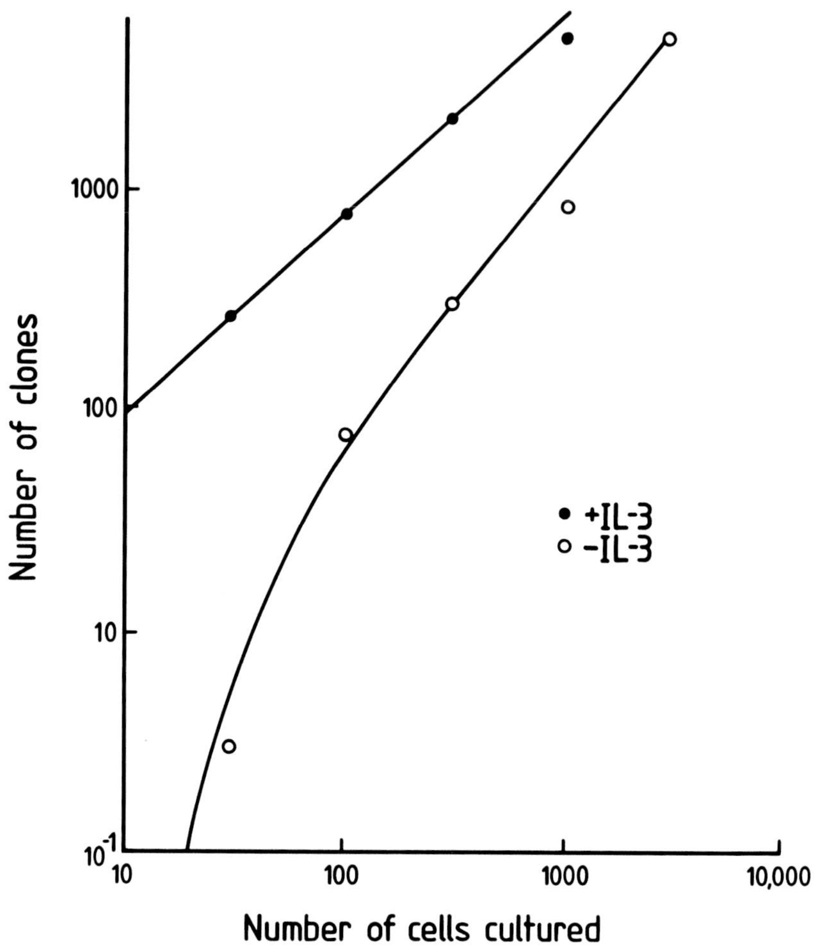

Figure 2: Clonability of tumor cells.

precursor cell lines and into fibroblasts at equal efficiency. Although gene transfer into hemopoietic stem cell lines is less efficient (10^{-4} compared to fibroblasts) this method is still applicable for these cells since helper cell lines transfected with MPSV based vectors produce up to 10^8 GTU.

We have previously reported that infection of GM-CSF dependent FDC-P1 cells with M3-GMV leads to G418 resistant cells which grow in the absence of externally added GM-CSF but which are dependent on their own produced and secreted GM-CSF. Most of these cells subsequently acquire autonomy from autocrine produced GM-CSF if cells are kept in culture for a few weeks. Infection of IL-3 dependent FDC-P2 cells with M3-MuV follows the same pattern and finally results in autonomous FDC-P2 cells. Injection of either autocrine or autonomous FDC-P2 cells caused tumour formation in nude mice. Tumour formation was much more efficient if cells were autonomous. Whether tumour formation after injection of autocrine FDC-P2 cells is due to the emergence of a subset of autonomous cells during the latent period in the mice is not clear. However, replating of tumour cells showed that not all tumour cells are independent (autonomous) of growth-factor. This suggests that autonomy is not necessarily a prerequisite for tumourigenicity.

Transplantation of hemopoietic stem cells (FDC-Pmix) infected with M3-MuV showed that these cells can cause a pattern very similar to chronic myeloid leukemia, with cells still capable of undergoing differentiation similar to their uninfected counterparts except for producing a decreased proportion of mixed/erythroid cells.

These data show that the efficient transfer of growth factor genes into hemopoietic precursor and stem cell lines offers the possibility to study the multistep process of tumourigenicity and the role activated growth factor genes play in leukemogenesis.

ACKNOWLEDGMENTS

We thank F. Hölzel for kindly providing the cell line EFO and G. Beck-Engeser for expert technical assistance.

REFERENCES

1. Metcalf D (1984) "The hemopoietic colony stimulating factors". Amsterdam: Elsevier.
2. Weinberg RA (1985) The action of oncogenes in the cytoplasm and nucleus. Science 230:770.
3. Chen SJ, Holbrook NJ, Mitchell KF, Vallone CA, Greengard JS, Crabtree GR, Lin Y (1985) A viral long terminal repeat in the interleukin 2 gene of a cell line that constitutively produces interleukin 2. Proc. Natl. Acad. Sci. 82:7284.
4. Duprez V, Lenoir G, Dautry-Varsat A (1985) Autocrine

growth stimulation of a human T-cell lymphoma line by interleukin 2. Proc. Natl. Acad. Sci. 82:6932.

5. Ymer S, Tucker WQJ, Sanderson CJ, Hapel AJ, Campbell HD, Young JG (1985) Constitutive synthesis of interleukin 3 by leukemia cell line WEHI-3B is due to retroviral insertion near the gene. Nature 317:255.
6. Lang RA, Metcalf D, Gough NM, Dunn AR, Gonda T (1985) Expression of a hemopoietic growth factor cDNA in a factor-dependent cell line results in autonomous growth and tumorigenicity. Cell 43:531.
7. Ostertag W, Stocking C, Johnson G, Franz T, Kluge N, Kollek R, Hess N (1987) Transforming genes and target cells of murine spleen focus forming viruses. Adv. Cancer Res. 48:193.
8. Laker C, Stocking C, Bergholz U, Hess N, DeLamarter JF, Ostertag W (1987) Autocrine stimulation after transfer of the granulocyte/macrophage colony-stimulating factor gene and autonomous growth are distinct but interdependent steps in the oncogenic pathway. Proc. Natl. Acad. Sci. U.S.A. 84:8452.
9. Stocking C, Kollek R, Bergholz U, Ostertag W (1985) Long terminal repeat sequences impact hematopoietic transformation properties to the myeloproliferative sarcoma virus. Proc. Natl. Acad. Sci. USA 82:5746.
10. Mann RC, Mulligan RC, Baltimore D (1983) Construction of a retrovirus packaging mutant and its use to produce helper free defective retroviruses. Cell 33:153.
11. Miller AD, Baltimore C (1986) Redesign of retrovirus packaging cell lines to avoid recombination leading to helper virus production. Mol Cell. Biol. 6:2895.
12. Dexter TM, Garland J, Scott D, Scolnick E, Metcalf D (1980) Growth of factor-dependent hemopoietic precursor cell lines. J. Exp. Med. 152:1036.
13. Martin P, Papayannopoulou Th (1982) HEL cells: A new human erythroleukemia cell line with spontaneous and induced globin expression. Science 216:1233.
14. Stocking C, Kollek R, Bergholz U, Ostertag W (1986) Point mutations in the U3 region of the long terminal repeat of Moloney murine leukemia virus determine disease specificity of the myeloproliferative sarcoma virus. Virology 153:145.
15. Spooncer E, Heyworth CM, Dunn A, Dexter TM (1986) Self renewal and differentiation of interleukin-3 dependent multipotent stem cells are modulated by stromal cells serum factors. Differentiation 31:111.

Gene Transfer and Gene Therapy, pages 89–101

RETROVIRAL TRANSFER OF THE HUMAN IL-2 RECEPTOR GENE INTO MURINE FIBROBLASTS AND T-LYMPHOID CELLS[1]

Gracia Kruppa[2], K. Pfizenmaier, and Barbara Seliger

Clincial Research Group BRWTI, Max-Planck-Society, University of Göttingen, 3400 Göttingen, FR Germany

We have compared the suitability of various retroviral vectors for gene transfer into murine fibroblasts and T-lymphoid cells, using the human IL-2 receptor (IL-2R) gene as a model. In the vectors employed, the Moloney leukemia virus(MoLV)-derived LTR promoted the selectable neomycin resistance gene. Expression of the IL-2 receptor gene was either controlled by the same retroviral promoter (double expression (DE) vector) or by an additional internal promoter (VIP) derived from thymidine kinase (Tk) and simian virus 40 (SV 40), respectively. Gene transfer and stable expression was obtained not only in fibroblasts, but also in a T-cell line. However, cells infected with distinct IL-2R vectors differed in IL-2R mRNA levels as well as in surface expression of the Tac antigen. The highest level of Tac antigen was observed, when the expression of IL-2R gene was directed by the internal SV 40 promoter, followed by the Tk promoter and the retroviral LTR, which was active only in fibroblasts. These data show that VIP vectors are superior to DE vectors with respect to both gene transfer efficiency and high expression of the gene product in lymphoid cells.

INTRODUCTION

Retroviral vectors are important tools for transfer and stable integration of heterologous genes in a variety of mammalian cells (1-5), including cells of the hematopoietic

---

1) This work was supported by Stiftung Volkswagenwerk.
2) in partial fulfilment of her PhD-thesis

system (6-9). However, the general application of these vectors is often limited due to low and variable virus titer and low efficiency of expression of the transduced gene in the infected cell (6,7,10). To date, most vectors used for gene transfer into cells of the hematopoietic lineage have been designed to promote the expression of the inserted DNA sequences from the transcriptional signal within the viral long terminal repeat (LTR) (6-9). In the case of low expression of the transduced genes tissue-specific factors and negative regulatory elements within the LTR might be responsible (11-13). To circumvent the problems of restricted tissue-specificity of heterologous gene expression controlled by retroviral LTRs, the use of vectors containing different transcriptional enhancer or promoter regions may be advantageous (14).

In order to investigate the suitability of vectors with internal promoters (VIP) for gene transfer into cells of hematopoietic lineage, we compared retroviral vectors carrying either the thymidine kinase (Tk) or the simian virus 40 (SV 40) promoter with a typical double expression vector, in which both a selectable marker (neo^R) and the gene of interest are controlled by the same retroviral promoter (MoLV-LTR). The human IL-2 receptor (Tac antigen) cDNA clone (15) was chosen as a reporter gene, because its expression is restricted to cells of hematopoietic origin, predominantly to lymphocytes and can be readily followed at both the RNA and the protein level.

MATERIAL AND METHODS

Retroviral vectors.

All vectors used for construction of recombinant IL-2R vectors were derived from the Moloney leukemia virus. The construction of N2, NTk, and NSV vectors (gift of T. von Rüden, Heidelberg) have been described elsewhere in detail (14,16,17).

The 2.335 base pair (bp) cDNA insert of pIL-2R3 (15, kindly provided by W. Greene, Durham, NC, USA) was inserted into the XhoI(N2), BglII(NTk), and HindIII(NSV) of the distinct vectors using standard techniques (18) (fig. 1). The recombinant plasmids containing the IL-2R cDNA in the desired orientation were identified by restriction enzyme analysis.

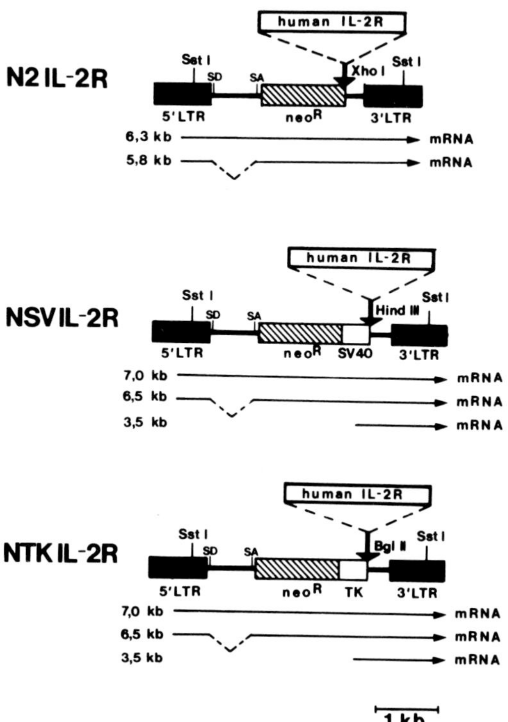

FIGURE 1. Schematic diagram of the N2 vector system used. The black boxes represent the LTR and the cross-hatched boxes the selectable neomycin-resistance (neo^R) gene. Tk: thymidine kinase promoter; SV: simian virus type 40 promoter.

Cell lines and viruses.

NIH3T3 fibroblasts and cell clones were maintained in MEM supplemented with 10 % FCS. IL-2R recombinant virus producer cells were generated from all three vectors by transfection using calcium phosphate precipitation (19). Neo^R colonies were isolated by selection in modified Eagle's medium (MEM) containing 400 µg/ml G418 (Gibco, BRL, Karlsruhe, FRG). To rescue neo^R IL-2R virus, Friend helper virus (20) was used for superinfection of NIH3T3 cells containing the provirus. One week later, cell-free culture supernatant from producer cells was assayed by limiting dilution analysis for infectious neo^R IL-2R virus titer on NIH3T3 cells; generation of neo^R colonies was monitored after 2 weeks, colonies were isolated and analyzed for IL-2R expression.

The T-lymphoid cell line BW5147 is a ouabain-resistant HPRT thymoma of the AKR mouse (21), which is maintained in Click's RPMI 1640 supplemented with 10 % FCS. BW5147 cells expressing the hIL-2R were generated by cocultivation of BW5147 with 5 x 10^4 irradiated (8000 R) neo^R IL-2R virus producing NIH3T3 cells in 96-well microculture plates as previously described (22). Culture medium was removed after one day and replaced by medium containing 2 mg/ml G418. Neo^R colonies were individually isolated after 2 weeks of culture in selection medium and analyzed for IL-2 receptor gene expression.

Northern blot analysis.

Total cellular RNA from 5 x 10^7 cells extracted according to Chirgwin et al. (23), size fractionated on 1 % denaturing formaldehyde-agarose gels and transferred to nitrocellulose filters as described (24). Blots were hybridized to cDNA probes labelled with ^{32}P by random priming (25), washed and exposed at -70°C to Kodak XAR film in the presence of intensifying screens.

Cytofluometric analysis of IL-2R expressing mouse cells.

Expression of IL-2R on transformed mouse cells was determined by indirect immunofluorescence analysis on Epics C flow cytometer (Coulter, Krefeld, FRG) with mouse anti-human IL-2R monoclonal antibody (anti-Tac) kindly provided by T. Waldmann (26,27), using FITC-conjugated goat-anti-mouse IgG antibodies (Medac, Hamburg, FRG) as a second reagent. The percentage of cells expressing Tac-antigens was calculated upon subtraction of the respective background fluorescence with isotype-matched control antibodies, using the Epics immuno program.

RESULTS

Efficient transfer and expression of the human IL-2 receptor gene in murine fibroblasts.

The various retroviral vectors N2 IL-2R, NTk IL-2R and NSV IL-2R (fig. 1) were transfected into NIH3T3 cells. $G418^R$ colonies, which developed with comparable efficiency for all three vectors, were screened for IL-2R expression by Tac-specific immunofluorescence. On the selected Tac-positive neo^R 3T3 clones, the level of IL-2R expression differed significantly. As shown in table 1, NSV IL-2R-transformed NIH3T3 clones express on average higher levels of Tac antigen on the cell surface as compared to clones transfected by NTk IL-2R, and by clones transfected with N2 IL-2R, which produced the lowest level of Tac expression.

TABLE 1
TAC ANTIGEN EXPRESSION ON MOUSE FIBROBLASTS TRANSFECTED WITH DISTINCT RETROVIRAL VECTORS

Cell lines (no. of clones analyzed)	mean specific fluorescence intensity[a]
3T3 IL-2R (3)	50 ± 13
3T3 Tk IL-2R (6)	84 ± 52
3T3 SV IL-2R (12)	120 ± 71

[a] mean ±SD of 3 independent experiments performed with each of the independently derived clones. Fluorescence on transformants with control antibody as well as on nontransformed 3T3 with anti-Tac ranged from 10 to 32 arbitrary units and has been subtracted. Tac-specific fluorescence intensity of HUT 102, expressing high levels of Tac antigen (554 ± 33), served as positive control.

To determine the steady state levels of specific retroviral RNAs, the above clones were examined by Northern blot analysis. NIH3T3 cells transfected with either NTk-IL-2R or NSV IL-2R DNA express LTR-directed RNAs in an unspliced (7.0 kb) and a spliced form (6.5 kb) as well as a third RNA species corresponding in size to the Tk- and SV 40-promoted IL-2R RNA transcript, respectively. This is the presumptive mRNA for the IL-2 receptor gene product (fig. 2). In contrast, only a single faint band of 6.3 kb was observed in NIH3T3 cells transfected with N-IL-2R (data not shown). As expected, neo^R-specific RNA was present in all clones analyzed (fig. 2).

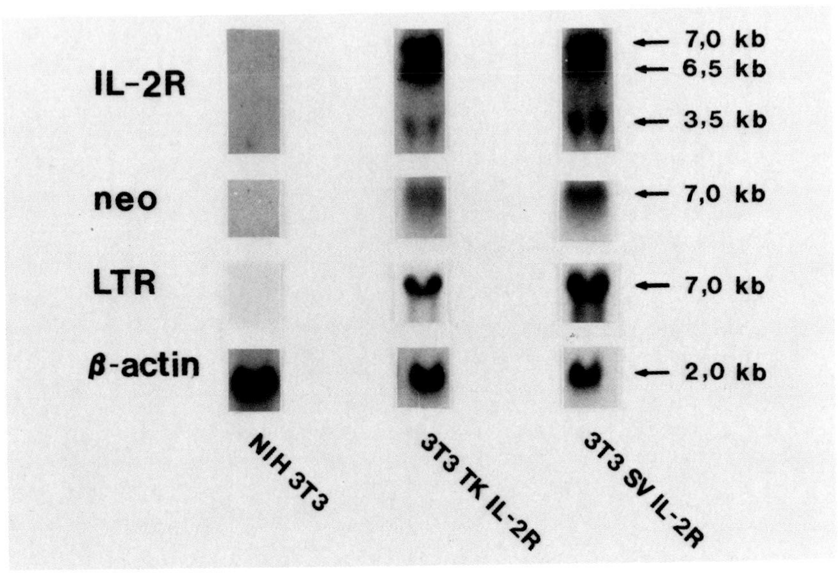

FIGURE 2. Northern blot analysis of NIH3T3 cells transfected with distinct neo IL-2R vectors. Total cellular RNA from neomycin-resistant NIH3T3 cells was probed with IL-2R (15), neo^R (28), LTR (29), and ß-actin (30) specific cDNAs.

Together, these data indicate that the neo^R gene can be efficiently transferred and expressed in NIH3T3 cells transformed by different retroviral vectors, but there exists a great variability in the levels of Tac expression.

Virus producing cell lines were generated by superinfection of neo IL-2R clones with Friend helper virus (20). To test the transfer efficiency of the 3 different vectors into fibroblasts, NIH3T3 cells were infected with serial dilutions of virus containing culture supernatants from three distinct producer clones and the number of G418 resistant colonies was determined. The neo^R Tk IL-2R and neo^R SV IL-2R vectors transferred G418 resistance at equally high efficiency (table 2). In contrast, a 1000-fold lower efficiency was observed for all of the neo IL-2R clones tested (table 2).

TABLE 2
TRANSFER EFFICIENCY OF NEO IL-2R VECTORS TO MURINE FIBROBLASTS

Vector (no. of clones analyzed)	GTU on NIH3T3[a]
neo IL-2R (3)	$1.2 \times 10^3 \pm 5.8 \times 10^2$
neo Tk IL-2R (6)	$2.4 \times 10^6 \pm 3.0 \times 10^5$
neo SV IL-2R (12)	$6.2 \times 10^6 \pm 3.0 \times 10^6$

[a]The selective G418^R transfer efficiency (GTU-G418 resistance transfer units) (5) to fibroblasts (NIH3T3) was determined as described in Materials and Methods for the indicated IL-2R vectors. Data represent mean ± SD of 3 experiments.

Expression of the human IL-2 receptor gene in murine T-lymphoid cells.

Based on the above finding of high infection rate by VIP-vectors of fibroblast cell lines and efficient expression of the IL-2 receptor gene in these cells, we next investigated whether the IL-2R gene could be expressed with similar efficiency in cells of the hematopoietic origin. In order to obtain high transduction frequencies of heterologous genes into hematopoietic cells, cocultivation with irradiated retrovirus-producing cells appears superior to other techniques (16,22). In accordance with these observations, we here show that coculture of BW5147 cells with the virus-producing cell lines 3T3Tk IL-2R and 3T3SV IL-2R, resulted in generation of neo^R clones with a frequency of 3.2×10^{-2} and 8.5×10^{-2}, respectively. In contrast, only a low frequency of infection ($\sim 10^{-4}$) was obtained with supernatants from 3T3 IL-2R cell lines.

Northern blot analysis of these clones revealed the expected specific IL-2R transcripts with higher steady state mRNA levels detected in BWSV IL-2R clones as compared to the BWTk IL-2R clones (fig. 3). No detectable IL-2R mRNA was found in 3 clones infected with the double expression vector (data not shown).

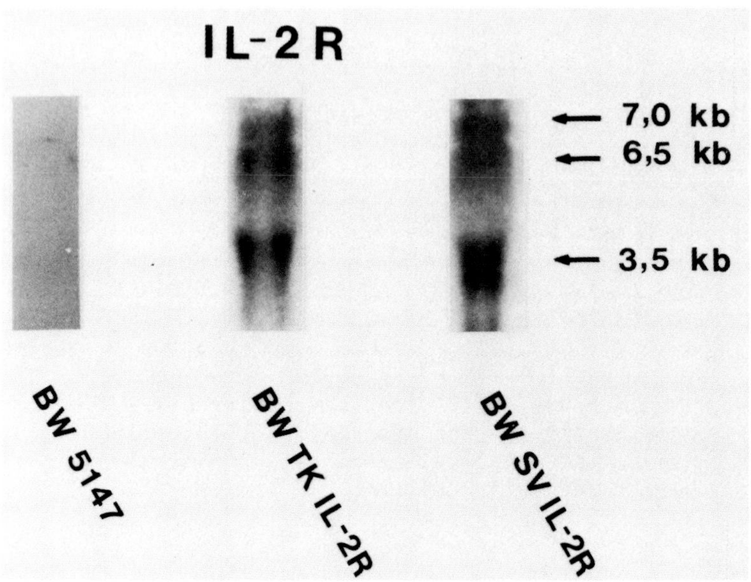

FIGURE 3. Northern blot analysis of murine T-lymphoid cells infected with distinct neo IL-2R virus. Total cellular RNA from neomycin-resistant BW5147 cells infected with neo IL-2R virus was analyzed by Northern blotting using a ^{32}P-labelled IL-2R cDNA probe (15).

This differential efficiency of IL-2R gene expression in BW5147 cells is reflected by the quantity of Tac surface antigen present in the various transformants with, on average, two- to threefold higher levels in BWSV IL-2R clones as compared to BWTk IL-2R clones (table 3).

TABLE 3
TAC ANTIGEN EXPRESSION OF T-LYMPHOID CELLS EXPRESSING THE HUMAN IL-2R GENE

Cell lines (no. of clones analyzed)	Specific fluorescence intensity[a]
BWIL-2R (3)	1 ± 1
BWTk IL-2R (5)	27 ± 9
BWSV IL-2R (9)	87 ± 47

[a]Mean ± SD of 2 independent experiments with the number of clones shown above. There was no Tac-specific fluorescence on parental BW5147.

DISCUSSION

In an attempt to delineate parameters influencing expression of genes transferred by recombinant retroviral vectors, we have explored various retroviruses containing the human IL-2R cDNA as a model gene. Two vectors with distinct constitutive promoters have been employed to internally control transcription of the IL-2R gene independent of the retroviral promoter. In addition, we have also assessed the direct influence of the Moloney virus-derived retroviral LTR on heterologous gene expression, since in other models, using different cell types, it has been implicated that a negative regulatory element within the LTRs may be responsible for tissue-specific activity of some retroviral enhancers (11-13).

Within all three vectors, we observed transfer and stable integration of the IL-2R gene in murine fibroblasts and lymphoid cells. However, the efficiency of gene transfer by virus infection and, in particular, expression of the IL-2R gene, as revealed from steady state mRNA levels and Tac antigen expression, was significantly different between these vectors. Thus, expression of the IL-2R gene by the DE-vector N2 IL-2R was restricted to fibroblasts, as neither Tac antigen on the cell surface nor IL-2R-specific mRNA was detectable in the few infected BW5147 cell clones obtained. In contrast, NSV IL-2R and NTk IL-2R are transferred with high efficiency and the IL-2R gene is expressed in fibroblasts and lymphoid cells. In both cell types, SV 40-directed expression of the IL-2R gene was found to be superior to that of Tk-promoted expression.

A substantial heterogeneity in Tac antigen expression of individual clones of transformed fibroblasts as well as T-cells was noted (tables 1,3). The clonal variability in IL-2R expression may reflect differences in transcriptional activity due to different chromosomal sites of vector integration. A more stringent control of promoters/enhancers may, however, reside in tissue-specific factors, selectively influencing the activity of some retroviral promoters, independent of their integration site into the cellular genome (11-13). For example, in a spleen colony formation assay, only NTk-infected, but not N2-infected spleen cells expressed the neo gene (16). Moreover, retroviral genomes are inefficiently expressed in various stem cells (11,31). Similarly, we here show that LTR-promoted IL-2R expression is inefficient in fibroblasts and virtually absent in T-lymphoid cells. In analogy to the embryonic stem cell system (13) it is possible that transacting-negative regulatory factors account for the lack of IL-2R gene expression in the N2 IL-2R infected BW cell clones.

Although we do not know at present which molecular processes interfere with Moloney virus-directed gene expression in the cell lines studied here, our data show that these inhibitory mechanisms can be bypassed with VIP-vectors, suggesting that such vectors are suitable for stable expression of foreign genes in cells of hematopoietic origin.

ACKNOWLEDGEMENTS

We thank T. v. Rüden for providing the vectors N2, NTk, and NSV, W. Greene and M. Krönke for providing the IL-2R cDNA. Technical help of M. Killian and secretarial assistance of G. Schmidt are gratefully acknowledged.

REFERENCES

1. Tabin CJ, Hoffmann JW, Goff SP, Weinberg RA (1982). Adaptation of retroviruses as a eucaryotic vector transmitting the herpes simplex virus thymidine kinase gene. Mol Cell Biol 2:426.
2. Joyner A, Bernstein A (1983) Retrovirus transduction: Segregation of the viral transforming function and the Herpes simplex virus tk gene in infectious Friend spleen

focus-forming virus thymidine kinase vectors. Mol Cell Biol 3:2191.

3. Miller AD, Joley DJ, Friedmann T, Verma JM (1983). A transmissable retrovirus expressing human hypoxanthine phosphoribosyltransferase (HGPRT): gene transfer into cells obtained from humans deficient in HPRT. Proc Natl Acad Sci USA 80:4709.
4. Cepko CL, Roberts BE, Mulligan RC (1984). Construction and applications of a highly transmissable murine retrovirus shuttle vector. Cell 37:1053.
5. Seliger B, Kollek R, Stocking C, Franz T, Ostertag W (1986). Viral transfer, transcription, and rescue of a selectable myeloproliferative sarcoma virus in embryonal cell lines: expression of the mos oncogene. Mol Cell Biol 6:286.
6. Dick JE, Magee MC, Huszar D, Phillips RA, Bernstein A (1985) Introduction of a selectable gene into primitive stem cells capable of long-term reconstitution of the hemopoietic system of W/W^{∇} mice. Cell 42:71.
7. Keller G, Paige C, Gilboa E, Wagner ET (1985). Expression of a foreign gene in myeloid and lymphoid cells derived from multipotent haematopoietic precursors. Nature 318:149.
8. Lemischka IR, Raulet DH, Mulligan RC (1986). Developmental potential and dynamic behavior of hematopoietic stem cells. Cell 45:917.
9. Eglitts M, Kantoff P, Gilboa E, Anderson WT (1985). Gene expression in mice after high efficiency retroviral gene transfer. Science 230:1395.
10. McIvor RS, Johnson MJ, Miller AD, Pitts SI, Williams SR, Valerio D, Martui DW, Verma JR (1987). Human purine nucleoside phosphorylase and adenosine deaminase: gene transfer into cultured cells and murine hematopoietic stem cells by using recombinant amphotropic retrovirus. Mol Cell Biol 7:838.
11. Stewart CL, Stuhlmann H, Jaehner D, Jaenisch R (1982) De novo methylation, expression, and infectivity of retroviral genomes introduced into embryonal carcinoma cells. Proc Natl Acad Sci USA 79:4098.
12. Stuhlmann H, Cone R, Mulligan RC, Jaenisch R (1984) Introduction of a selectable gene into different animal tissue by a retrovirus recombinant vector. Proc Natl Acad Sci USA 81:7151.
13. Gorman CM, Rigby PWJ, Lane DP (1985). Negative regulation of viral enhancers in undifferentiated embryonic stem cells. Cell 42:519.

14. Armentano D, Yu S-F, Kanthoff PW, von Rüden T, Anderson WF, Gilboa E (1987). Effect of internal viral sequences on the utility of retroviral vectors. J Virol 61:1647.
15. Leonard WJ, Depper RJ, Krönke M, Svetlik PB, Pepper NJ, Waldmann TA, Greene WC (1984). Molecular cloning and expression of cDNAs from the human interleukin-2 receptor. Nature 311:626.
16. Magli M-C, Dick JE, Huszar D, Bernstein A, Philipps RA (1987). Modulation of gene expression in multiple hematopoietic cell lineages following retroviral vector gene transfer. Proc Natl Acad Sci USA 84:789.
17. Yu SF, von Rüden T, Kantoff PW, Garber C, Seilberg M, Rüther K, Anderson WF, Wagner EF, Gilboa E (1986). Self-inactivating retroviral vectors designed for transfer of whole genes into mammalian cells. Proc Natl Acad Sci USA 83:3194.
18. Maniatis T, Frisch EF, Sambrook J (1982) Molecular cloning. Cold Spring Harbor Laboratory.
19. Graham FL, van der Eb AJ (1973) A new technique for the assay of infectivity with human adenovirus S DNA. Virol 52:456.
20. Pragnell IB, McNab A, Harrison PR, Ostertag W (1978) Are spleen focus-forming virus sequences related to xenotropic viruses and expressed specifically in normal erythroid cells? Nature 272:456.
21. Hyman R, Stallings V (1974). Complementation patterns of thy-1 variants and evidence that antigen loss variants "pre-exist" in parental population. J Natl Cancer Inst 52:429.
22. Seliger B, Kruppa G, Pfizenmaier K (1987) Stable expression of a selectable myeloproliferative sarcoma virus in murine T lymphocyte and monocyte cell lines. Immunobiol. 174:313.
23. Chirgwin JM, Przybyla AH, McDonald RJ, Rutter WJ (1979). Isolation of biologically active ribonuclei acid from sources enriched in ribonuclease. Biochem 18:5294.
24. Krönke M, Schlüter C, Pfizenmaier K (1987). Tumor necrosis factor inhibits MYC expression in HL-60 cells at the level of mRNA transcription. Proc Natl Acad Sci USA 84:469.
25. Feinberg HP, Vogelstein B (1983). A technique for radiolabelling DNA restriction endonuclease fragments to high specific activity. Analyt Biochem 132:6.
26. Uchiyama T, Broder S, Waldmann TA (1981). A monoclonal antibody (anti-Tac) reactive with activated and functionally mature human T cells. J. Immunol. 126:1393.

27. Leonard WJ, Depper JM, Uchiyama T, Smith KA, Waldmann TA, Greene WC (1982). A monoclonal antibody that appears to recognize the receptor for human T-cell growth factor: partial characterization of the receptor. Nature 300:267.
28. Colbère-Garapin F, Horodniceanu F, Kourilsky P, Garapin A-C (1981). A new dominant hybrid selective merker for higher eukaryotic cells. J Mol Biol 150:1.
29. Jähner D, Jaenisch R (1980). Integration of Moloney leukaemia virus into the germ line of mice: correlation between site of integration and virus activation. Nature 287:456.
30. Cleveland DW, Lopata MA, McDonald RJ, Cowan NG, Rutter WG, Kirschner MW (1980). Number and evolutionary conservation of alpha- and beta-tubulin and cytoplasmic beta- and gamma-actin genes using specific cloned cDNA probes. Cell 20:95.
31. Gautsch JW, Wilson MC (1983). Delayed de novo methylation in teratocarcinoma suggests additional tissue-specific mechanisms for controlling gene expression. Nature 301:32.

Gene Transfer and Gene Therapy, pages 103–116

REGULATION OF THE PROMOTER ACTIVITY OF HUMAN T-CELL LEUKEMIA VIRUS TYPE I[1]

Masataka Nakamura, Kiyoshi Ohtani, Kazuo Sugamura* and Yorio Hinuma

Institute for Virus Research, Kyoto University, Kyoto 606
and
*Department of Bacteriology, Tohoku University School of Medicine, Sendai 980, Japan

ABSTRACT We investigated the promoter activity of elements in the long terminal repeat (LTR) of human T-cell leukemia virus type I (HTLV-I). The HTLV-I LTR contains two distinct elements involved in maximum gene expression. One element is an enhancer responsible for _trans_-acting activation by p40. The other is present in the R region, and its function does not require p40 function. We also found that a tumor promoter, which is an activator of protein kinase C, and forskolin, which is an activator of adenyl cyclase, induced the enhancer function of the HTLV-I LTR independently of p40, and that a T-cell mitogen caused induction of the promoter activity, suggesting that replication of HTLV-I may be initiated by signals from immune responses.

INTRODUCTION

Human T-cell leukemia virus type I (HTLV-I) is a human retrovirus that is the etiological agent of clinically aggressive adult T-cell leukemia (ATL) (1,2). Infection

---

1 This work was supported in part by Grants-in-Aid for Cancer Research from the Ministry of Education, Science and Culture of Japan, and a grant from the Uehara Foundation.

with HTLV-I results in a low incidence of ATL *in vivo* and in transformation of normal T lymphocytes with the helper phenotype *in vitro*. The genome of HTLV-I, however, has no typical oncogene, unlike those of other animal acute transforming retroviruses (3), and is randomly integrated into the cellular genome unlike the genomes of chronic leukemia viruses (4). At present, the mechanism of transformation by HTLV-I is unknown. Nucleotide sequence analyses have indicated that the genome of HTLV-I contains a unique open reading frame, referred to as pX, between the *env* gene and the 3' long terminal repeat (LTR) (3), which encodes three distinct products (5).

Replication of retroviruses is initiated by transcription from the LTR in the provirus genome. Transcription initiation occurs at the 5' end of the R region of the LTR, and signals for transcription initiation are present in the U3 region. HTLV-I causes a great increase in the level of expression of genes directed by the promoter unit in the LTR of HTLV-I in cells productively infected with HTLV-I compared with that in uninfected cells (6,7). Recent studies have shown that this increase requires the function of one of the pX products, p40, which acts as a transcriptional activator in a *trans*-acting manner (8-10). A 21-base pair (bp) sequence, functioning as an enhancer in a *cis*-acting manner, confers susceptibility to this increase by p40 (11,12).

In studies on the transient expression of the bacterial chloramphenicol acetyltransferase (CAT) system, we noted that the enhancer containing the 21-bp sequence was not sufficient to induce the level of gene expression observed with the entire LTR as a promoter machinery, and found that the R region is also involved in maximum gene expression (13). In addition, we report here that extracellular stimuli cause activation of the HTLV-I enhancer. This p40-independent enhancer activation may well be an initial step in transcription initiation of HTLV-I.

MATERIALS AND METHODS

Cell Lines.

The cell lines used were the human erythroleukemic

cell line K562, the human B lymphoblastoid cell line LCL-Kan, and the human T cell line Jurkat.

Plasmids.

Two types of plasmids were used. The first is a plasmid expressing HTLV-I p40 (pMAXneo) (13). The frame shift mutant pMAXneo/M does not provide the function of p40. The second type of plasmid contains an assayable gene, the CAT gene, controlled by HTLV-I LTR promoter elements or chimeric promoter units from the HTLV-I LTR element and simian virus 40 (SV40) core promoter.

CAT Assay.

Cells transfected with plasmids by the DEAE-dextran method were harvested 48 hr after transfection. Cell extracts were prepared and CAT activity was assayed as described previously (13,14).

RESULTS

Promoter Activity of Deletion Mutants of the LTR.

To examine the ability of the HTLV-I LTR to activate gene expression, we studied the transient expression of the CAT gene under the control of the LTR, which is known to be active only in cells productively infected with HTLV-I or expressing the viral product p40 (15). We used a plasmid pMAXneo as a source of a p40 trans-acting factor. The plasmid pCHL4 contains the CAT gene directed by a 630-bp fragment from the LTR (13). We also constructed a series of plasmids containing LTR sequences with deletions on the 5' or the 3' side of the RNA start site (cap site), as shown in Figure 1. Mutant plasmids pCHL1, 2 and 3 have deletions in the sequence downstream from the cap site, and pdHE4 has a deletion in the sequence upstream from the cap site but contains the core promoter of HTLV-I. These plasmids were co-transfected into Jurkat cells, a human T cell line free of the HTLV-I sequence, together with pMAXneo and then CAT activity was measured.

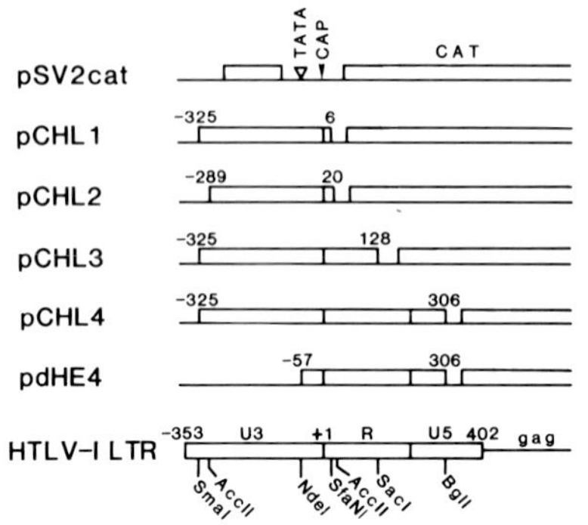

	CAT Activity Conversion (%) A pMAXneo	B pMAXneo/M	Activation index A/B
pSV2cat		0.10	
pCHL1 (-325 to 6)	0.90	0.11	8.2
pCHL2 (-289 to 20)	0.59	0.11	5.3
pCHL3 (-325 to 128)	3.1	0.10	31.0
pCHL4 (-325 to 306)	7.5	0.21	35.7
pdHE4 (-57 to 306)	0.11	0.15	0.7

HTLV-I LTR: -353 U3 +1 R U5 402 gag; SmaI, AccII, NdeI, SfaNI, AccII, SacI, BglI

FIGURE 1. Deletion effects of the LTR on the promoter activity

As the size of the deletion in the R region increased from the 3' end (pCHL4 to pCHL1), CAT activity gradually decreased in the presence of p40. *Trans*-activation by p40 was seen with mutants with a deletion in the R region, even with pCHL1 and pCHL2 which have only 6 and 20 nucleotides in the R region, respectively, and show less than 10% of the CAT activity of pCHL4. On the other hand, removal of the U3 region (pdHE4) also markedly reduced the activity, which was the basal level irrespective of p40. These observations suggest that at least two domains in the LTR affect gene expression.

Two Regulatory Elements in the LTR.

We examined the mechanism by which each domain in the LTR activated gene expression. For this, we constructed chimeric promoter units consisting of the core promoter of SV40 and elements from the HTLV-I LTR. A 230-bp fragment (*Acc*II to *Nde*I) in the U3 region (U3 fragment) and a 300-bp fragment (*Sfa*NI to *Bgl*I) in the R and the U5 regions (R fragment) were isolated and inserted separately into an enhancerless SV40-CAT plasmid (pSV1Ccat) to determine the activity of each element. Introduction of the U3 fragment enhanced the CAT activity irrespective of its position of insertion and orientation, but dependently on the presence of p40. These results demonstrate that the U3 fragment contains a cis-acting regulatory sequence that can be activated by the trans-activator, p40, and behaves like known enhancers. These results are consistent with observations by others (11,12). Our results also confirmed that the 21-bp sequence in the multi-repeated form is essential for trans-activation by p40 in K562 cells (Table 1) and in Jurkat cells (data not shown).

Similar experiments were performed to elucidate how the R fragment enhanced CAT gene expression. The R fragment was inserted into pSV2cat and pSV1Ccat in three different positions in the sense or anti-sense orientation. Transfection of these plasmids with pMAXneo or pMAXneo/M into Jurkat cells showed that the enhancing effect of the R fragment was independent of the presence of p40. In addition, enhanced CAT activity was observed after insertion of the R fragment at a site between the cap site and the translational initiation codon irrespective of its orientation (pdERH-1) (Figure 2). This fragment had no effect, however, when it was inserted at a site outside the promoter and the CAT gene.

	Jurkat			LCL-Kan		
	CAT Activity Conversion(%)		Activation index	CAT Activity Conversion(%)		Activation index
	A pMAXneo	B pMAXneo/M	A/B	A pMAXneo	B pMAXneo/M	A/B
pSV2cat	1.0	0.10	10	5.0	4.4	1.1
pSV1Ccat	0.1	0.07	1.3	0.06	0.05	1.2
pCHL4	15	0.60	25	1.8	0.23	7.8
pdEUC-1	1.1	0.10	11	0.71	0.10	7.1
pdERH-1	2.6	2.5	1.0	0.47	0.54	0.87
pHERH-1	14	1.3	11	3.8	0.69	5.5
pHERH-2	12	1.1	11	ND	ND	

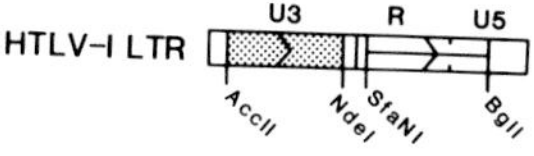

FIGURE 2. Reconstitution of the LTR promoter activity by the enhancer and R fragments.

The activity of the hybrid promoter unit containing each of two fragments was always much lower than that containing intact LTR (see Figure 2). Thus, we supposed that the association of the R fragment with the U3 fragment was necessary for full activity of the LTR, and examined this notion using plasmids including both fragments with the SV40 core promoter as an entire promoter machinery. These plasmids were co-transfected into Jurkat and LCL-Kan cells with either pMAXneo or pMAXneo/M and the levels of CAT activity were compared with that with pCHL4. Insertion of the U3 fragment (pdEUC-1) or the R fragment (pdERH-1) showed only 1.1% conversion in the presence of p40 and 2.5% conversion independent of the presence of p40, respectively. pCHL4 showed 15% conversion in the presence of p40 which is 25 times greater than that in the absence of p40 (Figure 2). The combination of both fragments with the SV40 core promoter (pHERH-1 and -2) resulted in a markedly enhanced level of CAT activity (14% conversion) which was equivalent to that seen with pCHL4. Essentially the same results were obtained with LCL-Kan cells.

Signals for Activation of the LTR Enhancer.

The activity of the HTLV-I enhancer region was induced by the own product p40, suggesting that cells carrying the entire HTLV-I provirus genome and expressing p40 produce the virus constitutively. However, expression of HTLV-I is rare in cells infected with HTLV-I, even in leukemic cells containing the entire provirus genome freshly isolated from patients with ATL, but culture of these cells _in vitro_ results in expression of virus antigens. Therefore, we assumed that stimuli from culture _in vitro_ induce signals for expression of the virus and that the stimuli can be transmitted through a sequence in the LTR.

The tumor promoter 12-O-tetradecanoylphorbol-13-acetate (TPA) is known to induce the expression of viral antigens of HTLV-I (1). We studied the molecular mechanism of this induction by TPA (16). Plasmid pCHL4 was transfected into K562 cells by the DEAE-dextran method. Then 12 hr after transfection TPA was added at a concentration of 5 ng/ml and the cells were cultured further for 24 hr. Treatment with TPA increased the CAT activity with pCHL4, indicating that the induction of HTLV-I by TPA could be due to activation of the promoter function in the LTR. The specificity of phorbol esters was examined by comparing the effect of TPA with that of phorbol-12, 13-didecanoate (4α-PDD), which does not induce viral expression or activate protein kinase C. The CAT activity from pCHL4 was increased 96-fold by treatment with TPA, but was not increased by 4α-PDD (data not shown). K562 cells were used because they showed the greatest response to TPA of the human hematopoietic cell lines tested. To determine the _cis_-acting element in the LTR that is responsive to TPA, we used chimeric promoter units that direct expression of the CAT gene. CAT plasmids with chimeric promoter units consisting of the SV40 core promoter and the enhancer element from the HTLV-I LTR were constructed by insertion of the enhancer fragment at the _Cla_I site of pSV1Ccat in the sense (pdEUC-1) or anti-sense (pdEUC-2) orientation, or at the _Bam_HI site in the sense orientation (pdEUB-1). K562 cells were transfected with these plasmids, and then CAT gene expression was determined with or without TPA treatment.

Little if any increase in CAT activity was seen with plasmids of pdHE4 lacking the enhancer element of HTLV-I, and pdERH-1 containing the R element with the SV40 core promoter. However, with plasmids (pdEUC-1, -2 and pdEUB-1) carrying the enhancer element with the SV40 core promoter, marked increases in CAT gene expression were observed upon TPA treatment (Table 1).

TABLE 1
EFFECT OF p40, TPA AND FORSKOLIN ON PROMOTER ACTIVITY IN K562

Plasmid	CAT activity, % conversion								
	p40			TPA			Forskolin		
	+	-	(index)	+	-	(index)	+	-	(index)
pCHL4	77.4	1.0	(77.4)	24.0	0.3	(80.0)	60.3.	5.1.	(13.0)
pdHE4	0.31	0.64	(0.5)	0.3	0.2	(1.4)	0.6	0.6	(1.0)
pdEUC-1	ND[a]			24.2	0.5	(52.4)	9.4	1.5	(6.3)
pdEUC-2	ND			67.6	3.0	(22.4)	38.0	6.5	(5.8)
pdEUB-1	ND			32.6	1.1	(28.7)	58.8	4.5	(13.1)
pdERH-1	ND			8.3	3.8	(2.2)	ND		
pdE21-1	0.7	0.3	(2.3)	0.4	0.3	(1.3)	0.3	0.3	(1.0)
pdE21-4	68.2	0.7	(97.4)	4.8	0.7	(6.9)	20.8	0.7	(29.9)
pSV1Ccat	0.5	0.4	(1.2)	0.2	0.2	(1.0)	0.6	0.6	(1.0)

[a]ND, not determined.

To determine whether the enhancement by TPA was dependent on the 21-bp sequence, which is involved in _trans_-activation by p40, plasmids containing the synthetic oligonucleotide sequence AAGGCTCTGACGTCTCCCCC, which is identical to that of one of three repeats in the U3 enhancer, were constructed. The synthetic sequence was flanked by the ClaI linker and inserted into the ClaI site of pSV1Ccat as an enhancer, generating pdE21-1 and pdE21-4 which have one and four repeats of the synthetic sequence in the sense orientation, respectively. These plasmids were transferred to K562 cells and CAT activity was measured with or without TPA treatment. Enhancement by TPA was seen with the synthetic 21-bp sequence in the repeated form, but not in the single form (Table 1). These results clearly demonstrate that the cis-acting element that responds to TPA is within the 21-bp enhancer of the HTLV-I LTR.

We also examined whether forskolin, which is known to activate adenyl cyclase, resulting in increase in the intracellular concentration of cyclic-AMP, induces function of the HTLV-I enhancer. As in studies with TPA, K562 cells transfected with CAT plasmids were cultured for 24 hr in the presence of forskolin at a concentration of 20 μM from 12 hr after transfection and the CAT activity of cell extracts was measured. Addition of forskolin caused marked increase in CAT gene expression from pCHL4, whereas no activation was observed in the absence of the U3 fragment (pdHE4). Experiments with the isolated U3 enhancer fragment showed that the enhancer conferred susceptibility to activation by forskolin, like known enhancers. Plasmid pdE21-4 showed increased CAT activity on treatment with forskolin, indicating that the enhancing action of forskolin is mediated through the 21-bp enhancer of HTLV-I.

TABLE 2

INDUCTION OF PROMOTER ACTIVITY BY PHA

		CAT activity % conversion		
		PHA		
Plasmid	PBL[a]	+	-	(Index)
pCHL4	-	4.6	1.8	(2.6)
	+	12.2	1.7	(7.2)

[a]peripheral blood leukocytes.

Protein kinase C and adenyl cyclase are known to be activated during T cell activation in the antigen or mitogen response. We, therefore, examined whether a T-cell mitogen, phytohaemagglutinin (PHA), induced promoter activity of the HTLV-I LTR. Jurkat cells were transfected with pCHL4 and cultured for 36 hr with or without PHA (0.25%) in the presence or absence of normal human peripheral leukocytes ($2X10^5$ cells/ml). Then cell extracts were prepared and their CAT activity was determined. The results showed that stimulation with PHA was enough to induce CAT gene expression from the HTLV-I LTR and that the presence of normal leukocytes was helpful in induction of promoter activity by PHA (Table 2).

DISCUSSION

We have shown that two independent domains are involved in the activity of the LTR to induce gene expression maximally in a *cis*-acting manner. One is present in the U3 region, is active only in the presence of an own viral product, a *trans*-acting activator p40, and functions like typical enhancers. We have also confirmed that a core sequence in the enhancer for p40-dependent *trans*-activation is the 21-bp sequence.

The extent of enhancement by the enhancer is, however, much lower than that with the intact LTR. The second regulatory element required for maximal gene expression is within a 300-bp fragment encompassing the unusually long R region and the 5' portion of the U5 region. Recently we found that this activity is within a 120-bp fragment in the R region (data not shown). This element is functional in both orientations but its effect is strictly dependent upon the position, and its enhancing effect is independent of p40. From results with deletion mutants of the LTR and examination of the activity of each isolated fragment from the LTR, we propose that the HTLV-I enhancer turns on transcription initiation in the presence of p40, and that the R region functions to maintain gene expression at a relatively high level. Thus, the association of these two elements which participate independently in positive gene expression is advantageous for production of the viral components directed by the promoter unit of the HTLV-I LTR.

Probably only cells expressing p40 contain the *trans*-acting factor(s) that turns on the enhancer function. It is unkown whether p40 itself interacts directly with the enhancer sequence. If p40 does not interact directly with the enhancer, it must induce or modulate a cellular factor that interacts with the enhancer and turns on the enhancer function. The latter possibility seems likely because our preliminary studies by the gel retardation assay suggested that nuclear extracts from several cells, even those not expressing p40, contain a factor(s) that binds to the synthetic 21-bp enhancer sequence.

The second domain in the LTR was functional only when it was placed downstream of the RNA start site and upstream of the translation initiation site. Insertion of an unrelated fragment of similar length into the same position had no effect, indicating that the effect is sequence-specific. Moreover, no special cellular environment or p40 was required for the activity of this fragment to facilitate gene expression. These structural and functional properties of the second domain suggest that it is a new control element for gene expression distinct from prototypic enhancers. In a recent study (17) the R region of the LTR of the related virus, bovine leukemia virus, was found to have enhancing effects on gene expression similar to those shown here. The mechanism by which the R element enhances gene expression is still unknown.

Figure 2 shows that p40 potentiates the function of the SV40 early promoter in Jurkat cells. Previously we analyzed this phenomenon in detail (18). Our results showed that p40 trans-activated the function of the enhancer from SV40 and that this trans-activation was seen in a limited repertoire of cell lines, unlike trans-activation of the HTLV-I enhancer by p40. p40 is also reported to induces expression of interleukin 2 and its receptor in a certain cell line (19,20). Expression of the interleukin 2 receptor is consistently associated with HTLV-I infection, although normal T lymphocytes express the receptor transiently. This abnormal expression of the receptor is thought to play a crucial role in the early stage of leukemogenesis by HTLV-I (21). In addition, introduction of the p40 gene caused induction of the protooncogene, c-fos (M. Fujii et al., personal communication). Furthermore, the enhancer in the LTR of human immunodeficiency virus was activated by HTLV-I p40 in a trans-acting manner (22). A pentanucleotide, CTGAC, or a closely related sequence is commonly found within sequences of enhancers or regions upstream from the promoter of these viral and cellular genes. (Table 3). This pentanucleotide sequence is also included in sequences recently determined as a TPA consensus sequence (23) and a consensus sequence of cyclic-AMP responsible element (CRE) (24).

TABLE 3
CTGAC MOTIF IN THE 5' REGION

AAGGCT	CTGAC	GTCTCC	HTLV-I	A
TAGGCC	CTGAC	GTGTCC		B
CAGGCG	TTGAC	GACAAC		D
TGGTTG	CTGAC	TAATTG	SV40	P
CCCTAA	CTGAC	ACACAT		A
AAAGTC	CTGAC	AAGATT	IL-2R	(human)
GGCTGC	CTGAC	CAGAAT		
TCCCGG	CTGAC	TCCTGA		
CAAAGA	CTGAC	TGAATG	IL-2	(human)
AACCTG	CTGAC	GCAGAT	c-fos	(human)
GAGCCC	GTGAC	GTTTAC		
GAACTG	CTGAC	ATCGAG	HIV-1	(BH-10)
	$^{C}_{G}$TGAC	T$^{C}_{A}$A	TPA consensus	
	TGAC	GTC	CRE consensus	

Little expression of HTLV-I products is seen in fresh leukemic cells from the peripheral blood of patients. Our finding that TPA and forskolin rather than p40 induce function of the enhancer of the 21-bp sequence could help to account for the initial step of transcription of HTLV-I in virus infected cells *in vivo*. Although TPA and forskolin are not natural stimuli, their biological activities are thought to be exerted through activation of protein kinase C and adenyl cyclase, respectively. These enzymes are important in transduction of signals for gene expression and cell proliferation, and are known to be activated during immune responses. In fact, the more natural immune stimulation by PHA induced gene expression from the HTLV-I promoter unit. Thus, we think that *in vivo* the activations of cellular enzymes which act as signals to induce expression of the genes necessary for immune reactions on antigen stimulation also act as signals to induce expression of the HTLV-I proviral genome.

REFERENCES

1. Hinuma Y, Nagata K, Hanaoka M, Nakai M, Matsumoto T, Kinoshita K, Shirakawa S, Miyoshi I (1981). Adult T-cell leukemia: Antigen in an ATL cell line and detection of antibodies to the antigen in human sera. Proc Natl Acad Sci USA 78:6476.
2. Kalyanaraman VS, Sarngadharan MG, Nakao Y, Ito Y, Aoki T, Gallo RC (1982). Natural antibodies to the structural core protein (p24) of the human T-cell leukemia (lymphoma) retrovirus found in sera of leukemia patients in Japan. Proc Natl Acad Sci USA 79:1653.
3. Seiki M, Hattori S, Hirayama Y, Yoshida M (1983). Human adult T-cell leukemia virus: Complete nucleotide sequence of the provirus genome integrated in leukemia cell DNA. Proc Natl Acad Sci USA 80:3618.
4. Seiki M, Eddy R, Shows TB, Yoshida M (1984). Nonspecific integration of the HTLV provirus genome into adult T-cell leukemia cells. Nature 309:640.
5. Kiyokawa T, Seiki M, Iwashita S, Imagawa K, Shimizu F, Yoshida M (1985). $p27^{X-III}$ and $p21^{X-III}$, proteins encoded by the pX sequence of human T-cell leukemia virus type I. Proc Natl Acad Sci USA 82:8359.
6. Sodroski JG, Rosen CA, Haseltine WA (1984). Trans-acting transcriptional activation of the long terminal repeat of human T lymphotropic viruses in infected cells. Science 225:381.
7. Fujisawa J, Seiki M, Kiyokawa T, Yoshida M (1985). Functional activation of the long terminal repeat of human T-cell leukemia virus type I by a trans-acting factor. Proc Natl Acad Sci USA 82:2277.
8. Felber BK, Paskalis H, Kleinman-Ewing C, Wong-Staal F, Pavlakis G (1985). The pX protein of HTLV-I is a transcriptional activator of its long terminal repeats. Science 229:675.

9. Sodroski J, Rosen C, Goh WC, Haseltine W (1985). A transcriptional activator protein encoded by the x-lor region of the human T-cell leukemia virus. Science 228:1430.
10. Seiki M, Inoue J, Takeda T, Yoshida M (1986). Direct evidence that $p40^x$ of human T-cell leukemia virus type I is a trans-acting transcriptional activator. EMBO J 5:561.
11. Fujisawa J, Seiki M, Sato M, Yoshida M (1986). A transcriptional enhancer sequence of HTLV-I is responsible for trans-activation mediated by $p40^x$ of HTLV-I. EMBO J 5:713.
12. Shimotohno K, Takano M, Teruuchi T, Miwa M (1986). Requirement of multiple copies of a 21-nucleotide sequence in the U3 regions of human T-cell leukemia virus type I and type II long terminal repeats for trans-acting activation of transcription. Proc Natl Acad Sci USA 83:8112.
13. Ohtani K, Nakamura M, Saito S, Noda T, Ito Y, Sugamura K, Hinuma Y (1987). Identification of two distinct elements in the long terminal repeat of HTLV-I responsible for maximum gene expression. EMBO J 6:389.
14. Gorman CM, Moffat LF, Howard BH (1982). Recombinant genomes which express chloramphenicol acetyltransferase in mammlian cells. Mol Cell Biol 2:1044.
15. Sodroski JG, Goh WC, Rosen CA, Salahuddin SZ, Aldovini A, Franchini G, Wong-Staal F, Gallo RC, Sugamura K, Hinuma Y, Haseltine WA (1985). trans-Activation of the human T-cell leukemia virus long terminal repeat correlates with expression of the x-lor protein. J Virol 55:831.
16. Fujii M, Nakamura M, Ohtani K, Sugamura K, Hinuma Y (1987). 12-O-Tetradecanoylphorbol-13-acetate induces the enhancer function of human T-cell leukemia virus type I. FEBS Lett 223:299.
17. Derse D, Casey JW (1986). Two elements in the bovine leukemia virus long terminal repeat that regulate gene expression. Science 231:1437.
18. Saito S, Nakamura M, Ohtani K, Ichijo M, Sugamura K, Hinuma Y (1988). trans-Activation of the simian virus 40 enhancer by a pX product of human T-cell leukemia virus type I. J Virol 62: in press.
19. Inoue J, Seiki M, Taniguchi T, Tsuru S, Yoshida M (1986). Induction of interleukin 2 receptor gene expression by $p40^x$ encoded by human T-cell leukemia virus type 1. EMBO J 5:2883.

20. Cross S, Feinberg MB, Wolf JB, Holbrook NJ, Wong-Staal F, Leonard WJ (1987). Regulation of the human interleukin-2 receptor α chain promotor: Activation of a nonfunctional promoter by the transactivator gene of HTLV-I. Cell 49:47.
21. Sugamura K, Fujii M, Ishii T, Hinuma Y (1986). Possible role of interleukin 2 receptor in oncogenesis of HTLV-I/ATLV. Cancer Rev 1:96.
22. Siekevitz M, Josephs SF, Dukovich M, Peffer N, Wong-Staal F, Green WC (1987). Activation of the HIV-1 LTR by T cell mitogens and the trans-activator protein of HTLV-I. Science 238:1575.
23. Lee W, Mitchell P, Tjian R (1987). Purified transcription factor AP-1 interacts with TPA-inducible enhancer elements. Cell 49:741.
24. Montminy MR, Sevarino KA, Wagner JA, Mandel G, Goodman RH (1986). Identification of a cyclic-AMP-responsive element within the rat somatostatin gene. Proc Natl Acad Sci USA 83:6682.

Gene Transfer and Gene Therapy, pages 117–127

THE TRANSCRIPTION OF THE T CELL RECEPTOR β-CHAIN GENE IS CONTROLLED BY A DOWNSTREAM REGULATORY ELEMENT*

Paul Krimpenfort[1], Yasushi Uematsu[2], Zlatko Dembic[3], Michael Steinmetz[3] and Anton Berns[1]

[1]Division of Molecular Genetics of The Netherlands Cancer Institute, and the Department of Biochemistry of the University of Amsterdam, Plesmanlaan 121, 1066 CX Amsterdam, The Netherlands.

[2]Basel Institute for Immunology, CH-4005 Basel, Switzerland.

[3]Central Research Units, Hoffmann-la Roche, Basel, Switzerland.

ABSTRACT To characterize cis-acting elements controlling the expression of T cell receptor β-chains we generated transgenic mice harboring a rearranged T cell receptor β chain gene with different extensions of 5' and 3' flanking sequences. Sequences located 5 kbp downstream from the poly-adenylation signal of the Cβ2 region appeared to be indispensible for expression in transgenic mice. Enhancer activity was further defined by transient CAT-assays. Oligonucleotide motifs characteristic for enhancer elements were identified in this region.

---

*This work was supported by The Netherlands Organization for the advancement of Pure Research (ZWO) through the Foundation for Medical Research (MEDIGON). The Basel Institute for Immunology was founded and is supported by F. Hoffmann-La Roche & Co, Limited, Basel, Switzerland.

INTRODUCTION

T cell receptors (TCR) recognize foreign antigens in the context of self major histocompatibility complex (MHC) molecules. Molecular genetic analysis has demonstrated that T lymphocytes generate diversity in receptor molecules in much the same way as B cells (1, 2). The genes encoding the chains of the T cell receptor are composed of segments which rearrange during T cell development thereby generating a T cell population with clonally distributed receptors. Somatic rearrangement of TCR and Ig genes seem to occur predominantly by deletion of intervening DNA. Two forms of T cell receptors have been described. The predominant form is composed of the α and β chain. TCR and Ig chains consist of N-terminal variable (V) and C-terminal constant (C) regions. The variable region of the TCR β chain is encoded by three gene segments, called variable (V), diversity (D) and joining (J) gene segments, which exist in multiple copies in germline DNA. A functional variable region gene is formed by an apparently random fusion of one of each of the three gene segment families. Analogous to immunoglobulin gene segments TCR gene segments are flanked by conserved nonamer and heptamer motifs presumably recognized by the recombinase (2, 3).

Expression of Ig genes is controlled by various regulatory elements. An enhancer element shown to be essential for Ig gene expression, is located within the intron in front of the Cμ and C-kappa genes (4-7). Elements regulating the expression of human T cell receptor α genes have recently been identified in front of the C_α gene at a similar location as in Ig genes (8). However, initial studies with TCR β genes suggested that some regulatory elements might be positioned differently (9, 10).

RESULTS

A Downstream Region is required for TCR β Gene Expression in Transgenic Mice.

The strategy to localize regulatory elements necessary for the expression of TCR β chain genes was based on the production of transgenic mice harboring a rearranged TCR β chain gene with different extensions of 5' and 3' flanking sequences. The cytotoxic T cell clone, B6.2.16, served as a source for a functionally rearranged TCR β chain gene. From this clone cDNA and genomic libraries were made in lambda gt11 and EMBL3 vectors, respectively (10). Functional rearranged genes were reconstituted from these clones and germline sequences from BALB/c liver DNA. Different fragments, carrying the complete β gene and varying amounts of 5' and 3' flanking sequences were used to generate transgenic mice. These fragments are depicted in Figure 1. The transgenic mice contained from 1 to 50 copies of the transgene, mostly integrated in head-to-tail concatemers (data not shown).

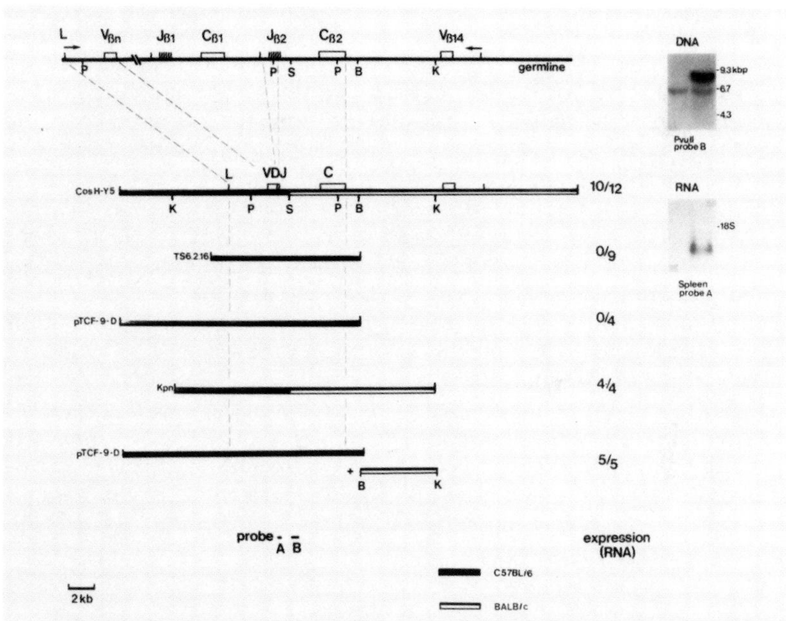

FIGURE 1. The TCR β gene fragments used to generate transgenic mice. The upper line shows the germline organization of the TCR β gene locus. Lines below represent the gene fragments that were injected to produce transgenic mice. On the right typical examples of DNA and RNA analysis are shown.

To determine whether the transgenic TCR β gene was transcribed, Northern blot analysis was performed on RNA isolated from the spleens of founder mice. A ^{32}P end-labeled 30-mer covering the VDJ join and therefore specific for the transgene was used as probe.

Ten out of twelve transgenic mice carrying the insert of cos HYβ9-1.14-5 showed a high, although variable level of mature transgenic mRNA (1.3 kb) in splenic lymphocytes. A similar high level of transgene transcripts was observed in splenocytes from all mice that harbored the 20 kbp KpnI fragment from cos HYβ9-1.14-5. No transgene-specific mRNA could be detected in splenocytes from any of the transgenic mice containing either the 13 kbp insert of clone 8 or the 20 kbp SalI insert of clone 9. These results strongly suggest that a regulatory element located between 2 and 7 kbp downstream of the Cβ2 gene is absolutely required for the expression of the TCR β gene in transgenic mice. However, in the constructs which showed transcriptional activity, both the Cβ2 and 3' flanking sequences were of BALB/c origin. Therefore, one could argue that the smaller C57BL/6-derived clones were inactive because of a mutation in the region replaced by BALB/c sequences in the expressing clones. To check this possibility transgenic mice were raised harboring a mixture of the 20 kbp SalI fragment of clone 9 and the 5.5 kbp BamHI/KpnI fragment located downstream from Cβ2 (Figure 1). Coinjected DNA fragments usually integrate at the same chromosomal location in host cell DNA. Five transgenic mice containing both fragments within the same chromosomal locus were obtained. All five expressed the transgene at a significant level. These results indicate that the C57BL/6-derived clones lack an essential cis-acting sequence, which is present in the 5.5 kbp BamHI-KpnI fragment.

The Transgenic TCR β Gene is expressed in a Tissue-specific manner.

To determine the tissue-specific expression of the different transgenes, Northern blot analysis was performed with RNA obtained from tissues of offspring mice heterozygous for the different transgenes. The results show that both the large cosmid clone and the 20 kbp KpnI fragment are expressed with high preference in thymus and spleen. Monitoring of expression by the monoclonal antibody F23.1, which recognizes TCR β-chains of the V8 family, indicated that normal levels of the transgene encoded TCR β chain was expressed on all T cells in association with the CD3 complex (10). To determine whether the TCR β transgene was also expressed in B cells, splenic B and T cells were stimulated for 4 days with LPS and ConA, respectively, RNA was isolated and analyzed by Northern blot analysis. The expression pattern varied among the different transgenic strains. The most pronounced expression was generally found in T cells. The expression in B cells varied from very low levels to levels comparable with the levels seen in T cells. In one of the strains analyzed also some aberrant expression was observed in non-lymphoid tissues. Apparently the site of integration or the structural organization of the transgenic copies affects the expression patterns. No difference was found between the transcription patterns of cosmid HYβ9-1.14-5 and the smaller 20 kbp KpnI fragment. Therefore, the transgenic TCR β gene carrying 5 kbp of 5' and 7 kbp of 3' flanking sequences is expressed in a lymphoid specific manner with a preference for T cells.

The 3' Region contains an Enhancer which can activate Promoters in a Transient Assay.

To determine whether the region downstream of Cβ2 responsible for the expression in transgenic mice harbored enhancer sequences, various subfragments were tested for their capacity to stimulate transcription from different promoters in a transient CAT assay (11). We used a promoter derived from the _pim_-1 gene (12), the TCR β promoter present in the transgene, both fused to the CAT gene, and δMoCAT (13). The fragments were inserted either downstream (_pim_-1 and TCR β promoter) or upstream (δMoCAT) of the CAT gene in both orientations. The YAC T cell lymphoma cell line and the 2M3 pre-B cell line served as recipients. The _pim_-1 promoter was chosen because of its low background and responsiveness to various enhancers in lymphoid cell lines (M. Van Lohuizen and A. Berns, unpublished results). A physical map of the Cβ2 downstream 5.5 kbp BamHI/KpnI fragment and the location of the fragments and their enhancer activity is visualized in Figure 2. The highest activity was observed with the BglII-NcoI fragment, whereas the overlapping HpaI/BglII fragment was somewhat less active. The most relevant enhancer sequences appeared to be contained within the overlapping 550 bp HpaI/NcoI fragment. The latter fragment showed a similar high activity both in T and B cells, whereas no activity was observed in NIH3T3 fibroblast (data not shown). The different promoters responded similarly to the various fragments, although the absolute level differed considerably, the Moloney promoter being less active.

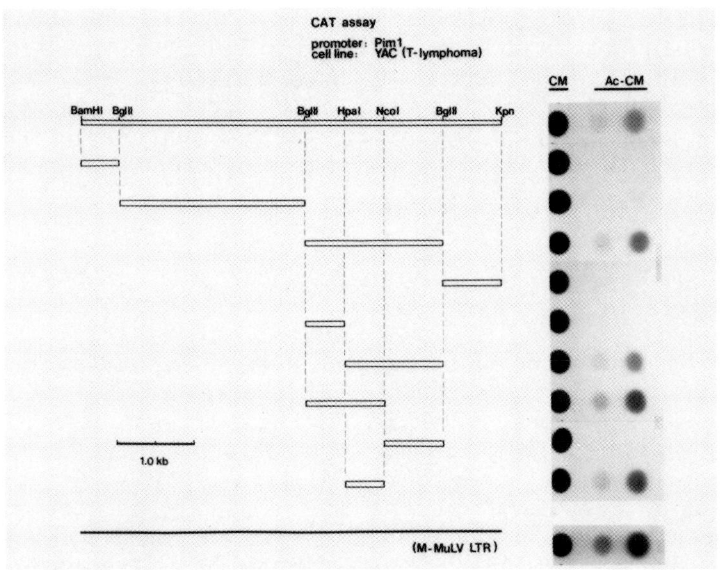

FIGURE 2. Activation of pim-1/CAT fusion genes by fragments located downstream of the TCR β transcription unit. YAC-1 cells were transfected with 30μg of plasmid DNA and cultured for 42 hrs. Protein extracts were prepared from the cells and analyzed for CAT activity. The amount of protein used in the reaction was the same for all samples. The upper line represents the 5.5 kbp BamHI/KpnI fragment, which is located downstream of the TCR β gene.

Figure 3 shows the nucleotide sequence of the 913 bp BglII/NcoI fragment. This fragment contains a number of typical enhancer consensus motifs. Four stretches show significant homology to the enhancer core sequence of SV40 (14), immunoglobulin heavy chain gene and polyoma, as indicated in Figure 3. Two motifs are found which are homologous to the lymphoid specific octamer sequence ATGCAAAT present in the promoters of immunoglobulin heavy and light chain genes and in the IgH enhancer (4, 15, 16).

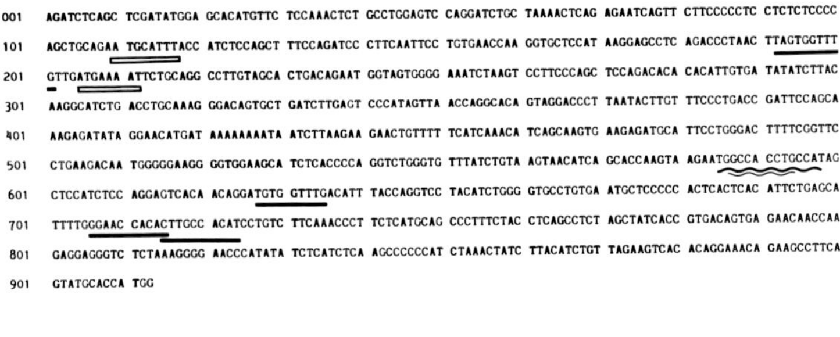

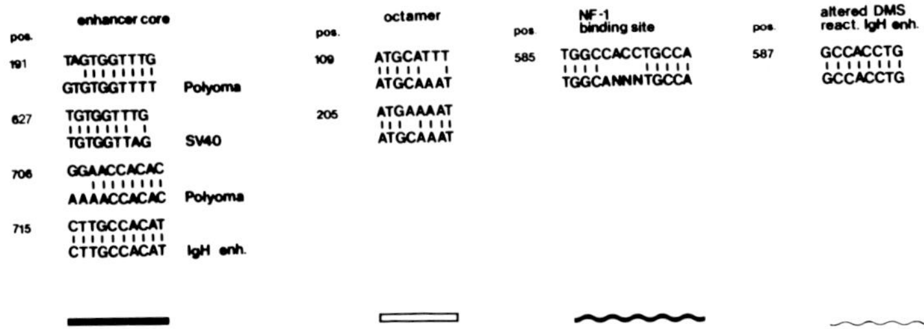

FIGURE 3. Relevant Sequences within the Enhancer Fragment. The nucleotide stretches that show significant homology with known enhancer motifs are presented below with their position (pos.) next to the most homologous enhancer motif.

A slightly modified version of this sequence is also found in the promoter of the TCR β chain gene (17). A nuclear factor-1 (NF-1) binding site, consisting of the inverted repeat TGGCA, with a 3 nucleotide-long spacer is present (18, 19). Within the NF-1 binding site, an 8 nucleotide long stretch (CAGGTGGC) is found, which is identical to a motif in the immunoglobulin enhancer, which is protected from chemical modification by dimethylsulphate in B lymphocytes (20). Therefore, the sequences downstream of Cβ2 that are essential for expression in transgenic mice, conform to an enhancer activity rather than a sequence that is only required for transgenic expression, e.g. by forming the boundery of a transcriptional active region (Grosveld, Cell, in press).

DISCUSSSION

We have shown that all the regulatory elements required for lymphoid specific expression of a transgenic TCR β gene are present in a 20 kbp KpnI fragment that contains 5 kbp of 5' and 7 kbp of 3' flanking sequences (28). An element essential for the expression of the TCR β chain gene in transgenic mice was localized to a 5.5 kbp fragment downstream of the poly-adenylation signal. Also Ig kappa genes contain an important downstream regulatory element which is required for their expression in transgenic mice (21). High expression levels were observed in T cells. A variable, although generally much lower, expression was observed in B-cells, while other cells showed a very low expression in only some transgenic lines. The absence of strict T cell specificity in transgenic mice is reminiscent of the expression behaviour of immunoglobulin heavy chain genes in transgenic mice, which show significant expression in T cells (22). The element downstream of TCR β shows all the characteristics of an enhancer element. Lymphoid specificity was also observed in a transient CAT assay which was used to map the enhancer to a 550 bp fragment downstream of the Cβ2 gene segment. Sequence analysis showed the presence of typical enhancer motifs also found in immunoglobulin genes and SV40. In the transient CAT assay the 550 bp enhancer element is as active in T cells as in B cells. However, when the larger BamHI/NcoI fragment is used, expression in B cell lines is strongly inhibited, suggesting the presence of a silencer element between the Cβ2 region and the TCR β enhancer. For a number of enhancer regions it has been shown that tissue-specificity can be imposed by flanking silencer elements, which can restrict the activity of the enhancer to distinct cell types (23-27).

Most enhancers are found upstream of transcriptional units. A few enhancers have been identified within introns and a few downstream of transcriptional units. Especially for genes which undergo diverse somatic rearrangements in their 5' regions like immunoglobulin genes and TCR genes the positioning of an enhancer into their 3' regions makes sense. For immunoglobulin heavy and kappa light chain genes and the TCR α gene an enhancer element is found in the intron between the J and C gene segments in a region which is not deleted during V-D-J or V-J joining or heavy chain switching. For TCR β genes one would predict that, if a single enhancer element is used, it should be located either within the intron between Jβ2 and Cβ2 or downstream of Cβ2. Our experiments identify an important enhancer element downstream of Cβ2.

References

1. Tonegawa S (1983). Somatic generation of antibody diversity. Nature 302:575.
2. Kronenberg M, Siu G, Hood LE, Shastri N (1986). The molecular genetics of the beta-chain genes of the T-cell antigen receptor and T-cell antigen recognition. Ann Rev Immunol 4:529.
3. Hood L, Kronenberg M, Hunkapillar T (1985). T cell receptors and the immunoglobulin supergene family. Cell 40:225.
4. Banerji J, Olson L, Schaffner W (1983). A lymphocyte-specific cellular enhancer is located downstream of the joining region in immunoglobulin heavy chain genes. Cell 33:729.
5. Gillies SD, Morrison SL, Oi VT, Tonegawa S (1983). A tissue-specific transcription enhancer element is located in the major intron of a rearranged immunoglobulin heavy chain gene. Cell 33:717.
6. Queen C, Baltimore D (1983). Immunoglobulin gene transcription is activated by downstream sequence elements. Cell 33:741.
7. Picard D, Schaffner W (1984). A lymphocyte-specific enhancer in the mouse immunoglobulin kappa gene. Nature 307:80.

8. Luria S, Gross G, Horowitz M, Givol D (1987). Promoter and enhancer elements in rearranged alpha gene of the human T cell receptor. EMBO J 6:3307.
9. McDougall S, Calame KL (1987). Transcriptional regulation of the T cell receptor beta chain locus. in: Abstracts UCLA symposia on molecular and cellular biology. J Cell Biochem sup 11D:280.
10. Uematsu Y, Ryser S, Dembic Z, Borgulya P, Krimpenfort P, Berns A, von Boehmer H, Steinmetz M (1988). In transgenic mice the introduced functional T-cell receptor beta gene prevents expression of endogenous beta genes. Cell:in press
11. Gorman CM, Moffat LF, Howard BH (1982). Recombinant genomes which express chloramphenicol acetyl transferase in mammalian cells. Mol Cell Biol 2:1044.
12. Selten G, Cuypers HT, Boelens W, Robanus-Maandag E, Verbeek J, Domen J, van Beveren C, Berns A (1986). The primary structure of the putative oncogene pim-1 shows extensive homology with protein kinases. Cell 46:603,
13. Linney E, Davis B, Overhauser J, Chao E, Fan H (1984). Non-function of a Moloney murine leukaemia virus regylatory sequence in F9 embryonal carcinoma cells. Nature 308:470.
14. Schirm S, Jiricny J, Schaffner W (1987). The SV40 enhancer can be dissected into multiple segments, each with a different cell type specificity. Genes & Development 1:65.
15. Falkner FG, Zachau HG (1984). Correct transcription of an immunoglobulin kappa gene requires an upstream fragment containing conserved sequence elements. Nature 310:71.
16. Parslow TG, Blair DL, Murphy WJ, Granner DK (1984). Structure of the 5' ends of immunoglobulin genes. A novel conserved sequence. Proc Natl Acad Sci USA 81:2650.
17. Royer HD, Reinherz EL (1987). Multiple nuclear proteins bind upstream sequences in the promoter region of a T-cell receptor beta-chain variable-region gene: evidence for tissue specificity. Proc Natl Acad Sci USA 84:232.
18. Nowock J, Sippel AE (1982). Specific protein-DNA interaction at four sites flanking the chicken lysozyme gene. Cell 30:607.
19. Rawlins DR, Rosenfeld PJ, Wides RJ, Challberg MD, Kelly Jr. TJ (1984). Structure and function of the adenovirus origin of replication. Cell 37:309.

20. Gimble JM, Max EE (1987). Human immunoglobulin kappa gene enhancer: Chromatin structure analysis at high resolution. Mol Cell Biol 7:15.
21. Storb U, Pinkert C, Arp B, Engler P, Gollahon K, Manz J, Brady W, Brinster R (1986). Transgenic mice with mu and kappa genes encoding antiphosphorylcholine antibodies. J Exp Med 164:627.
22. Grosschedl R, Weaver D, Baltimore D, Costantini F (1984). Introduction of a mu immunoglobulin gene into the mouse germ line: Specific expression in lymphoid cells and synthesis of functional antibody. Cell 38: 647.
23. Imler JL, Lemaire C, Wasylyk C, Wasylyk B (1987). Negative regulation contributes to tissue specificity of the immunoglobulin heavy-chain enhancer. Mol Cell Biol 7:2558.
24. Nir U, Walker MD, Rutter WJ (1986). Regulation of rat insulin 1 gene expression: Evidence for negative regulation in nonpancreatic cells. Proc Natl Acad Sci USA 83:3180.
25. Laimins L, Holmgren-Koenig M, Khoury G (1986). Transcriptional "silencer" element in rat repetitive sequences associated with the rat insulin 1 gene locus. Proc Natl Acad Sci USA 83:3151.
26. Baniahmad A, Muller M, Steiner C, Renkawitz R (1987). Activity of two different silencer elements of the chicken lysozyme gene can be compensated by enhancer elements. EMBO J 6:2297.
27. Kuhl D, De la Fuente J, Chaturvedi M, Parimoo S, Ryals J, Meyer F, Weissmann C (1987). Reversible silencing of enhancers by sequences derived from the human IFN-alpha promoter. Cell 50:1057.
28. Krimpenfort P, de Jong R, Uematsu Y, Dembic Z, Ryser S, von Boehmer H, Steinmetz M, Berns A (1988). Transcription of T cell receptor beta-chain genes is controlled by a downstream regulatory element. EMBO J :in press.

Gene Transfer and Gene Therapy, pages 129–149

REGULATION OF FOS GENE: A PARADIGM FOR NUCLEAR ONCOGENES

I.M. Verma, J. Visvader, W.W. Lamph, P. DeTogni
J. Barber and P. Sassone-Corsi

The Salk Institute
P.O. Box 85800
San Diego, California 92138

ABSTRACT Proto-oncogene fos is an inducible gene. Expression is invariably very rapid but transient. A cyclic AMP responsive element (CRE) has been identified to be localized between positions -57 to -63 upstream of the 5'-cap site. In DNaseI footprint analysis, purified CRE binding protein protects this region. The c-fos cDNA is able to induce transformation if an A-T rich stretch located downstream of the coding domain is removed. Both the viral and cellular fos protein are extensively modified with serine phosphorylation as the predominant modification. Fos monoclonal antibodies have been generated. Regulation of the fos gene is complex, but appears to involve controls at the level of transcription, post-transcription and post-translation.

INTRODUCTION

With nearly two-scores of proto-oncogenes capable of malignant transformation, it is a miracle that most cells maintain their normal growth and differentiation program. Why has the cell not eliminated such potentially lethal genes during the course of evolution? A glance at Table I would indicate that many of oncogenes are essential for the normal metabolic function of the cell (1-12). Clearly there is a delicate balance between the normal and malignant activities of the products of proto-oncogenes. We have, for the past 5 years, been interested to study the regulation of proto-oncogene fos whose viral cognate is the resident transforming gene of FBJ-murine osteosarcoma virus. The viral and cellular fos proteins share identities in the first 332 a.a. (with 5 single a.a. changes), but the remaining 48 a.a. of c-fos protein are in a different reading frame as compared to 49 a.a. of the v-fos protein. Despite altered C-termini, both proteins can induce cellular transformation and are localized in the nucleus (13,14).

TABLE 1

PROTO-ONCOGENES WITH STRUCTURAL HOMOLOGIES TO NORMAL CELLULAR PROTEINS

ONCOGENE	CELL HOMOLOGUE
c-sis	A chain of PDGF
c-erbB	EGF Receptor
c-fms	CSF-1 Receptor
c-erbA	Thyroxine (T3) Receptor
c-jun	Transcriptional factor (AP1)
Int-1	Wingless in Drosophila
Int-2	FGF-related
c-rel	Dorsal in Drosophila

A remarkable property of the c-fos gene is its induction in response to a large array of diverse agents ranging from growth factors, differentiation agents, stress, pharmacological reagents, immunomodulators, etc. (15, 16). Induction is invariably very rapid, but transient (17). Thus, we are faced with a situation where the same c-fos protein capable of inducing cellular transformation is also induced in nearly every cell type during growth, differentiation, and development. How does the cell prevent from being transformed? Clues for this delicate balance can be discerned by studying regulation of the fos gene. It appears proto-oncogene fos is regulated at the transcriptional, post-transcriptional and post-translational level.

RESULTS

Cyclic AMP responsive elements in the c-fos promoter:

Since diverse agents induce c-fos gene, we would have suspected that its promoter might contain multiple DNA motifs (16). Briefly, there are two types of elements, (a) those required for basal level of transcription and (b) inducible elements. The dyad symmetry elements (DSE) responsive to serum, TPA, EGF, etc. has been extensively analyzed by a number of laboratories. Here we describe a cyclic AMP response element (-60 element) located between -57 to -63 in the human, mouse and chicken c-fos gene. Figure 1A shows the sequence of the -60 element which has strong homology to the consensus TGACGT present in promoter elements required for cAMP regulation of the rat somatostatin gene (18), the human glycoprotein hormone α-subunit gene, α-CG (19), and others. In PC12 cells, expression of the c-fos gene is induced by dibutyryl-cAMP, forskolin and choleratoxin (20-22). This induction is rapid and transient, similar to that observed with other inducing agents (17, 23).

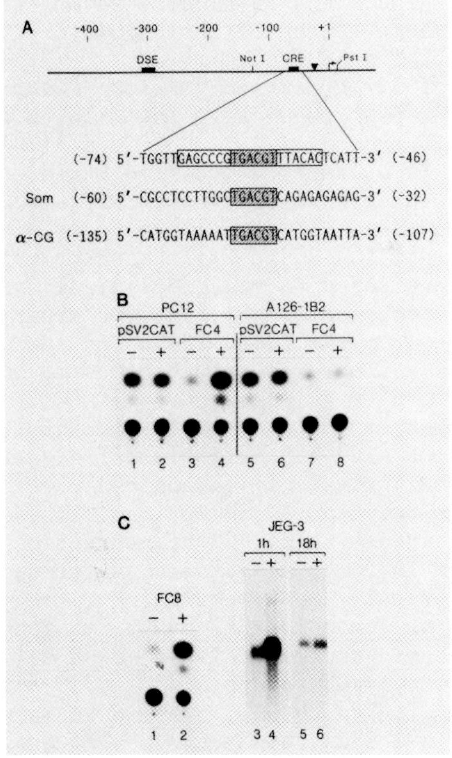

fig. 1

FIGURE 1. cAMP responsive element (CRE) in the human c-fos gene. **A)** Structure of the c-fos promoter. The positions of the dyad-symmetry element (DSE), CRE, TATA-box (filled triangle) and the transcription start-site are indicated. The NotI and PstI restriction sites were used to prepare a 5' end-labeled fragment as probe in the DNA footprinting experiments. The sequence around the c-*fos* CRE is indicated. The shadowed sequence 5' TGACGT 3' is the consensus sequence common to cAMP responsive genes such as the rat somatostatin (Som) and the human glycoprotein hormone α-subunit (a-CG). In the case of the α-CG gene, the conserved sequence is repeated in tandem; in the Figure only the most downstream repeat is shown. The sequence of the c-*fos* promoter protected in the DNaseI footprinting experiments (see Fig. 3) is boxed (positions -51 to -69).
B) Induction of c-fos-CAT by forskolin in PC12 cells. pSV2CAT contains the entire SV40 promoter linked to the CAT bacterial gene. FC4 is a deletion mutant of the c-*fos* promoter where 402 bp of the regulatory region are linked to the CAT gene (26).
Transfection of 10 μg plasmid DNA into PC12, A126-1B2 and JEG-3 cells was performed using the calcium phosphate co-precipitation technique (27). PC12 and A126-1B2 cells were passaged in Dulbecco's modified Eagle's medium (DMEM) supplemented with 10% fetal calf serum (FCS) and 5% horse serum (HS). JEG-3 cells were grown in DMEM supplemented with 10% FCS. Cells were transfected at 70% confluence and exposed to the precipitate for 12 hrs. Cells were harvested after 24 hrs; for experiments with forskolin, the drug was added 16-20 hrs. prior to harvesting, to a final concentration of 20 μm.
The plasmids were transfected into rat PC12 pheochromocytoma cells (lanes 1-4) and into the PC12 mutant cell line A126-1B2 which is deficient in cAMP-dependent protein kinase II (lanes 5-8). + and - indicate cells treated (+) or not (-) with forskolin. The forskolin-induction of FC4 transcription in PC12 cells is about 15-fold, as measured after densitometric scanning of several autoradiograms.
C) Induction of the c-fos-CAT and c-fos gene by forskolin in human choriocarcinoma JEG-3 cells: Induction of the transfected FC8 (containing a -220 deletion of the c-*fos* promoter linked to CAT) plasmid is shown in lanes 1 and 2. Lanes 3-6 show the Northern blot analysis of endogenous c-*fos* RNA induction by forskolin after 1 hr.

To characterize cAMP-responsive sequences within the c-fos promoter, we fused several deletions of the c-fos 5' flanking region to the bacterial chloramphenicol acetyltransferase (CAT) gene. Promoter activity was determined by transient assay after calcium-phosphate transfection of the fos-CAT fusion constructs into PC12 cells. To assess non-specific effects of cAMP on gene expression, a plasmid (pSV2CAT) containing the SV40 promoter fused to the CAT gene was used as a control (24). We tested whether the regulation of the transfected c-fos promoter by cAMP would be analogous to that observed for the endogenous gene by treating the cells with forskolin, 12 hours after transfection. As shown in Figure 1B (lanes 3-4); a 15-fold induction of the c-fos-CAT fusion gene is observed after addition of forskolin, whereas expression from the SV40 promoter is unchanged (lanes 1-2).

To determine if induction of the c-fos promoter by cAMP requires cAMP-dependent protein kinase activity, as previously reported for the somatostatin gene, we examined the expression of the c-fos-CAT fusion gene in the PC12 mutant cell line A126-1B2. These cells have normal levels of cAMP-dependent protein kinase I activity, but are markedly deficient in type II activity (25). In contrast to the 15-fold stimulation observed in wild-type PC12 cells, CAT activity in A126-1B2 cells is unchanged or minimally induced (less than 2-fold, lanes 7-8). The pSV2CAT control shows the same profile as in PC12 wild-type cells (lanes 5-6).

We also examined c-fos induction by cAMP in another endocrine cell line, JEG-3, a placental trophoblast cell line derived from a human choriocarcinoma tumor (19). A 12-fold induction by forskolin is also observed when the c-fos-CAT fusion gene is transfected into JEG-3 cells (Fig. 1C; lanes 1-2). Since induction of the c-fos gene in this cell line has not been reported previously, we studied the effect of cAMP on the endogenous gene by Northern blot analysis of RNAs purified 1 and 18 hours after forskolin treatment. As shown in Figure 1C there is a marked increase in c-fos mRNA after 1 hour of forskolin treatment (lanes 3-4). Little induction is observed after 18 hours (lane 5-6) indicating the transient nature of the effect. These results

---

and 18 hrs. of treatment. The probe used in the hybridization was a full-length human c-fos cDNA clone.

demonstrate that the stimulation of c-fos expression by forskolin in JEG-3 cells is similar to that described in PC12 cells and that in either cell type, the episomal conformation of the gene does not alter cAMP responsiveness.

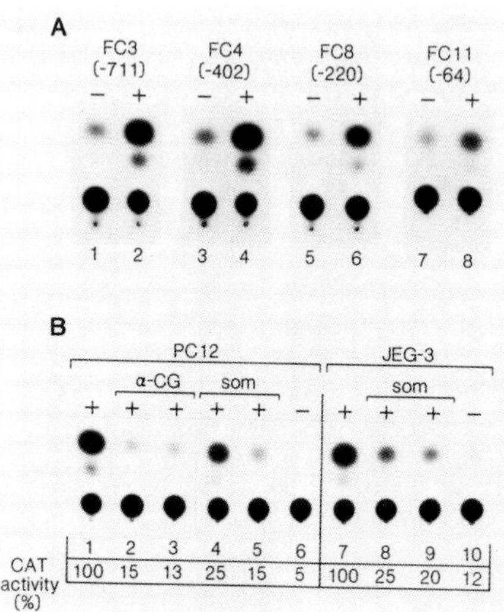

FIGURE 2. Sequence characterization of the c-fos cyclic-AMP responsive element. **A)** Analysis of forskolin induction of the deletion mutants of the c-fos promoter. Several deletions were linked to the CAT gene and the plasmid DNA was transfected in PC12 cells as described in Figure 1. The data shown is representative of a series of experiments. Quantification of CAT activity was performed by densitometric scanning of several autoradiograms. Induction of c-fos transcription by forskolin for FC3 and FC4 is about 15-fold; induction of FC8 is about 10-fold. Induction of FC11 is about 4-fold. Lanes 5 to 8 are from autoradiograms that have been exposed three times longer than lanes 1 to 4. + and - indicate cells treated (+) or not (-) with forskolin. **B)** Competition experiments in PC12 and JEG-3 cells. Lane 1, CAT activity after transfection of 4 μg of FC8 DNA in forskolin-stimulated PC12 cells. Increasing amounts of α-CG competitor (15 and 25 μg; lanes 4 and 5 respectively) show the same effect. Lane 6 represents

To delineate the sequences in the c-*fos* promoter responsible for the induction by cAMP, we progressively deleted sequences from the 5'-flanking region of the c-*fos* gene (26). CAT activity derived from FC3, FC4, and FC8 plasmids was induced 10- to 15-fold in response to forskolin (Fig. 2A; lanes 1 to 6), whereas the induction observed with the FC11 plasmid was reduced to 3-4-fold (Fig. 2A; lanes 7-8). Since the end-point of the promoter deletion in FC11 is at position -64 (see Fig. 1), this result indicates that the sequences flanking the CRE consensus 5' TGACGT 3' are probably required for complete induction.

To demonstrate that this putative CRE in the proximal upstream region of the c-*fos* promoter is responsible for cAMP induction by interaction with a nuclear *trans*-acting factor(s), we performed *in vivo* competition experiments (27, 28) in which increasing amounts of plasmid DNAs containing CREs from the somatostatin and α-CG genes were co-transfected with a fixed amount of a c-*fos*-CAT gene. For this study we used FC8, a c-*fos*-CAT recombinant in which only 220 bp of the c-*fos* promoter are linked to the CAT gene. This fusion gene is still efficiently induced by cAMP, as shown by the deletion analysis (Fig. 2A; see also Fig. 2B, compare lanes 1 and 6 for PC12 cells and

---

the FC8 non-induced basal level. Similar results were obtained in JEG-3 cells with 15 and 25 μg of the somatostatin competitor (compare FC8 forskolin induced level, lane 7, and lanes 8-9). Analogous results were obtained with the α-CG competitor (not shown). Lane 10 represents the FC8 non-induced level. CAT activity was quantified by densitometric scanning of autoradiograms from several experiments. The relative ratio of the observed activity of the various samples to the maximal activity recorded (lanes 1 and 7) is expressed as a percentage.
The α-CG competitor contains the original double CRE present in the promoter of the gene (positions -152 to -100) cloned into the GEM 4 plasmid. The somatostatin competitor contains the CRE sequence (positions -32 to -61) cloned into the GEM 4 plasmid. DNA transfections, forskolin treatments and CAT assays were performed as described in the legend to Figure 1. + and - indicate cells treated (+) or not (-) with forskolin. The total amount of DNA in each transfection was the same, being adjusted to 30 μg with GEM 4 DNA when necessary.

lanes 7 and 10 for JEG-3 cells). Increasing amounts of the α-CG and somatostatin CRE-containing plasmids drastically reduced the cAMP induction of the c-fos promoter in PC12 cells (Fig. 2B; lanes 2-3 and 4-5) and JEG-3 cells (lanes 8-9; data with the α-CG competitor are not shown). These results indicate the presence of a trans-acting factor(s) for the cAMP response of the c-fos promoter in both PC12 and JEG-3 cells. Furthermore, this factor(s) is probably common to other CREs since cAMP induction of the c-fos gene can be drastically reduced by using heterologous competitors.

To further delineate the CRE factor-binding site, DNase I footprinting assays using nuclear extracts of JEG-3 and PC12 cells were carried out. Increasing amounts of the extracts were incubated with a 5' end-labeled c-fos promoter fragment (see Fig. 1; NotI (-120) to PstI (+15)). Samples were treated with DNase I and the digestion products resolved on denaturing polyacrylamide gels. We observed a single protected region which covered 19 nucleotides, extending from -51 to -69 relative to the transcription initiation site. To demonstrate that the c-fos CRE blinding factor is actually the same protein that binds to other CREs, we performed DNA footprints using the PC12-derived affinity-purified 43 kd nuclear protein that has been shown to specifically interact with the rat somatostatin CRE (CREBP, 29). As shown in Figure 3, the purified 43 kd CREBP binds to the c-fos CRE, protecting from DNaseI digestion the same -51/-69 region observed in the footprint experiments performed with nuclear extracts. Furthermore, the binding of the purified protein is also decreased by competitors containing somatostatin and α-CG CREs (lanes 4 and 5). It is interesting to note that the promoter region containing the CRE appears to be required for basal level transcription and for full TPA induction. This observation would suggest possible multiple functions for the same promoter region.

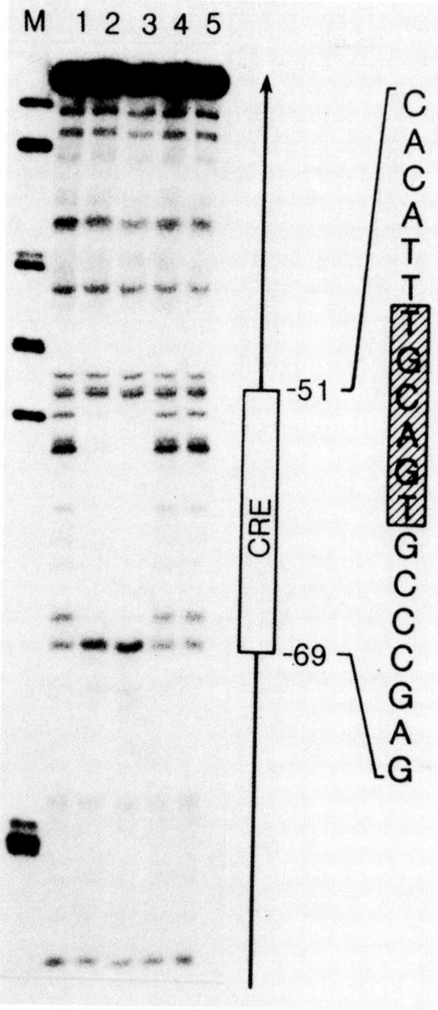

FIGURE 3. DNaseI footprint of the c-fos CRE by the affinity-purified CREB protein from PC12 cells. Lane 1 is the control DNaseI digest of the probe without binding protein. Lanes 2 and 3 show the footprint of the -51/-69 element by increasing amounts of the CREB protein. Lanes 4 and 5 show the competition for binding when an excess of somatostatin (lane 4) and α-CG (lane 5) CREs were added to the reaction.

Transformation by c-fos cDNA:

We have previously demonstrated that c-fos gene can induce cellular transformation if sequences at the 3'-end of the non-coding domain are deleted (30). Removal of a 67 bp A-T rich region located some 500 nt downstream of the chain terminator, and 150 nt upstream of the poly (A)-addition signal was sufficient to induce transformation (31). These experiments were carried out with either hybrid viral-mouse c-fos genes or total mouse c-fos gene (containing introns). To rule out the influence of non-coding sequences other than those located at the 3'-non-coding domain, we repeated the

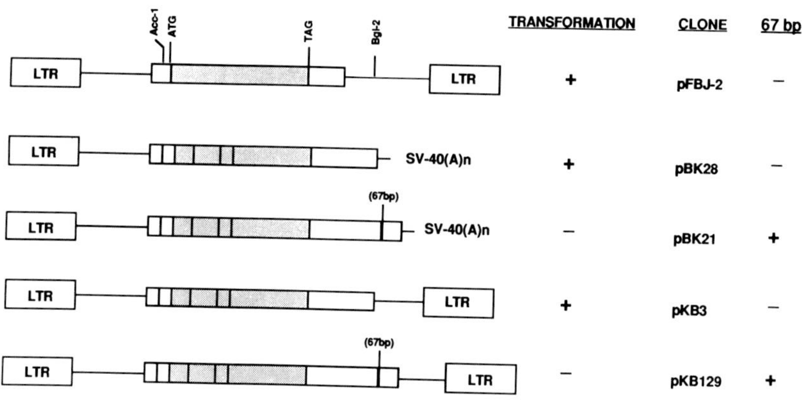

FIGURE 4. C-fos cDNA constructs: The structure and the transforming potential of the various cDNA constructs (with or without the 67 bp region) is shown.

experiments with human c-fos cDNA clone. Figure 4 shows that only those constructs where the 67 bp A-T rich region has been removed are capable of transformation. We are now pursuing studies to determine if the A-T rich region influences the stability of the fos mRNA.

Fos proteins are post-translationally modified:

Cells transformed by either the c-fos or v-fos DNA constructs constitutively produce the corresponding fos proteins. In Figure 5, fos proteins were identified by metabolic labeling and immunoprecipitation of denatured lysates using fos-specific antisera (32). Figure 5A shows the profile of c-fos and v-fos proteins. In addition to the fos specific proteins, in the c-fos protein lane there are other proteins ranging from 35-45 kda. None of them appears to be related to p39, a protein which has been shown to be non-covalently bound to fos protein (33). Pulse-chase analysis indicates that the primary translation products of c-fos and v-fos begin as forms with relatively high electrophoretic mobility (Fig. 1 lanes 2 and 6) which are rapidly modified to forms with lower mobility (lanes 4 and 8).

Since it had previously been shown that [^{32}P]phosphate could be incorporated into both c-fos and v-fos proteins (34) and since phosphorylation often affects the electrophoretic mobility of proteins in SDS gels, it was possible that phosphorylation of the proteins might account for part of this post-translational processing. To test this, c-fos and v-fos proteins were purified by immunoprecipitation from cells metabolically labeled with [^{32}P]phosphate or L[^{35}S]methionine. Both proteins were then treated for various times with the enzyme, alkaline phosphatase (AP), which hydrolyzes organic phosphoesters. The data in Figure 5B indicates that enzymatic removal of metabolically incorporated [^{32}P]phosphate was very efficient for both c-fos and v-fos proteins. Enzymatic phosphoester hydrolysis was inhibited by the specific phosphatase inhibitor, para-nitrophenylphosphate (pNPP) (see Fig. 5B, lanes 4 and 8). This demonstrates that the removal of radioactivity from the proteins was due to phosphatase activity alone and not due to spontaneous hydrolysis or to the presence of contaminating enzymes in the preparation. We conclude from this that most, if not all, of the protein phosphate groups are in the form of phosphoesters.

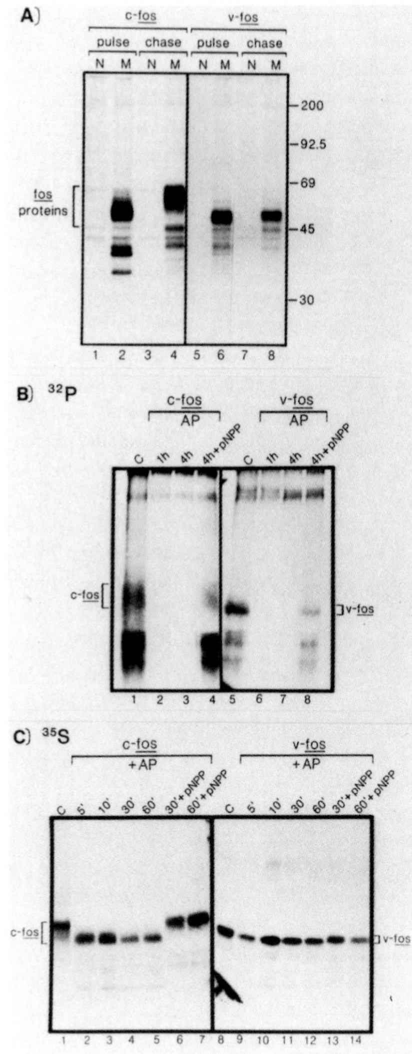

FIGURE 5. Analysis of fos proteins. **A)** Immunoprecipitation and pulse-chase analysis of c-fos and v-fos proteins. Cells were pulse-labeled for 10 min with L-[^{35}S]methionine and then either lysed immediately (lanes 2 and 6) or first "chased" for an additional 20

To determine the effect of phosphoester removal on the electrophoretic mobility of the proteins a similar experiment was performed using c-fos and v-fos proteins immunoprecipitated from cells metabolically labeled with L-[^{35}S]methionine. As shown in Figure 5C hydrolysis of protein phosphoesters results in a dramatic increase in the electrophoretic mobility of the c-fos protein as well as more subtle, but reproducibly observed, shift for the v-fos protein. It is thus likely that covalent modification by phosophorylation is solely responsible for the electrophoretic heterogeneity of both proteins in SDS gels.

Fos monoclonal antibodies:

Paucity of a good and reliable source of fos antibodies lead us to generate fos monoclonal antibodies

---

min by the addition of a 1000-fold molar excess of unlabeled L-methionine (lanes 4 and 8). C-fos and v-fos proteins were then immunoprecipitated and analyzed by SDS gel electrophoresis and autofluorography. Odd lanes (N) - normal preimmune serum; even lanes (M) - αM2 fos antiserum. Arrows on the side indicate molecular weight size markers.

B) Treatment of c-fos and v-fos proteins with alkaline phosphatase. Cells were metabolically labeled for 5 hr. with [^{32}P]phosphate, denatured lysates were prepared, and c-fos (left) and v-fos (right) proteins were immunoprecipitated. The proteins were removed from the final Pansorbin pellet, concentrated solutions of Nonidet P-40, deoxycholate, and NaCl were added to a final concentration of 1%, 1%, and 150 mM, respectively, the samples were then aliquoted and incubated at 37°C as follows: 1 and 5, no incubation; 2 and 6, 1 hr. with alkaline phosphatase (AP); 3 and 7, 4 hr. with AP; 4 and 8, 4 hr. with AP plus 2 mM P-nitrophenylphosphate (p-NPP). The samples were then analyzed by SDS polyacrylamide gel electrophoresis, and the radioactive bands were visualized by autofluorography.

C) The experiment was essentially as described in B above, but cells were metabolically labeled with L-[^{35}S]methionine for 1 hr. and the resultant immunoprecipitated proteins incúbated at 37°C with AP as follows: 1 and 8 no incubation; 2 and 9, 5 min; 3 and 10, 10 min; 4 and 11 30 min; 5 and 12, 60 min; 6 and 13, 30 min plus 10 mM p-NPP; 7 and 14, 60 min plus 10 mM pNPP.

against synthetic peptides. We synthesized synthetic peptides corresponding to the predicted human fos protein amino-terminal (4-17 with one serine deletion at position 15), carboxy-terminal (359-378 aa.) and to M peptide region 132-153 aa) using modifications of the solid phase synthesis as described.

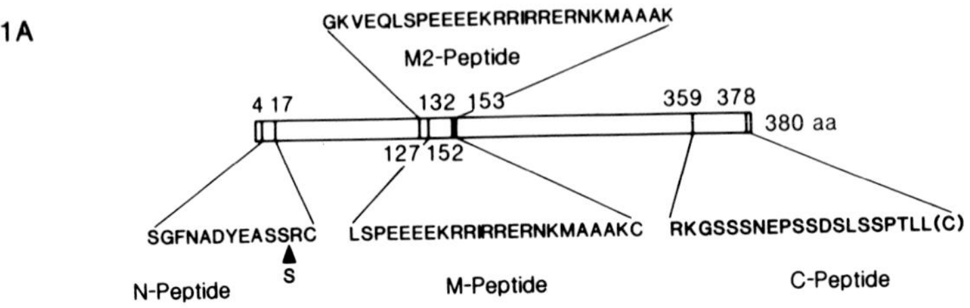

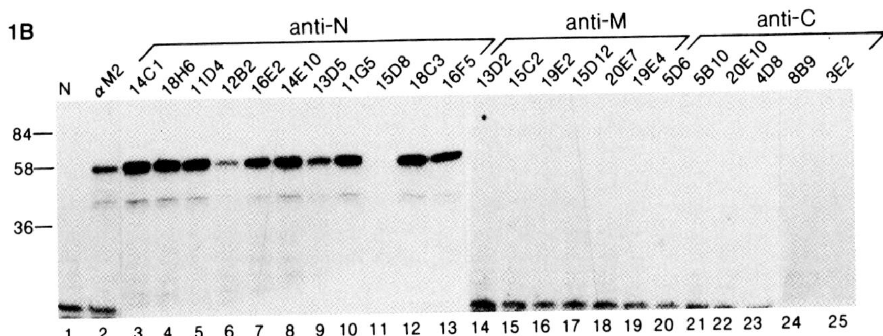

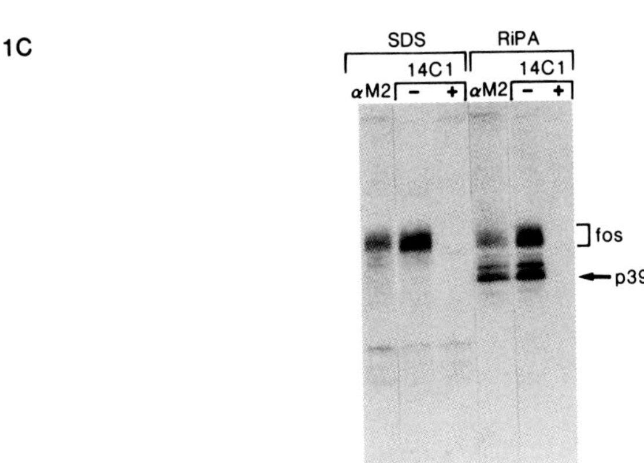

fig. 6

Figure 6 shows that monoclonal antibodies raised against NH_2-terminal synthetic peptide can immunoprecipitate ^{35}[S]-labeled c-fos protein from Hela cells induced with TPA. Of the 11 monoclonal ab's directed against N-terminal peptide, ten immunoprecipitated a 55 kda protein analogous in size to that observed with M2 antisera (Fig. 6B). We wanted to further characterize some of the monoclonal antibodies and chose 14C1 because it gave the strongest signal (Fig. 6B lane 3). Figure 6C (lane 5) shows that 14C1 antibodies are also capable of recognizing p55-p39 cellular protein complex similar to that observed with M2-antisera. The p39 is a cellular protein which is non-covalently bound to p55 and is not observed when the samples are treated with SDS prior to immunoprecipitation (Fig. 6C). No p55 kda protein can be immunoprecipitated if the samples are preincubated with N-peptide indicating that the 14C1 monoclonal antibody is specific for fos protein and is directed against amino-terminal region of the protein.

FIGURE 6. Immunoprecipitation of $p55^{c\text{-}fos}$ with various monoclonal antibodies: **A)** The location and the primary sequence of the fos peptides used to raise the antibodies is indicated. The numbers indicate amino acid residues in the protein. The arrow indicates the serine residue which was deleted in the synthesis of the N-peptide. The sequence of the M2 peptide (127-152) used for the preparation of the rabbit fos antiserum (αM2) is also shown.
B) Immunoprecipitation of p55 c-fos from Hela cell nuclear extracts: The monoclonal Ab's were grouped on the bases of their specificity for the N-peptide (lanes 3-13) the M-peptide (lanes 14-19) and the C-peptide (lanes 20-25). The control with normal serum is shown in lane 1. The products were analyzed by electrophoresis on a 10% SDS-polyacrylamide gel. The numbers on the left indicate the molecular weights ($x10^3$) of marker proteins.
C) Nuclei from [^{35}]methioninein-labeled Hela cells were extracted in SDS lysis buffer (lanes 1-3) or in RIPA buffer (lanes 4-6). Aliquots from each nuclear extract were immunoprecipitated with affinity purified αM2 fos antiserum or with the monoclonal antibody 14C1. The symbols + and - indicate if 14C1 was preabsorbed or not with 100 μg/ml of N-peptide respectively.

DISCUSSION

The c-fos promoter has been extensively analyzed by a number of laboratories and its salient features are depicted in the figure. The promoter contains three types of regulatory elements:

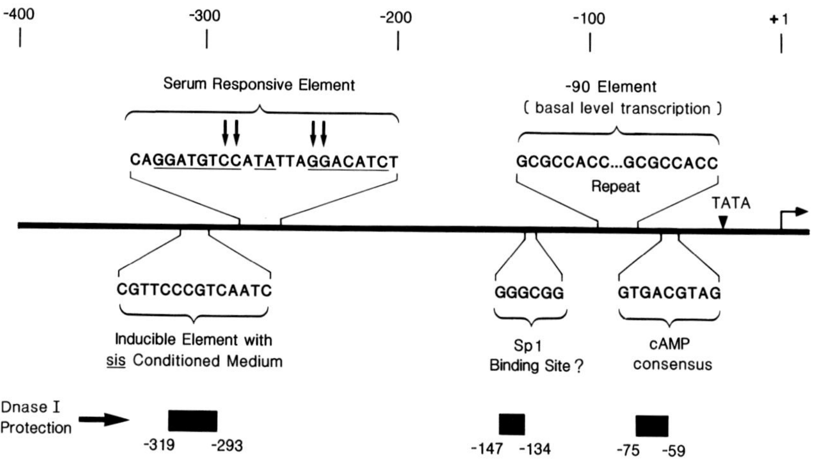

FIGURE 7. The c-fos promoter. Asterisks indicate protein-DNA contact sites determined by DNA methylation interference assays. The element responsive to sis-conditioned medium has been experimentally identified in the human c-fos gene; an analogous sequence is present in the mouse gene.

(i) The TATA box (-25/-32). As in other promoters, this sequence is required for proper initiation and orientation of transcription.

(ii) Two upstream elements for basal transcription. The first element, containing the consensus sequence TGACGT , is located between -57 to -63 in the human, mouse, and chicken c-fos genes. Deletion of this element reduces the basal transcription level by 10-fold (35). This sequence in the mouse c-fos gene is protected from DNAase I digestion following incubation with nuclear extracts (36). Interestingly, This element resembles the consensus sequence of cAMP-regulated promoters (29). An identical sequence is also found in the upstream promoter of several adenovirus genes. A second element required for basal transcription of the human c-fos gene is a direct repeat located between -76/-97; deletion of this element decreases transcription by 5- to 10-fold (35). In the mouse c-fos gene there are two analogous GC-rich regions (-130/-150 and -90/-112) that are also required for the basal level of transcription.

(iii) Inducible elements. The upstream enhancer region, located between -297 and -317, encompasses a sequence element with dyad symmetry, termed DSE. The DSE is required for induction of the c-fos gene with serum, the phorbol ester TPA, and EGF (37, 38). Additionally, binding to nuclear factors is completely abolished by competition with a synthetic oligonucleotide containing the DSE. Heterologous promoters linked to DSE also respond to induction with serum or EGF, but require the -60 element for maximal induction. The DSE is conserved between chicken, mouse, and human genes and is similar to sequences in the enhancers of polyoma, SV40, and Moloney murine leukemia virus LTR.

Since induction of the c-fos gene is not affected by protein synthesis inhibitors, it would appears that trans-acting factor(s) required for transcription preexist in the cell. We have previously shown that c-fos gene expression is regulated by both negative and positive cellular factors (27). In quiescent or serum-starved cells, transcription of the c-fos gene can be induced by addition of excess copies of the upstream c-fos promoter element, which suggests that some negative factor(s) is being competed out. Conversely, addition of the same c-fos promoter element in serum-induced cells diminishes c-fos transcription, indicative of competition with positive factor(s).

On the basis of these observations we have proposed the following model. In uninduced cells, a negative factor(s) is associated with the c-fos promoter-transcriptional complex, resulting in transcriptional shut-off. Upon induction, this factor(s) is modified and its affinity for the regulatory sequences decreases (alternatively, the positive trans-acting factor is modified and its affinity for the regulatory sequences increases). In this model, the intracellular balance of the two classes of factors is crucial as their relative affinities for c-fos regulatory sequences change upon induction. Post-translational modification of transcriptional factors has also been proposed for control of immunoglobulin genes (39) and the TPA-inducible human metallothione II_A gene.

Evidence to date indicates that two cellular pathways are involved in c-fos induction - one involving protein kinase C activation, the other involving adenylate cyclase.

Growth factors increase intracellular degradation of membrane phosphoinositides, resulting in the production of diacylglycerol (DAG) and IP-3, the former activating protein kinase C and the latter mobilizing release of CA^{2+} from the endoplasmic reticulum. Down-regulation of protein kinase C by sustained treatment with TPA (which mimics DAG in activating the kinase), or inhibition of the kinase, precipitously depresses c-fos gene induction. Furthermore, TPA-resistant variants of HL-60 cells do not express the c-fos gene upon induction with TPA (40). TPA induction of c-fos involves the DSE binding protein, whereas other TPA-inducible genes require different transcriptional factors like AP-1 and NF-$_K$B. Thus, TPA can activate genes that respond to different nuclear transcriptional factors but use common cellular signaling pathways.

The c-fos gene is also inducible with agonists of adenylate cyclase like dibutyryl-cAMP, forskolin, or cholera toxin (20-22). The -60 promoter element, which resembles the consensus sequence for cAMP regulation, binds to a nuclear factor cAMP-binding protein (Fig. 3). Thus the c-fos gene can be induced by two diverse pathways that may "cross-talk", as suggested by the observation that TPA induces phosphorylation of the catalytic subunit of adenylate cyclase.

Despite extensive knowledge of the structure and expression of the fos gene, its role in the cell remains largely conjectural. The rapid induction of the fos gene in a wide variety of cell types in response to different stimuli would suggest a generalized role in signal transduction. Induction of the fos gene in PC12 pheochromocytoma cells in response to NGF. In spinal cord neurons in response to sensory stimulation, and in specific subsets of neurons in the central nervous system following administration of convulsants suggests that the fos protein may be involved in processing information, generated upon short-term stimulation, in the nucleus. A role for fos protein has also been postulated in the establishment of long-term memory.

The ability of the fos protein to form a complex with DNA would suggest a role in the regulation of gene expression (41, 42). This view is strengthened by the observations that the v-fos protein trans-activates the α1-collagen promoter, and that the c-fos protein directly participates in the nucleoprotein complexes regulating gene expression during adipocyte differentiation (43). The evidence to date, however, indicates that fos protein may not bind to DNA directly, but may require the formation of a complex with other nuclear proteins. Modification of the c-fos protein could play a crucial role in the formation of this complex.

ACKNOWLEDGEMENTS

We would like to thank Dr. Marc Montminy for the generous supply of CREB protein. We also thank Drs. L. Ferland and P. Mellon for their interest in this work. This work was carried out through funds from NIH and ACS grants. J.V. is in receipt of a Jane Coffin Childs Fellowship, and P.S.-C. is on leave from CNRS, France.

REFERENCES

1. Bishop JM: Trends Genet 1:245, 1985.
2. Van Beveren C, Verma IM: Curr Top Microbiol Immunol 123:73, 1985.
3. Bishop JM: Cell 42:23, 1985.
4. Doolittle RF, Hunkapiller MW, Hood LE, Devare SG, Robbins EC, Aaronson SA, Antoniades HN: Science 221:275, 1983.
5. Waterfield MD, Scrace GT, Whittle N, Stroobant P, Johnsson A, Wasteson A, Westermark B, Heldin C-H, Huang JS, Deuel TF: Nature 304:35, 1983.
6. Ullrich A, Coussens L, Hayflick JS, Dull TJ, Gray A, Tam AW, Lee J, Yarden Y, Libermann TA, Schlessinger J, Downward J, Mayes ELV, Whittle N, Waterfield MD, Seeburg PH: Nature 309:418, 1984.
7. Scherr CJ, Rettenmier CW, Sacca R, Roussel MF, Look AT, Stanley ER: Cell 41:665, 1981.
8. Weinberger C, Thompson CC, ONG ES, Lebo R, Gruol DJ, Evans RM: Nature 324:641, 1986.
9. Sap J, Munoz A, Damm K, Goldberg Y, Ghyduel J, Leutz A, Beug H, Vennstrom B: Nature 324:635, 1986.
10. Bohmann D, Bos TJ, Admon A, Nishimura T, Vogt PK, Tjian R: Science 238:1386, 1987.
11. Rijsewijk F, Schuermann M, Nagenaar E, Parren P, Wiegel D, Nusse R: Cell 50:649, 1987.
12. Steward R: Science 238:692, 1987.
13. Van Beveren C, van Straaten F, Curran T, Muller R, Verma IM: Cell 32:1241, 1983.
14. Verma IM, Graham WR: Adv Can Res 49:29, 1987.
15. Verma IM: Trends Genet 2:93, 1986.
16. Verma IM, Sassone-Corsi P: Cell 51:513, 1987.
17. Greenberg ME, Ziff EB: Nature 311:433, 1984.
18. Montminy MR, Sevarino KA, Wagner JA, Maudel G, Goodman RH: Proc Natl Acad Sci USA 83:6682, 1986.
19. Delegeane AM, Ferland LH, Mellon PL: Mol Cell Biol 7:3994, 1987.
20. Curran T, Morgan JT: Science 229:1265, 1985.
21. Greenberg ME, Greene LA, Ziff EB: J Biol Chem 260:14101, 1985.
22. Kruijer W, Schubert D, Verma IM: Proc Natl Acad Sci USA 82:7330, 1985.
23. Kruijer W, Cooper JA, Hunter T, Verma IM: Nature 312:711, 1984.

24. Angel P, Imagawa M, Chiu R, Stein B, Imbra RJ, Rahmsdorf HJ, Jonat C, Herrlich P, Karin M: Cell 49:729, 1987.
25. Van Buskirk R, Corcoran T, Wagner JA: Mol Cell Biol 5:1984, 1985.
26. Deschamps J, Meijlink F, Verma IM: Science 230:1174, 1985.
27. Sassone-Corsi P, Verma IM: Nature 326:507, 1987.
28. Scholer HR, Gruss P: Cell 36:403, 1984.
29. Montminy MR, Bilezikjian L: Nature 328:175, 1987.
30. Miller AD, Curran T, Verma IM: Cell 36:51, 1984.
31. Meijlink F, Curran T, Miller AD, Verma IM: Proc Natl Acad Sci USA 82:4987, 1985.
32. Curran T, Van Beveren C, Ling N, Verma IM: Mol Cell Biol 5:167, 1985.
33. Curran T, Teich N: Virology 116:221, 1982.
34. Barber JR, Verma IM: Mol Cell Biol 7:2201, 1987.
35. Fisch TM, Prywes R, Roeder RG: Mol Cell Biol 7:3490, 1987.
36. Gilman MZ, Wilson RN, Weinberg RA: Mol Cell Biol 6:4305, 1986.
37. Prywes R, Roeder RG: Cell 47:777, 1986.
38. Greenberg ME, Siegfried Z, Ziff EB: Mol Cell Biol 7:1217, 1987.
39. Sen R, Baltimore D: Cell 47:921, 1986.
40. Mitchell RL, Henning-Chubb C, Huberman E, Verma IM: Cell 45:497, 1987.
41. Sambucetti LC, Curran T: Science 234:1417, 1986.
42. Renz M, Verrier B, Kurz C, Muller R: Nucl Acids Res 15:277, 1987.
43. Distel RJ, Ro H-S, Rosen BS, Groves DL, Spiegelman BM: Cell 49:835, 1987.

Gene Transfer and Gene Therapy, pages 151–162

ONCOGENE EFFECTS ON DIFFERENTIATION AND TRANSFORMATION OF MAMMARY EPITHELIAL CELLS IN TRANSGENIC MICE

A.-C. Andres, M.A. van der Valk[1], C.-A. Schoenenberger, F.Flueckiger and B. Groner

Ludwig Institute for Cancer Research, Inselspital, 3010 Bern, Switzerland

ABSTRACT We describe the effect of the tissue-specific and lactogenic hormone-dependent expression of the Wap-ras and the Wap-myc oncogenes on differentiation and transformation of mammary epithelial cells in transgenic mice. Ras expression impairs functional differentiation but does not prevent terminal differentiation. Mammary adenocarcinomas were rarely detected in the Wap-ras transgenic females. In contrast, myc expression may block differentiation and allows continuous proliferation of the mammary epithelial cells. About 80% of the females developed mammary adenocarcinomas. Cooperation of these two oncogenes was analyzed in double-transgenic mice bearing the Wap-ras and the Wap-myc oncogene. Co-expression of both genes synergistically affects the differentiation leading to an almost non-functional mammary gland. The high number of neoplastic foci in mammary glands of double transgenic females indicates cooperative effects in the transformation process. However, the tumor latency suggests that expression of the myc and the ras oncogenes is still not sufficient to cause full tumor formation.

INTRODUCTION

Transgenic mice provide an excellent means to study the role of activated oncogenes in transformation in vivo. They allow the effects to be studied in the context of different

---

[1]The Netherlands Cancer Institute, CX1066 Amsterdam, NL

cell types and tissue organization. The finding that gene sequences around the promoter region are usually sufficient to direct tissue-specific expression of transferred genes makes it possible to analyze oncogene effects on defined cell types in the animal (1). So far, most information has been obtained from transgenic animals bearing the ras or the myc oncogene under the control of the elastase gene promoter, the immunoglobulin enhancer and the long terminal repeat (LTR) of the mouse mammary tumor virus (MMTV). Investigation of tumor formation in the oncogene-bearing animals revealed a different tumorigenic potential of the oncogenes, depending on the cell type in which they are expressed, and confirmed the view of carcinogenesis as a multi-step process (2). Tumor formation in vivo possibly requires the activation of several cooperating oncogenes. The ability of the myc and the ras oncogenes to cooperate in the transformation in vivo was analyzed in transgenic females bearing the two oncogenes under the control of the LTR of MMTV. Despite the widespread coexpression of myc and ras, cooperation in tumor formation was detected only in a few tissues (3). This indicates that the ability of oncogenes to cooperate is cell type-specific.

Our interest in transformation of mammary epithelial cells led us to study the effect of oncogenes on this particular cell type in vivo. The differentiation of the mammary epithelium is dependent on the hormonal status of the individual, and complex cell interaction is thought to be crucial for the developmental process (4,5). Neoplastic diseases are considered to be the escape of the tumor cells from this complex growth control. Therefore, to understand the involvement of oncogenes in this process, their mechanism of action has to be characterized by a concerted view of the effects on cellular growth, differentiation and transformation.

Several oncogenes and different mechanisms of their activation are thought to contribute to carcinogenesis in the mammary gland. Activation of members of the ras gene family by point mutations has been observed in mammary tumor cell lines (6) and in chemically induced mammary adenocarcinomas in rodents (7). Amplification of the c-myc oncogene was detected in about one-third of human breast tumors (8). These results suggest that the myc and the ras oncogenes may play an important role in the carcinogenesis of the mammary gland.

We investigated the effect of the Ha-ras and the c-myc oncogenes on the differentiation and transformation of the mammary gland in transgenic mice (9,10). To target the expression of the oncogenes they were subjected to the control of the murine Whey acidic protein (Wap) gene. The Wap is a milk protein of rodents and the gene is exclusively expressed in the mammary epithelial cells during pregnancy and lactation (11). The specific expression of the ras or myc oncogenes in the transgenic mice exerted a different effect on differentiation as well as on transformation of the mammary epithelium. In order to analyze if the ras and myc oncogenes act synergistically in the transformation of the mammary epithelial cells in vivo, we established double-transgenic mice bearing both oncogenes. Our results suggest that cooperative effects of ras and myc may be confined to impaired differentiation and that additional events are required for tumor formation.

RESULTS

Establishment of Single and Double Transgenic Mice.

Transgenic mice provide the means to study the effect of the ras and the myc oncogenes on the mammary epithelium in vivo. To target the expression, we subjected the oncogenes to the control of the murine Wap gene promoter. We isolated a 2.5 kb fragment from the Wap gene, which spans the untranslated RNA leader sequence, the RNA cap site and about 2.4 kb of additional 5' flanking sequences (12). This fragment was ligated either to the coding part of the activated human Ha-ras gene (13) or to exon 2 and 3 of the murine c-myc gene (14). The resulting hybrid genes were introduced into the germline of mice by microinjection of fertilized eggs. We established five transgenic lines with the Wap-ras gene, and four with the Wap-myc gene (9,10). To determine the expression pattern of the transgenes we analyzed RNA from various tissues by RNase protection assays. The probe to detect Wap-ras RNA protects a fragment of 143 nucleotides (N) which is indicative for exon 1. Wap-myc RNA was analyzed with a probe protecting 198 N of exon 1. The probe specific for the endogenous Wap RNA comprises exons 1 and 2 and protects two fragments of 110 N (exon 1) and 135 N (exon 2) (9,10). Expression of the Wap-ras gene is found in mammary glands of lactating females of two independent lines indicated by the protection of a RNA fragment of 143 N (figure 1A, lane 1). Wap-ras RNA is absent in the mammary gland of the same individual after

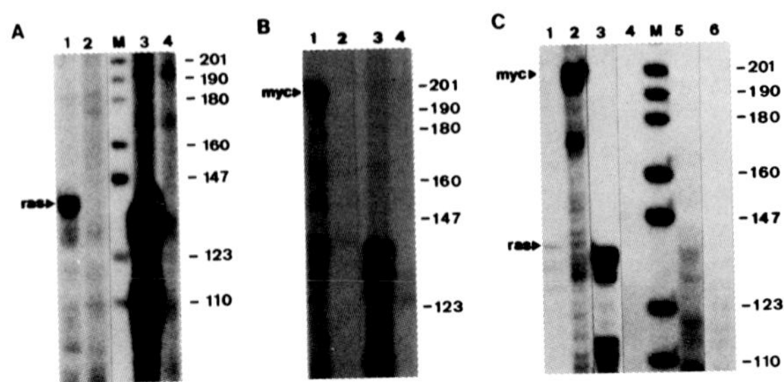

FIGURE 1. A: Transgene expression in Wap-ras transgenic females. B: Expression in Wap-myc transgenic females. Lanes 1+3: lactating, lanes 2+4: regressed mammary glands. 10 µg RNA were hybridized to the Wap-ras (A) or to the Wap-myc probe B (lanes 1+2), 5 µg RNA were hybridized to the Wap probe (lane 3+4). C: Expression in double transgenic females. Lanes 1-3: lactating, lanes 4-6: regressed mammary glands. 10 µg RNA were hybridized to the Wap-ras (lanes 1+4) or to the Wap-myc probe (lanes 2+5). 15 µg RNA were hybridized to the Wap probe (lane 3+6). M: HpaII digested ^{32}P end-labeled pBR322 DNA (indicated in nucleotides).

lactation (lane 2). This parallels the expression of the endogenous Wap gene (lanes 3+4). The hormone-dependent expression of the transgene was also found in three out of four Wap-myc transgenic lines. The fragment of 198 N indicative for Wap-myc RNA is only found in mammary glands during lactation (lane 1) but not after lactation (lane 2), as is seen for the endogenous Wap gene (lane 3+4). We detected no transgene expression in any organ other than the mammary gland. This indicates that the Wap gene promoter conferred not only hormone-dependent but also tissue-specific expression to the linked genes.

To analyze the ability of the ras and the myc oncogenes to cooperate in the transformation of the mammary epithelial cells in vivo, we established mice bearing the Wap-ras and the Wap-myc oncogenes by crossing animals of the respective single transgenic lines. The resulting double-transgenic

TABLE 1
HISTOPATHOLOGY OF THE MAMMARY GLANDS

	wt	ras	myc	myc-ras
end-pregnant	100% of fat pad replaced by alveoli	75% of fat pad replaced by alveoli	50% of fat pad replaced by alveoli	25% of fat pad replaced by alveoli
	Alveoli dilated	100% of fat pad filled with smaler sized alveoli	Fat pad not completely replaced	40% of fat pad replaced by immature alveoli
lactating	Secretion No mitosis	Secretion No mitosis Epithelial cells atypic	Poor secretion Mitosis Abnormal nuclei and nucleoli	+/- no secretion Mitosis Abnormal nuclei and nucleoli Epithelial cells atypic
			Hyperplastic foci	Hyperplastic foci
		Main ducts normal	Main ducts normal	Main ducts normal
after lactation	Lobular-alveolar part regressed	Lobular-alveolar part regressed	Moderate number of hyperplastic foci	Many hyperplastic foci or carcinomas in situ
			Proliferation	No or low proliferation

females showed the coordinate expression of the Wap-ras (figure 1C, lane 1) and the Wap-myc genes (lane 2) in the mammary gland during lactation. Again both genes were silent after lactation (lane 4+5) and no expression was detected in the other organs. Even though the expression of the endogenous Wap gene is weak in lactating mammary glands of these females (lane 3) we detect its normal hormonal control (lane 6).

Oncogene Expression Affects the Differentiation of the Mammary Gland.

Since the Wap promoter is only activated when the mammary gland is differentiating to the lactating state, oncogene effects on a defined differentiation process can be analyzed in our transgenic animals. We analyzed the histological appearance of mammary glands from transgenic and wild type mice at different stages of development. These results are summarized in Table 1. Wap-ras expression results in a retarded development of the alveolar cells at end-pregnancy, but during lactation the entire fat pad is replaced by secretory alveoli. Atypia in the epithelium such as increased cell volume and poor lateral cell contact suggest an impaired functional differentiation. This is supported by the reduced ability of these glands to express the milk protein genes (data not show). Even though the epithelial cells are impaired in function, they reach an end-differentiated state during lactation. This is indicated by the cessation of cell division during lactation and the normal regression after lactation.

Wap-myc expression exerts different effects on the differentiation of the mammary epithelium. The replacement of the fat pad remains incomplete also during lactation. The increase in milk protein gene expression at the onset of lactation which is typical for wild type females is not observed in Wap-myc transgenic mice (data not shown). The poor secretional activity and continuing cell division suggest that the epithelial cells may be blocked in their differentiation. Some hyperplastic foci can be detected during lactation. At least some of these foci remain in the tissue after lactation and show considerable mitotic activity.

The analysis of mammary glands from double-transgenic females revealed a limited development of the alveolar part and an immature appearance of the epithelial cells during

TABLE 2
TUMORFORMATION IN TRANSGENIC MICE

Strain	Line	Incidence	Latency
Wap-ras	3	3/100	12-14 mos*
	58	1/50	7 mos*
Wap-myc	33	10/12	
	19	2/3	2-5 mos*
	21	1/1	
Wap-ras/Wap-myc	58x33	4/4	3-4 mos*

* after first pregnancy

lactation. Double transgenic females show only residual levels of milk protein gene expression (Figure 1C). The almost complete lack of secretional activity indicates that the glands of double-transgenic females are not functional. In contrast to the single transgenic mice these females are not able to nurse their progeny. Compared to the mammary glands of Wap-myc expressing females we detect a significantly increased number of neoplastic foci, and even some carcinomas in situ are found. These lesions are also seen in the regressed mammary gland. However, in contrast to the Wap-myc transgenic females, these lesions show no or only low proliferation. We conclude from the histological appearance and from the non-functional state of the mammary glands that coexpression of the ras and myc oncogenes synergistically affects the differentiation process of the mammary epithelium.

Oncogene Expression Predisposes the Mammary Epithelium to Transformation.

Even though we find single and double oncogene expression in the mammary gland of transgenic mice, the animals initially appeared normal and healthy. Palpable mammary tumors started to appear only after several months. The tumor occurrence in each transgenic line is listed in Table 2. In the Wap-ras-expressing females only 4 mammary adenocarcinomas were observed after a latency of about one year. The tumors showed an increased expression of the Wap-ras gene compared to the corresponding normal tissue of the same individual (9). Wap-myc expression exerts a completely different tumorigenic potential on the mammary epithelium. About 80% of the females develop mammary adenocarcinomas and the latency can be as short as 2 months. Myc expression has to be considered as a potent predisposing factor for transformation of the mammary epithelial cells. Analysis of gene expression revealed that not only the Wap-myc gene but also the endogenous milk protein genes are expressed in the tumors independent of lactogenic hormones. Wap-myc expression contributes to overcoming the hormonal control of the mammary epithelial cells (10).

Apparently normal mammary glands of double-transgenic females show an increased number of neoplastic foci compared to animals expressing only the Wap-myc gene. This may indicate oncogene cooperation in the transformation process, leading to the rapid development of mammary gland tumors. However, the latency of palpable tumors is comparable to that found in the single Wap-myc transgenic females. We assume that ras may interfere with the myc-induced course of transformation. The fact that proliferation is only found in lactating mammary glands but not after lactation suggests that ras may interfere with the hormonal abrogation exerted by the myc oncogene. This is supported by the lack of hormone-independent expression of milk protein genes in tumors of double transgenic females. Interestingly, these tumors show increased expression of the Wap-ras gene compared to the corresponding normal tissue of the same individual (data not shown). This increase in ras expression was typical for the tumors of single transgenic animals and seems to be maintained in double transgenic females.

DISCUSSION

The development of the mammary epithelium to the end-differentiated lactational state involves cellular proliferation as well as the commitment of the cells to express the milk protein genes. Our studies revealed that these processes are differently and specifically impaired by the expression of the ras or the myc oncogenes. Even though mammary epithelial cells expressing the Wap-ras oncogene show an impaired functional differentiation, the cells reach an end-differentiated state and regress normally after lactation. Differentiation promoting effects as well as proliferative effects have been attributed to the ras oncogene in different experimental systems (15,16,17). This indicates that the phenotypic consequences of ras expression are strictly dependent on the cell type in which it is expressed. More consistent effects were observed with the myc oncogene. Experiments in cultured primary cells indicate that deregulated myc expression induces growth transformation but is accompanied by the preservation of differentiated functions (18,19). In transgenic animals myc expression targeted to the lymphoid compartment by the E enhancer favored proliferation over maturation of developing pre-B cells (20). We made a similar observation in the Wap-myc transgenic females. The differentiation of the mammary epithelial cells is blocked at an intermediate stage and they continue to proliferate during lactation. The different effects of myc or ras expression on the differentiation explains the different tumorigenic potentials. Since ras does not prevent the cells from reaching an end-differentiated, non-dividing state they remain susceptible to the signals governing postlactational regression. In contrast, the proliferative effect of myc may prevent their end-differentiation. Additional events may occur in the population of cells which escape regression leading to their transformation. In all three transgenic lines the abnormalities are confined to the secretory alveolar cells. The secretory alveolar cells develop from stem cells located in the main ducts, which are present in all developmental stages (21). By using the Wap promoter we expect oncogene expression in the alveolar cells only. Thus, the observed abnormalities concern exclusively the oncogene expressing cells in the transgenic females.

The concept gained from studies in cultured primary cells suggest that the ras and the myc oncogenes cooperate in the transformation (14). The high number of neoplastic foci in regressed mammary glands of double transgenic females suggests that these oncogenes may also cooperate in transformation of mammary epithelial cells in vivo. In contrast to the neoplastic foci in the regressed mammary gland of females expressing only the Wap-myc gene, the lesions in double-transgenic animals show a very low mitotic activity. Expression of both oncogenes initially does not result in increased cell proliferation. We assume that ras mainly interferes with the hormonal abrogation exerted by the myc oncogene, which leads to the cessation of cell division in the absence of lactogenic hormones. Conceivably, cooperation may be confined to early stages of carcinogenesis in our experimental system. Our finding that ras and myc do not cooperate in all stages of carcinogenesis is in contrast to the results obtained in transgenic mice bearing these two oncogenes under the control of the LTR of MMTV (3). The LTR-ras gene alone is highly tumorigenic in the mammary epithelium and a clear cooperation with the LTR-myc gene in transformation is found. This is illustrated by a significantly shorter tumor latency and by the appearance of mammary gland tumors in males. The different effects of the ras gene may be due to the different properties of the LTR and the Wap promoter. First, the LTR promoter allows expression already during embryonic development. The tumorigenic potential of ras may not only be dependent on the cell type but also on the differentiation state of a cell. Second, the LTR promoter allows higher ras expression than the Wap promoter. Transformation by ras may be dependent on the level of expression in addition to the point mutation. It was shown that myc oncogene expression could not be substituted for the high levels of ras expression required to transform baby rat kidney cells (22). Similarly, the increase in Wap-ras expression, which was found in tumors of single transgenic mice, is also a characteristic of the mammary tumors of double-transgenic females. Our results suggest that increasing the ras expression is still a prerequisite for the development of the fully malignant phenotype despite the coexpression of the myc oncogene.

REFERENCES

1. Palmiter RD, Brinster RL (1986). Germline transformation of mice. Ann Rev Genet 20:465.
2. Groner B, Schoenenberger C-A, Andres A-C (1987). Targeted expression of the ras and the myc oncogenes in transgenic mice. TIG 3:306.
3. Sinn E, Muller W, Pattengale PK, Tepler I, Wallace R, Leder P (1987). Coexpression of MMTV/v-Ha-ras and MMTV/c-myc genes in transgenic mice: Synergistic action of oncogenes in vivo. Cell 49:465.
4. Banerjee MR (1976). Responses of mammary cells to hormones. Int Rev Cytol 47:1
5. Levine JF, Stockdale FF (1985). Cell interactions promote mammary epithelial cell differentiation. J Cell Biol 100:1415.

6. Kozma S, Bogaard ME, Buser K, Saurer SM, Bos JC, Groner B, Hynes NE (1987). The human c-Ki-ras gene is activated by a novel mutation in codon 13 in the breast carcinoma cell line MDA-MB231. Nucl Acids Res 15:5963.
7. Zarbl H, Sukumar S, Arthur AV, Martin-Zanca D, Barbacid M (1985). Direct mutagenesis of Ha-ras-1 oncogenes by N-nitroso-N-methylurea during initiation of mammary carcinogenesis in rats. Nature 315:382.
8. Varley JM, Swalloa JE, Brammar WJ, Whittaker JL, Walker RA (1987). Alterations to either c-erbB-2 (neu) or c-myc proto-oncogenes in breast carcinomas correlate with poor short term prognosis. Oncogene 1:423.
9. Andres A-C, Schoenenberger C-A, Groner B, Hennighausen L, LeMeur M, Gerlinger P (1987). Ha-ras oncogene expression directed by a milk protein gene promoter: Tissue specificity, hormonal regulation and tumor induction in transgenic mice. Proc Natl Acad Sci USA 84:1299.
10. Schoenenberger C-A, Andres A-C, Groner B, van der Valk MA, LeMeur M, Gerlinger P (1988). Targeted c-myc gene expression in mammary glands of transgenic mice induces mammary tumors with constitutive milk protein gene expression. EMBO J 7:169.
11. Hobbs AA, Richards DA, Kessler DJ, Rosen JM (1982). Complex hormonal regulation of rat casein gene expression. J Biol Chem 257:3598.
12. Campell SM, Rosen JM, Hennighausen L, Strech-Jurk U, Sippel AE (1984). Comparison of the whey acidic protein genes of the rat and the mouse. Nucl Acids Res 12:8685.
13. Capon DJ, Chen EY, Levinson AD, Seeburg PH Goeddel PV (1983). Complete nucleotide sequence of the T24 human bladder carcinoma oncogene and its normal homologue. Nature 302:33.
14. Land H, Parada LF, Weinberg RA (1983). Tumorigenic conversion of primary embryo fibroblasts requires at least two cooperating oncogenes. Nature 304:596.
15. Bar-Sagi D, Feramisco JR (1985). Microinjection of the ras oncoprotein into PC12 cells induces morphological differentiation. Cell 42:841.
16. Bell JC, Jardine K, McBurney MW (1986). Lineage-specific transformation after differentiation of multipotential murine stem cells containing a human oncogene. Mol Cell Biol 6:617.

17. Payne PA, Olson EN, Hsian P, Roberts R, Perryman MB, Schneider MD (1987). An activated c-Ha-ras allele blocks the induction of muscle specific genes whose expression is contingent on mitogenic withdrawal. Proc Natl Acad Sci USA 84:8956.
18. Alema S, Tato F, Baettiger D (1985). myc and src oncogenes have complementary effects on cell proliferation and expression of specific extracellular matrix components in definitive chondroblasts. Mol Cell Biol 5:538.
19. Casalbore P, Agostini E, Alema S, Falcone G, Tato F (1987). The v-myc oncogene is sufficient to induce growth transformation of chick neuroretina cells. Nature 326:188.
20. Langdon WY, Harris AW, Cory S, Adams JM (1986). The c-myc oncogene perturbes B lymphocyte development in Eμ-myc transgenic mice. Cell 47:11.
21. Taylor-Papadimitriou J, Lane BE, Chang SE (1983). Cell lineages and interactions in neoplastic expression in the human breast. In: Rich MA, Hager JC, Furmanski P (eds): "Understanding breast cancer" M Dekker INC NY
22. Kelekar A, Cole M (1987). Immortalization by c-myc, Ha-ras and Ela oncogenes induces differential cellular gene expression and growth factor responses. Mol Cell Biol 7:3899.

Gene Transfer and Gene Therapy, pages 163–177

INDUCIBLE ABLATION OF A SPECIFIC CELL TYPE BY THE HERPES THYMIDINE KINASE GENE EXPRESSION

Emiliana Borrelli, Richard Heyman,
Mary Hsi and Ronald M. Evans

Howard Hughes Medical Institute
Gene Expression Laboratory
The Salk Institute
La Jolla, California 92037

ABSTRACT The introduction of the Herpes virus 1 thymidine kinase (HSV1-TK) gene, into cultured cells in concomitance with a nucleoside analog treatment, leads to a drug dependent inhibition of DNA replication and, consequently, cell death. Expression of HSV1-TK in the immune system of transgenic animals leads to a pronounced drug induced degeneration of lymphoid tissue.

INTRODUCTION

Gene transfer methodology has provided a valuable approach with which to impart new phenotypic properties on cultured cells (1), as well as in transgenic animals (2,3). The possibility of introducing a desired cloned gene into the mouse germ line has inspired the approach of expressing negative selectable markers into cells, which would either inhibit cell proliferation or lead to cellular degeneration. The knowledge that the functional cis-elements which specify gene expression, in a single cell type, reside in the promoter/enhancer regions of all genes (4) is of great utility. These elements are regulated during development, and once activated, insure an efficient transcription of the linked gene. This property restricts the design of a toxic vector, as the ablation of a cell line during embryonal life could lead to lethal phenotypes.

A valuable alternative would be the manifestation of a toxic phenotype that is both cell specific and inducible.

One candidate of potential use in such an approach is the herpes virus 1 thymidine kinase (HSV1-TK) (5). This enzyme alone is not harmful to cells and in thymidine kinase negative cells it can allow cell survival (6). The pathogenic effects of herpes virus infection has led to the development of nucleoside analogs that selectively block viral spread as a consequence of their metabolism by the herpes enzyme but not its cellular counterpart (7-15). One analog, acyclovir (acycloguanosine; ACV) (12-13) has demonstrated clinical efficacy for the treatment of herpes infections in humans. Furthermore, in non-infected cells, ACV, FIAU (1-(2-deoxy-2-fluoro-β-D-arabinofuranosil)-5-iodouracil) (14) and related compounds display little cytotoxicity (12,14-15). These observations lead us to examine whether the expression of the cloned HSV1-TK gene either in transfected cells or in a fibroblast cell line having a resident HSV1-TK gene would convert ACV and FIAU to cytotoxic intermediates. In this paper we show that the introduction of the HSV1-TK gene in cultured cells and in the immune system of transgenic animals, in conjunction with nucleoside analog treatment, leads to a drug concentration dependent cytotoxicity and, consequently, to cell death.

MATERIALS AND METHODS

Gene constructs.

The structure of the recombinant DNAs used for transfection of cos cells (16) is shown in Figure 1a. Briefly, paPS1 contains the mouse mammary tumor virus long terminal repeat (MTV-LTR) linked to the HSV1-TK gene (17). pRShGRα contains the Rous sarcoma virus long terminal repeat (RSV-LTR) linked to the human glucocorticoid receptor (18). pSV$_2$CAT contains the Simian virus 40 promoter region linked to the chloramphenicol acetyltransferase gene (19). pKHTK has been used to obtain transgenic mice. The structure is shown in Figure 3a. pKHTK plasmid is derived from pKCATH (20) in which the CAT gene was substituted with the HSV1-TK gene. Thick lines denote a 682 bp XbaI-EcoRI fragment of the μ-heavy chain enhancer (21,22) linked to a 1.1 kb fragment of the K-light chain promoter (23) (hatched box) directing the transcription of a 1.5 kb fragment of the HSV1-TK gene (open box) from position +2 relative to the TK transcription startsite to 236 bp downstream of the TK polyadenylation site.

Drugs.

Two nucleoside analogs have been used for the purpose of this paper: Acyclovir (acycloguanosine; ACV Burroughs-Wellcome) (12,13) and FIAU (1-(2-deoxy-2-fluoro-β-D-arabinofurenosil)-5-iodo-uracil Bristol Meyers) (14). Both were a a generous gift from Dr. D. Richman.

Tissue Culture Transfection.

Fifty percent confluent cos cells (16) were transfected by the calcium phosphate technique (24), with 1 ug of pSV_2CAT, 5 ug of paPS1 and 1 ug of pRShGRα. The cells were grown in Dulbecco media containing 5% FCS. Before each experiment Dex, at a concentration of 10^{-7} M, was added. Where the efficiency of the nucleosides analogs were tested either ACV or FIAU were added to the media at a concentration of 10 uM. The cells were trypsinized and washed after 24 hours then replated, the new media contained the same concentration of Dex and ACV or FIAU than before. 12 hours later the cells were harvested and he plasmid DNAs were recovered by the Hirt method (25). The recovered plasmid DNAs were digested by BamHI and separated on a 0.8% agarose gel and then transferred on nitrocellulose paper. The immobolized DNA was then hybridized (26) to a nick-translated probe spanning the polyadenylation site of SV40 (BamHI-EcoRI fragment from SV40 position 2533-1782) or with a pBR322 probe.

Cell Growth Assay.

The cytotoxicity of the nucleoside analogs ACV and FIAU, was determined by using a cell growth assay. Rat fibroblast 208 (27), or Rat2#3B clone (28), a TK^- cell line stably transformed in TK^+ by an MTV-TK construction, were seeded at low density ($3x10^4$) in

Dulbecco media containing 10% FCS. Cells were incubated with varying concentration of drug incorporated into the media 12 hours after the cells were plated either in the presence or in the absence of 10^{-7} M Dex. After a total of 72 hours, cells were detached by trypsin treatment and resuspended in medium. Viable cells were recognized by Trypan blue exclusion. Cell numbers were determined with a Neubauer Hemocytometer.

Transgenic Mice.

Several transgenic lines were obtained by the injection of the KHTK hybrid construct (see Figure 3a). In this paper we describe #686 whose pattern of RNA expression is similar to the other lines obtained. Transgenic and non-transgenic littermates were treated for one week with 50 mg/ml solution of FIAU in PBS released by a Miniosmotic pump, implanted subcutaneously (Model 2002, Alzet, Alza Corp., Palo Alto, CA).

RNA Analysis.

Northern analysis was performed on total RNA isolated from various tissues of the transgenic line #686, using the LiCL and urea method (29), separated on 1% agarose-formaldehyde gel, transferred to nitrocellulose and hybridized under stringent conditions (26) using the whole HSV1-TK gene as probe. The filter was autoradiographed at -70°C with an intensifying screen for 24 hours. S1 nuclease analysis was performed on 10μg of total RNA from different tissues of the transgenic mice; the RNAs were hybridized with a double strand probe (30) spanning from the AvaI site (+350) in the TK gene to the EcoRI site (-1100) indicated in the picture (Figure 3a). AFter S_1 nuclease treatment, the protected fragment, corresponding to RNA transcribed from the K light chain promoter, was 380 nucleotides in length in agreement with the expected size.

RESULTS

Inhibition of Plasmid Replication by Synergistic Action Between the HSV1-TK and ACV or FIAU.

It is presumed that the toxic efficacy of anti-herpetic drugs is a consequence of the metabolism of these nucleoside analogs (such as ACV and FIAU) by HSV1-TK to phosphorylated derivatives that are incorporated into DNA by herpes DNA polymerases. This, in turn, leads to chain termination of the replicating viral template (7-15 and ref. therein). The inability of the virus to replicate its genome appears to be directly related to the clinical benefit of these drugs. We wished to explore the possibility that HSV1-TK expressed from a transfected plasmid could metabolize anti-herpetic nucleosides to substrates for the cellular DNA polymerase in uninfected cells. In principle, the incorporation of these activated drugs would lead to a net decrease in cellular DNA replication. Accordingly, a plasmid capable of replicating in cos cells (16), pSV2CAT (Fig. 1a) (19), was cotransfected with a plasmid, pAPS1 (17), expressing HSV1-TK gene under the control of the glucocorticoid inducible promoter of the mouse mammary tumor virus (MTV) (Fig. 1a). Here, the expression of TK is dependent on both the presence of glucocorticoids, such as the synthetic dexamethasone (Dex), and the glucocorticoid receptor to activate the MTV-promoter. For this reason, a human glucocorticoid receptor (hGR) expression vector, pRShGR*α* (18), was also cotransfected to insure high levels of this <u>trans</u>-acting factor (Fig.1a). pRShGR*α* itself contains the origin of replication of SV40 and therefore is able to replicate in cos cells.

Semi-confluent cos cells (16) were cotransfected by the calcium phosphate technique (24) with pSV_2CAT, paPS1 and pRShGR*α* and analyzed 36 hours later. The effect of ACV or FIAU on the replication of pSV_2CAT was gauged by the reduction in the number of molecules of this plasmid extracted by the Hirt method (25) from transfected cells. Southern blot analysis (26) of BamHI digests was performed using as probe a nick-translated DNA fragment that specifically recognizes the polyadenylation site of SV40, present in both the plasmids pSV_2CAT and pRShGR*α* (Fig. 1a). The results of this analysis are shown in Fig.1b; in lane 1 the intensity of the band is indicative of the amount of linearized pSV2CAT recovered from transfected cos cells, when no drugs are added to the media (see also Fig. 1c, lane 1). The intensity of the band corresponding to the pSV_2CAT replication in the absence of drug (Fig. 1b, lane 1) is normalized to 100%

of relative value in the densitometric scanning analysis of the samples (see Fig. 1c). When 10μM of ACV is added to the culture media, under the same experimental condition,the total plasmid copy number is reduced by 40% relative to no drug treatment (see Fig. 1b, lane 2 and Fig. 1c, lane 2 (pSV2CAT); the reduction is even greater (70%) when 10μM of FIAU is used (see Fig. 1b, lane 3 and Fig. 1c, lane 3 (pSV2CAT).

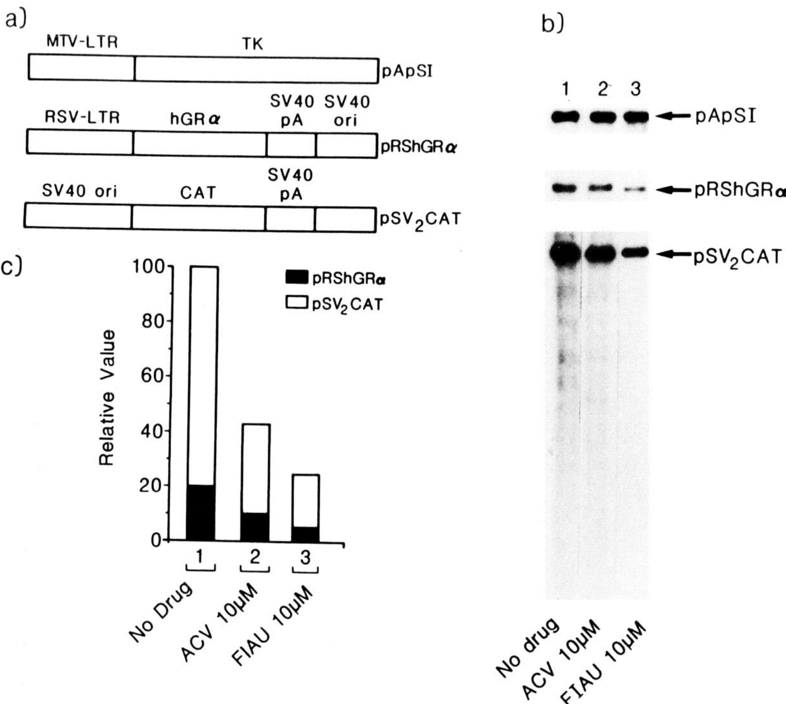

FIGURE 1. HSV1-TK inhibition of plasmid DNA replication in transiently transfected cells. a) Structure of the recombinants used for transfection of cos cells. b) Replication of pSV_2CAT and pRShGRα in cos cells in the presence of ACV or FIAU. The presence or the absence of drugs is as indicated at the bottom of the lanes. The bands corresponding to each plasmid are indicated by the arrows on the right side. c) Densitometric scans of autoradiograms of Southern blot analysis from the Hirt extracts. pSV_2CAT is represented by the white columns and the intensity of

Replication of pRShGRα (see Fig. 1b) is also sensitive to the addition of ACV and FIAU to the media. The reduction is similar to that observed with pSV2CAT. [We note that pRShGRα never reached the same copy numbers as pSV_2CAT; we have no explanation for this difference.] Thus, with ACV and FIAU there is 40% and 70% decrease in the replication of these plasmids. The plasmid pApS1 does not replicate and thus serves as an internal control. As expected, the number of copies was unchanged by any treatment [see Figure 1b (pApS1)]. Finally, when either the nucleoside analogs were absent or the plasmid expressing TK was absent, no reduction in the amount of pSV_2CAT replication was observed (data not shown). These experiments indicate that the expression of the thymidine kinase gene is required for the conversion of the nucleoside analogs to toxic precursors.

Cytotoxic effect of ACV and FIAU on a cell line expressing the HSV1-TK gene. To examine the effect on cellular DNA replication we used the rat fibroblast cell line #3B (28) that stably expresses the glucocorticoid inducible MTV-HSV1-TK gene. Addition of Dex to these cells leads to a 50-fold increase in HSV1-TK activity. We next examined the sensitivity of these and control cells to various levels of ACV and FIAU.

The effect of the drugs on the 3B cells was compared to the one obtained on the control rat 208 fibroblast cell line (27),not expressing HSV1-TK. The experiments were performed in the presence or in the absence of Dex at a concentration of 10^{-7}M. The efficacy of the drugs was monitored as reduction of viable cells. At low concentrations of ACV ranging from 0.1 to 1.0μM, no effect of the drug was observed (see Fig. 2). ACV at 3.0μM appeared to be slightly toxic (20% cell death) but only to 3B cells, in the presence of Dex. Its toxicity to 3B in the presence of Dex was greater at higher concentrations: at 10μM we observed 60% cell death, and 98% cell death at 100μM. At 100μM we observed non-specific toxicity of ACV that was independent of the presence or absence of Dex. When FIAU was used similar but more dramatic cytotoxic effects were observed (see Fig. 2).

---

its band in the absence of drug, was taken as the 100% value, pRShGRα is represented by the black columns.

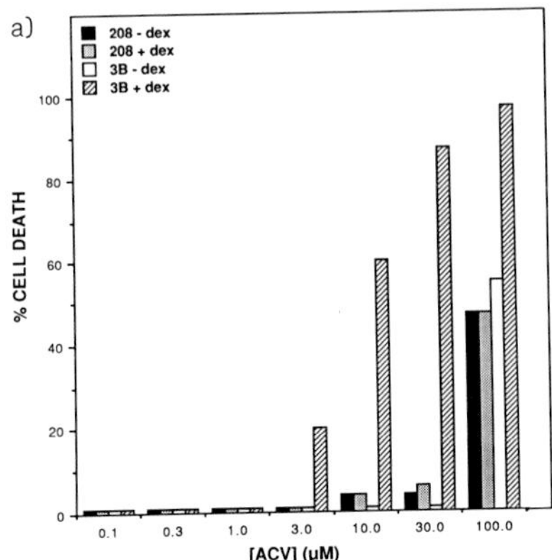

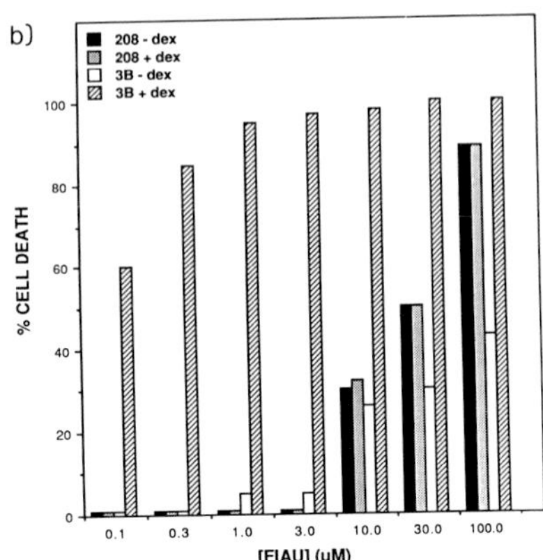

FIGURE 2. Cytotoxic effect of ACV (a) and FIAU (b) on growth of 208 versus 3B cells. The symbols on the top left side indicate the conditions and the cell line used.

Remarkably, concentrations as low as 0.1μM reduced viability by more than 60% and at 3.0μM 98% cell death was achieved in the absence of severe non-specific toxicity. Thus, both drugs could reduce cell viability at doses that were apparently not harmful to the 208 cells. These results indicate that the expression of the HSV1-TK gene confers upon cells a toxic potential which can be revealed by the addition of appropriate levels of drugs.

Specific cell-killing in transgenic mice. We wished to establish whether the cell-specific expression of HSV1-TK gene in transgenic mice would confer drug-dependent cytotoxicity as observed in cultured cells. As model system we chose to express the HSV1-TK gene into cells of the immune system. Accordingly, the immunoglobulin heavy chain enhancer (21-22) and the light chain promoter (23) in front of the HSV1-TK gene was constructed. This promoter-enhancer combination has been previously shown, in vitro, to be the most efficient to insure high level of expression to a linked gene in a myeloma cell line (28).

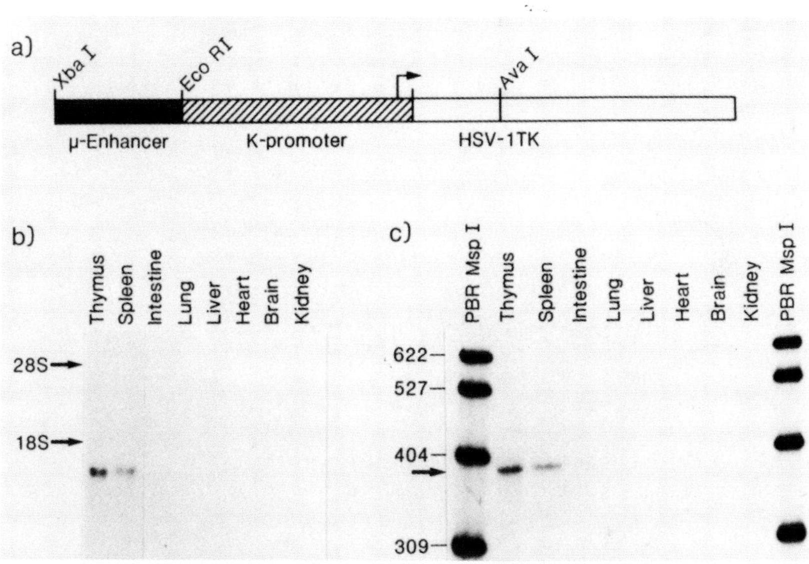

FIGURE 3. Expression in transgenic mice of the KHTK hybrid construct. a) Structure of the KHTK plasmid. b) Northern blot analysis of HSV1-TK mRNA in transgenic mice. 20 μg of total RNA was used in all lanes. Migration of the ribosomal RNAs are indicated as size markers. c) S1 nuclease analysis of the KHTK mRNA from transgenic mice.

The KHTK (Fig.3a) was injected into fertilized mouse eggs, giving rise to a transgenic founder that established the 686 pedigree. RNA analysis that line revealed transcripts corresponding to the correct tk RNA in spleen and thymus (Fig.3b). Although the expression of the heavy chain enhancer-light chain promoter, *in vitro*, is preferential to B-cells, in transgenics the presence of the HSV1-TK in the thymus indicate that this hybrid promoter is also active in the T-cell lineage. Other reports on transgenic mice obtained by the injection of either the entire mu-gene (31) or of a construct containing a similar promoter-enhancer combination (32) showed similar results. Very low level of expression of the transgene could be observed in other tissues including liver, intestine, lung and brain following long exposure of the blot (not shown). The expression in lung and intestine may result from the presence of lymphnodes in these tissues while the significance of tk transcripts in liver and brain is unclear. To further verify the location of the transcription start site in the transgenic tissues, an S1 analysis was performed that reveals a correct initiation of transcription from the hybrid promoter in the same tissues already shown to contain detectable levels of tk RNA by Northern analysis (Fig. 3c). Furthermore, to test if the protein was biologically active, a tk assay was performed on extracts from these tissues. The results show a detectable enzyme activity only in thymus and spleen (data not shown).

To evaluate the efficacy of our system *in vivo*, an experiment was performed in which transgenic and non-transgenic littermates were implanted subcutaneously with a mini-osmotic pump, containing FIAU at a concentration of 50 mg/ml. FIAU was chosen because it was most efficient in blocking the DNA replication *in vitro*. The dose was calculated to obtain serum levels of the drug to concentrations shown to be effective in cell culture. One week after the pump was implanted the animals were sacrificed. Autopsy revealed a completely normal anatomy in the control mouse, with no apparent toxic effects after one week of FIAU treatment (Fig. 4). In the transgenic mouse, most organs were also normal with the dramatic exception of the spleen and thymus which were visually atrophied (Fig.4). This was confirme by weight and cytometric analysis revealing a reduction of 83% of the cell number in the thymus and 57% reduction in the spleen.

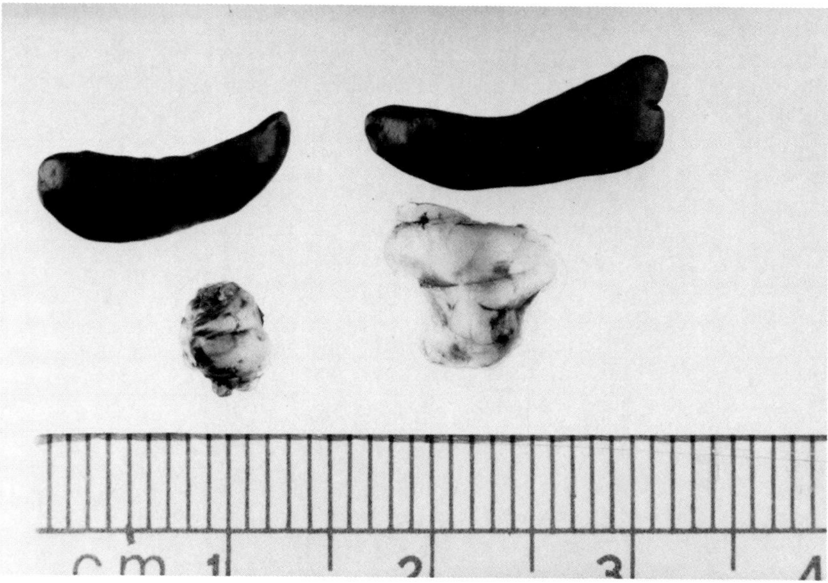

Figure 4. Comparison of transgenic versus non-transgenic TK expressing tissues after treatment with the nucleoside analog FIAU. Left: transgenic tissues (spleen, top left; thymus bottom left). Right: non-transgenic tissues (spleen, top right; thymus bottom right).

DISCUSSION

We have described the generation of a toxic or suicide vector which is based on the controlled expression of the herpes thymidine kinase gene product. The expression of this product alone seems to have no beneficial or obvious deleterious effects on cultured cells. The rational of this approach is based on the ability of this enzyme to convert ACV, FIAU and related compounds to toxic nucleotide intermediates. The activation of these compounds as a consequence of the HSV1-TK expression disrupts cellular DNA replication leading subsequently to cell death. It is not clear the extent of chromosomal damage that must be absorbed to manifest cytotoxic effects, but it is clear that the toxicity is proportional to both the level of expression of the HSV1-TK and the concentration of ACV or FIAU to which the cells are exposed. The toxic effects exerted by this system on cultured cells was so striking that it seemed likely that this technique would be useful for studying cell lineage in transgenic animals. Such a system would exploit the ability of tissue-specific promoters to target restricted expression of HSV1-TK to desired cell types. Since the expression of HSV1-TK alone is not harmful one can generate stable transgenic pedigrees prior to conducting ablation studies. The ability to control drug dose and delivery should allow manipulation of the relative extent of the toxic effects and thus, not only look at lineage ablation but also to look at the plasticity of residual stem cells and their capacity for regeneration.

Recently, another toxic system with the same purpose has been described (33-34), in which a toxic gene, the diphteria toxin, under the control of a specific promoter induces cell death in transgenic mice. An advantage of our system is due to the inducibility of the toxic function compared to the costitutive one achieved by the diptheria toxin expression.

Although it was possible to achieve complete cytotoxicity in the _in vitro_ cultured cell, the toxicity in animals, although severe, was not absolute. It is likely that the protocol used in this study to induce toxicity in animals, was not optimal. In future studies, we should be able to achieve greater toxicity by increasing the relative concentration of FIAU by extending the time of treatment or by a combination of both. By controlling these parameters it might be possible to progress from mild cellular degeneration to near complete destruction of a specific cell and thus provide valuable animal models to study lineage formation and cell function.

ACKNOWLEDGEMENTS

We wish to acknowledge the valuable technical expertise of Deborah Anderson, Drs. Bart Sefton, Geoff Wahl and Douglas Richman for helpful discussion, Drs. Keith Yamamoto, Vincent Giguere and Cary Queen for the gift of the plasmids, pAPS1, pRShGRα and pkCATH, respectively and Dr. M. Pfahl for the rat 2 cell line

#3B. Kevin Murakami for helpful assistance and Elaine Stevens for typing the manuscript. E.B. is on leave from the Unite 184 de Biologie Moleculaire et de Genie Genetique de l'I.N.S.E.R.M., Strasbourg, France. RME is an Investigator of the Howard Hughes Medical Institute. This work was supported by the Howard Hughes Medical Institute and by grants from the National Institutes of Health and the Mathers Foundation.

REFERENCES

1. Gluzman Y (ed) (1982). Eukaryotic Viral Vectors, Cold Spring Harbor Laboratory, Cold Spring Harbor, NY.
2. Palmiter RD, Brinster RL (1985). Transgenic mice. Cell 41:343.
3. Rosenfeld MG, Crenshaw III EB, Borrelli E, Heyman R, Lira SA, Swanson L, Evans RM (1988). Transgenic mice: applications to the study of the nervous system. Ann Rev Neurosci 11:353.
4. Sassone-Corsi P, Borrelli E (1986). Transcriptional regulation by transacting factors. TIG 2:215.
5. McKnight SL (1980). The nucleotide sequence and transcript map of the herpes simplex virus thymidine kinase gene. Nucl Acids Res 8:5949.
6. Wigler M, Silverstein S, Lee L, Pellier A, Cheng Y, Axel R (1977). Transfer of purified herpes virus thymidine kinase gene to cultured mouse cells. Cell 11:223.
7. Cheng Y-C, Dutschman G, Fox JJ, Watanabe KA, Machida H (1981). Differential activity of potential antiviral nucleoside analogs on herpes simplex virus-induced and human cellular thymidine kinases. Antimicrobial Agents and Chemotherapy 20:420.
8. St. Clair MH, Miller WH, Miller RL, Lambe CU, Furman A (1984). Inhibition of cellulr DNA polymerase and herpes simplex virus-induced DNA polymerases by the triphosphate of BW759U. Antimicrobial Agents Chemotherapy 25:191.
9. Fyfe JA, Keller PM, Furman PA, Miller RL, Elion GB (1978). Thymidine kinase from herpes simplex virus phosphorylates the new antiviral compound, 9-(2-Hydroxyethoxymethyl)guanine. J Bio Chem 253:8721.

10. Chen MS, Amico LA, Speelman DJ (1984). Kinetics of the interaction of monophosphates of the antiviral nucleosides 2'-fluoro-1-β-D-arabinofuranosyl-pyrimidine and (E)-5-(2-bromovinyl)-2'-deoxyuridine with thymidylate kinases from vero cells and herpes simplex virus types 1 and 2. Antimicrobial Agents and Chemotherapy 26:778.
11. Grant AJ, Feinberg A, Chou T-C, Watanabe KA, Fox JJ, Philips FS (1982). Incorporation of metabolites of 2'-fluoro-5-iodo-1-β-D-arabinofuranosylcytosine into deoxyribonucleic acid of neoplastic and normal mammalian tissues. Biochem Pharm 31:1103.
12. Elion GB (1982). Mechanism of action and selectivity of acyclovir. Am J Med 73:7.
13. Tucker WE Jr (1983). Preclinical toxicology studies with acyclovir - preface. Fund Appl Toxicol 3:559.
14. McLaren C, Chen MS, Barbhaiya RH, Buroker RA, Olsen FB (1985). Preclinical investigations of FIAU, an anti-herpes agent. in Kono, R. (ed) "Herpes Virus and Virus Chemotherapy," Elsevier Science Publishers, p. 57.
15. Mansuri MM, Ghazzouli I, Chen MS, Howell HG, Brodfuehrer PR, Benigni DA, Martin JC (1987). 1-(2-eoxy-2-fluoro-β-D-arabinofuranosyl)-5-ethyluracil. A highly selective antiherpes simplex agent. J Med Chem 30:867.
16. Gluzman Y (1981). SV40-transformed simian cells support the replication of early SV40 mutants. Cell 23:175.
17. Chandler VL, Maler BA, Yamamoto KR (1983). DNA sequences bound specifically by glucocorticoid receptor in vitro render a heterologous promoter hormone responsive in vivo. Cell 33:489.
18. Giguere V, Hollenberg S, Rosenfeld MG, Evans RM (1986). Functional domains of the human glucocorticoid receptor. Cell 46:645.
19. Gorman CM, Moffat LF, Howard BH (1982). Recombinant genomes which express chloramphenicol acetyl transferase in mammalian cells. Mol Cell Biol 2:1044.
20. Victor Garcia J, Le thi Bich-Thuy, Stafford J, Queen C (1986.) Synergism between immunoglobulin enhancers and promoters. Nature 322:383.
21. Banerji J, Olson L, Shaffner W (1983). A lymphocyte-specific cellular enhancer is located

downstream of the joining region in immunoglobulin heavy chain genes. Cell 33:729.

22. Gillies SD, Morrison SL, Oi VT, Tonegawa S (1983). A tissue-specific transcription enhancer element is located in the major intron of a rearranged immunoglobulin heavy chain gene. Cell 33:717.
23. Queen C, Baltimore D (1983). Immunoglobulin gene transcription is activated by downstream sequence elements. Cell 33:741.
24. Graham FL, van der Eb AJ (1973). A new technique for the assay of infectivity of human adenovirus 5 DNA. Virology 52:456.
25. Hirt B (1967). Selective extraction of polyoma DNA from infected mouse cell cultures. J Mol Biol 26:365.
26. Meinkoth JL, Wahl GM (1984). Hybridization of nucleic acids immobilized on solid supports. Anal Biochem 138:267.
27. Quade K (1979). Transformation of mammalian cells by avian myelocytomatosis virus and avian erythroblastosis virus. Virology 98:461.
28. Pfahl M, Payne J, Benbrook D, Wu KC (1987). Proc UCLA Symp Conf Steroid Hormone Action, in press.
29. Auffray C, Rougeon F (1980). Eur J Biochem 107:303.
30. Maniatis T, Fritsh EF, Sambrook J (1982). "Molecular Cloning," Cold Spring Harbor Laboratory, NY.
31. Grosschedl R, Weaver D, Baltimore D, Costantini F (1984). Introduction of a μ immunoglobulin gene into the mouse germ line: specific expression in lymphoid cells and synthesis of functional antibody. Cell 38:647.
32. Gerlinger P, LeMeur M, Irrmann C, Renard P, Wasylyk E, Wasylyk B (1986). B-lymphocyte targeting of gene expression in transgenic mice with the immunoglobulin heavy-chain enhancer. Nucl Acids Res 14:6565.
33. Palmiter RD, Behringer RR, Quaife CJ, Maxwell F, Maxwell IH, Brinster RL (1987). Cell lineage ablation in transgenic mice by cell-specific expression of a toxin gene. Cell 50:435.
34. Breitmen ML, Clapoff S, Rossant J, Tsui L-C, Glode LM, Maxwell, IH, Bernstein A (1987). Genetic ablation: targeted expression of a toxin gene causes microphthalmia in transgenic mice. Science 238:1563.

Gene Transfer and Gene Therapy, pages 179–188

HEMATOPOIETIC ABNORMALITIES INDUCED BY ECTOPIC EXPRESSION OF THE THY-1 ANTIGEN IN TRANSGENIC MICE[1]

Carlisle P. Landel, Shizhong Chen, Florence Botteri, Herman van der Putten[2] and Glen A.Evans

Cancer Biology and Gene Expression Laboratories, The Salk Institute, P.O. Box 85800, San Diego, CA 92138.

ABSTRACT Expression of the Thy-1 glycoprotein was induced on several populations of bone marrow cells which are normally Thy-1^{-} by introducing a hybrid Thy-1 gene containing an immunoglobulin enhancer into the germline of transgenic mice. These animals express the Thy-1 antigen on over 80% of bone marrow cells and develop a chronic B lymphocyte hyperplasia characterized by the expansion of several populations of cells in the bone marrow and lymph nodes. An additional population of cells expressing substantially lower surface levels of Thy-1 is observed in the bone marrow of some transgenic strains and has the antigenic characteristics of Thy-1^{lo} early hematopoietic progenitor cells. The unusual characteristics of this disorder can be transmitted to

[1]This work was supported by grants from the National Institutes of Health (HD18012,GM33868) and funds from the G. Harold and Leila Y. Mathers Charitable Foundation. S. Chen is a predoctoral trainee from the Department of Biology, University of California, San Diego. G.A. Evans is a Pew Scholar in the Biomedical Sciences.

[2]Present address: Department of Biotechnology, Ciba-Geigy A.G., CH-4002 Basel Switzerland.

lethally irradiated host animals by transplantation of transgenic bone marrow cells.

INTRODUCTION

The differentiation of mature hematopoietic cells from pluripotent stem cells in the bone marrow can be traced by following the expression of lineage-specific cell surface antigens (1). In the mouse, the Thy-1 antigen is a well characterized surface marker of T lymphocyte differentiation that is also expressed on fibroblasts, neurons (2), and a rare population of bone marrow cells with the characteristics of hematopoietic stem cells (3). Murine Thy-1 exists in two allelic forms, Thy-1.1 and Thy-1.2, differing by only a single amino acid substitution. The structure and expression of the mouse and human Thy-1 genes have been well characterized, and it is thought that the cell-type specificity of expression is regulated not by its two alternate promoters but rather by downstream enhancer elements (4,5).

The function of the Thy-1 glycoprotein and its significance for hematopoietic differentiation and T cell function are less well understood. Anti-Thy-1 antibodies can trigger a rapid increase in intracellular [Ca^{2+}] in T cells and transfected cell lines consistent with a role in transmembrane signalling events (6). However, the initiation of subsequent events in lymphocyte activation, including IL-2 secretion, is dependent on the presence of the T-cell receptor-CD33 complex and does not occur in B cells transfected with the Thy-1 gene(7). These data, and the observation that Thy-1 is absent from T cells of many species including man (2), suggest that the Thy-1 antigen is unlikely to play a critical, evolutionarily conserved role in T cell activation. However, the Thy-1 molecule could potentially interact with an endogenous ligand *in vivo* leading to intracellular [Ca^{2+}] activation and subsequent cellular differentiation.

To test this concept, we attempted to redirect Thy-1 gene expression in transgenic mice to an inappropriate cell type and determine the developmental consequences of this mutation. Transgenic mice were produced carrying a hybrid Thy-1 transgene containing an immunoglobulin heavy chain enhancer, Thy-1/Eμ, which directs expression of this T cell antigen to the B cell lineage (8). These mice express the Thy-1/Eμ transgene at high levels on mature B cells and their progenitors and develop a heritable lymphoid hyperplasia characterized by the appearance of several cell populations expressing large amounts of Thy-1 in the bone marrow and lymph nodes. Previous characterization of this abnormality demonstrated the presence of a population of large cells expressing both Thy-1 and the pre-B cell antigen B220, and a population of cells expressing lower amounts of Thy-1 but lacking most other cell-surface markers characteristic of the hematopoietic lineage. Because of the similarities in the cell surface characteristics between hematopoietic stem cells and the novel bone marrow cell population found in the transgenic mice, we have examined the ability of transgenic bone marrow to reconstitute the hematopoietic system of lethally irradiated mice. We report here that transgenic bone marrow is capable of completely repopulating the hematopoietic system of lethally irradiated mice and induces a hematopoietic phenotype indistinguishable from that of the transgenic donors.

METHODS

Transgenic mice were previously produced carrying one or more copies of the Thy-1/Eμ gene (Figure 1). This construction contains a 300 bp deletion in the exon encoding the 3' untranslated region of the mRNA, shortening the length of the RNA transcript such that the transgene product can be easily detected on Northern blots. A 997 nucleotide restriction fragment containing the mouse immunoglobulin heavy chain enhancer was substituted for a 1085 nucleotide region of the Thy-1 intron downstream of the promoters (8).

Seven founder lines of transgenic mice were established in the C57BL/6 x BALB/c background (Thy-1.2) and the transgene introduced into a Thy-1.1 background by backcrossing with the AKR/J (Thy-1.1) strain and the B6.PL-Thy-1^a/Cy (Thy-1.1) congenic strain. All transgenic mice have been bred through four or five successive generations and display a consistent phenotype of mild anemia and lymphoid hyperplasia manifested primarily in the bone marrow and lymph nodes. Older transgenic mice have been found to develop benign tumors but no evidence of increased frequency of malignant tumors has been observed.

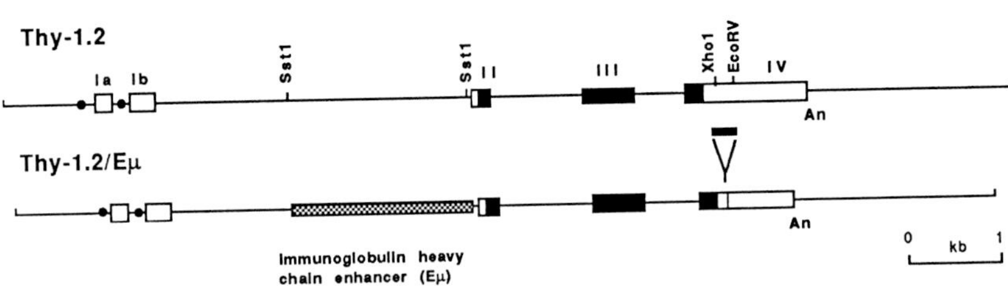

FIGURE 1. Thy-1/Eμ hybrid gene construction used for the production of transgenic mice.

Bone marrow transplantation was performed using selected BALB/c hosts 12 hours after receiving 680 R from a ^{60}Co source. H-2 haplotypes of individual donor and recipient mice were determined by restriction fragment length polymorphisms in the mouse major histocompatibility class I genes (S. Chen and B.E.Rothenberg, unpublished data) and transplantation performed only into H-2 matched hosts. 10^3-10^5 bone marrow cells were injected into the tail veins of H-2 matched host animals and the transplanted mice were then maintained in a sterile environment. After 3 months, the surviving mice were sacrificed and bone marrow cell suspensions prepared. Cells were stained with FITC or phycoerythrin-labeled antibodies and multiparameter flow cytometry was carried out as described (8).

RESULTS

Transgenic mice carrying the Thy-1/Eμ gene develop a lymphoproliferative disease with massive expansion of Thy-1^+,$B220^+$ cells in the bone marrow, lymph nodes and spleen. The abnormality is most evident in the bone marrow where more than 80% of the cells express the Thy-1 antigen (Figure 2) and can be classified into several cell populations. About 45% of the bone marrow consisted of large, $B220^+$ cells expressing as much cell surface Thy-1 antigen as normal mature T lymphocytes. This population is shifted towards the G2/M phases of the cell cycle and most likely includes activated B lymphocytes and B cell progenitors (8). A second major population of cells expresses Thy-1 at about 10% of the level normally found on thymocytes and does not express other surface antigens indicative of T cell, B cell, macrophage or granulocyte cell lineages. These cells lack the B220 B cell surface marker. This population is remarkably similar in surface characteristics to the Thy-1^{lo},$TBMG^-$ stem cell described by Muller-Sieburg, et al. (3) and in some Thy-1/Eμ transgenic strains represents up to 40% of the bone marrow cells. To characterize the stem cell potential of the bone marrow of Thy-1/Eμ transgenic mice, transplantation of transgenic cells into normal recipients was carried out.

Transplantation of transgenic or control bone marrow cells allowed long term survival of the irradiated recipient and apparent complete repopulation of all hematopoietic lineages. At 12 days after transplant, initial bone marrow repopulation was seen with the appearance of the characteristic Thy-1^{hi},$B220^+$ hyperplastic cells (data not shown). Three months after transplantation, all of the hematopoietic cell lineages were repopulated and long term survival of the recipients assured.

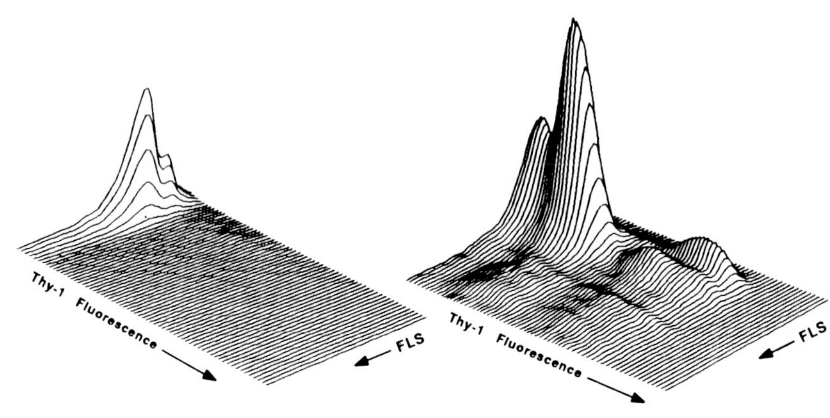

FIGURE 2. Thy-1^{hi} and Thy-1^{lo} hematopoietic cell populations in the bone marrow of Thy-1/Eμ transgenic mice. FLS indicates the forward angle light scatter signal.

Flow cytometry was carried out on the repopulated bone marrow to assess the extent of hematopoietic lineage replacement. Bone marrow analysis demonstrated the presence of B and T lymphocytes, pre-B cells, granulocytes and macrophages. Microscopic examination confirmed the presence of erythroid progenitors cell as well. Interestingly, this analysis demonstrated that the majority of cells in the reconstituted bone marrow expressed Thy-1 and that both Thy-1^{hi} and Thy-1^{lo} populations were present (Figure 3). The distribution of cells in transplanted bone marrow was, in general, remarkably similar to that of Thy-1/Eμ transgenic donors. In particular, repopulation studies demonstrated the presence of both Thy-1^{hi},B220^{+} cells and Thy-1^{lo},TBMG^{-} cells in proportions equivalent to that of the transgenic donors. Additional analysis suggests that the transplanted recipients demonstrated B lymphoid hyperplasia with lymph node involvement similar to that seen in Thy-1/Eμ transgenic mice.

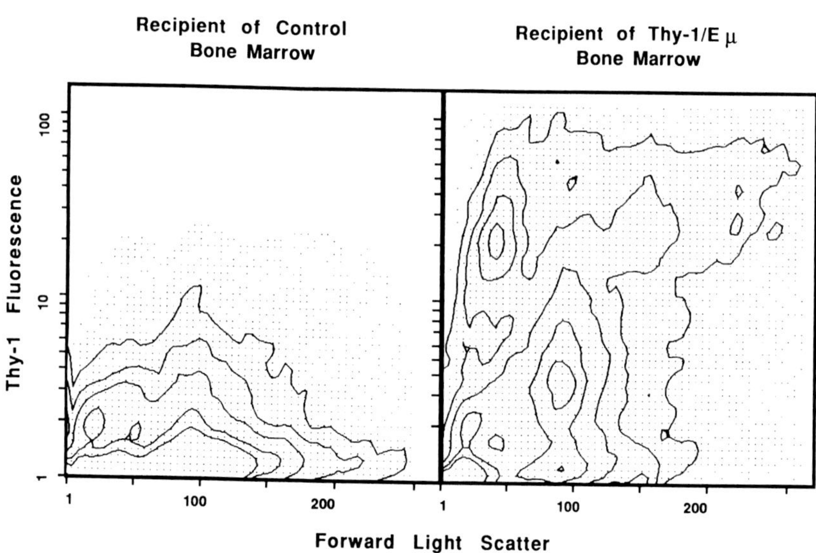

FIGURE 3. FACS analysis of bone marrow cells from recipients of bone marrow transplants from Thy-1/Eμ mice.

DISCUSSION

Recently, the mouse and human Thy-1 antigen genes have been shown to be efficiently and appropriately expressed as transgenes in the mouse genome with nearly correct developmental and cell-lineage specific expression (8-10). Thus, the *cis*-acting regulatory sequences responsible for the complex multi-tissue regulation may be located within a relatively small region near the Thy-1 gene. However, inappropriate developmental regulation, manifested as inappropriate or ectopic expression in several organ systems, has been found to produce proliferative, non-malignant abnormalities (8-9). In the hematopoietic system, expression of the Thy-1 molecule on early B progenitors leads to B cell hyperplasia. Expression of the mouse Thy-1 gene in the epithelial cells of the kidney, induced by the expression mouse-human Thy-1 fusion genes, leads to hyperplasia of the proximal tubule epithelium with frequent progression to malignancy (9). Table 1 summarizes the results of a number of transgenic studies on the tissue distribution and proliferative effects of the expression of Thy-1 in transgenic mice. These data also indicate that, consistent with previous *in vitro* studies, the tissue-specific regulatory elements are separable into neural and hematopoietic cell specific domains, and these elements map downstream of the promoters at the 3' flanking region of the Thy-1 gene (4,9; S. Chen, G. Andreason, unpublished data).

The association of ectopic Thy-1 expression with non-malignant proliferation suggests that the physiological function of this enigmatic molecule may involve cellular self-renewal. This is consistent with *in vitro* studies demonstrating the capacity of Thy-1, singly or in association with other cell surface molecules, to induce biochemical changes consistent with cell activation (6,7). In the hematopoietic system, Thy-1 is normally distributed on various mature cells in different species, but in man, mouse, rat and dog, it appears to be consistently expressed on early lymphoid progenitors. The Thy-1^{lo} phenotype appears to be a characteristic surface phenotype of early hematopoietic progenitor cells. Thy-1^{lo},$TBMG^{-}$, cells comprise about 0.1% or less of bone marrow cells (3). This suggests that the Thy-1 molecule expressed on hematopoietic progenitor cells is involved in the activation of these cells for differentiation or self-renewal.

A significant aspect of this study is that, in addition to B cell hyperplasia, a second cell population of cells is present in bone marrow which expresses only low amounts of surface Thy-1. This large population of Thy-1^{lo}, TBMG$^-$ cells in the transgenic bone marrow suggests the intriguing possibility that, in addition to B cell proliferation, a primary effect of the Thy-1/Eμ phenotype is expansion of the hematopoietic stem cell population. Further analysis of Thy-1/Eμ transgenic mice will help determine if the Thy-1 antigen functions as a regulator of hematopoietic cell differentiation.

TABLE 1.
EXPRESSION OF THY-1 IN TISSUES OF TRANSGENIC MICE

Gene(ref.)	Lung	Kidney	Brain	Thymus	Spleen	BM
Thy-1 (8)	-	-	+++	+++	+	-
Thy-1 (8)	-	-	+++	+++	+	-
Thy-1δX (8)	-	-	+++	+++	-	-
Thy-1δS (8)	-	-	+/-	+++	-	-
Thy-1/Eμ (8)	-	-	+/-	+++	+++	+++*
mThy-1 (10)	-	-	+++	+++	+	-
mhThy-1 (9)	-	++*	+++	-	+/-	-
hThy-1 (9,10)	-	++*	+++	++	-	-

+ indicates expression
* indicates proliferative abnormality

ACKNOWLEDGEMENTS

We thank R. Hyman for reagents and advice on their use, J. Trotter for assistance with flow cytometry and the use of his data analysis software, J. Zhao and K.A. Lewis for expert technical assistance, and B. Rothenberg, K. Pischel, J. H. Eubanks, R. Heyman, and R. Evans for advice and discussions.

REFERENCES

1. Williams, AF (1985). Immunoglobulin-related domains for cell surface recognition. Nature 314:579.
2. Williams, AF, and Gagnon, JJ (1982). Neuronal cell Thy-1: homology with immunoglobulin. Science 216:696.
3. Muller-Sieburg, CE, Whitlock, CA, and Weissman, IL (1986). Isolation of two early B lymphocyte progenitors from mouse marrow: a committed pre-pre-B cell and a clonogenic Thy-1^{lo} hematopoietic stem cell. Cell 44:653.
4. Ingraham, HA, and Evans, GA (1986). Characterization of two atypical promoters and alternate mRNA processing in the mouse Thy-1.2 glycoprotein gene. Mol. Cell. Biol. 6:2923.
5. Giguere, V, Isobe, KI, and Grosveld, F (1985). Structure of the murine Thy-1 gene. EMBO J. 4:2017.
6. Kroczek, RA, Gunter, KC, Seligmann, B, and Shevach, EM (1986). Thy-1 functions as a signal transduction molecule in T lymphocytes and transfected B lymphocytes. Nature 322:181.
7. Gunter, KC, Germain, RN, Kroczek, RA, Saito, T, Yokoyama, WM, Chan, C, Weiss, A, and Shevach, EM (1987). Thy-1-mediated T-cell activation requires co-expression of CD3/Ti complex. Nature 326: 505.
8. Chen, S, Botteri, F, van der Putten, H, Landel, CP, and Evans, GA (1987). A lymphoproliferative abnormality associated with inappropriate expression of the Thy-1 antigen in transgenic mice. Cell 51:7.
9. Kollias, G, Evans, DJ, Ritter, M, Beech, J, Morris, R, and Grosveld, F (1987). Ectopic expression of Thy-1 in the kidneys of transgenic mice induces functional and proliferative abnormalities. Cell 51: 21.
10. Gorden, JW, Chesa, PG, Nishimura, H, Rettig, WJ, Maccari, JE, Endo, T, Servalli, E, Seki, T, and Silver, J (1987). Regulation of Thy-1 gene expression in transgenic mice. Cell 50:445.

Gene Transfer and Gene Therapy, pages 189–204

TARGETED CELL SUICIDE BY TOXIN GENE EXPRESSION[1]

Ian H. Maxwell[2], L. Michael Glode[2], Françoise Maxwell[2], Gail S. Harrison[2], Martin L. Breitman[3], Hector Rombola[3], Dawn-Marie Coulson[3], and Alan Bernstein[3]

[2]Division of Medical Oncology, University of Colorado Health Sciences Center, Denver, CO 80262

[3]Mount Sinai Hospital Research Institute, 600 University Avenue, Toronto and Department of Medical Genetics, University of Toronto, Toronto, Ontario, Canada

ABSTRACT The A polypeptide of diphtheria toxin (DT-A) is extremely toxic when introduced into the cytoplasm of eukaryotic cells. We have explored the use of regulated expression of the DT-A coding sequence as a means of selectively killing specific cell types. We constructed DT-A expression plasmids and obtained evidence for lethality resulting from their transfection into cultured mammalian cells. The coding sequence for an attenuated DT-A mutant, *tox* 176, was also cloned and, as expected, this showed a diminished level of toxicity when expressed in cultured cells. Specificity for B-lymphoid cells in culture was

---

[1]This work was supported by grants from the National Institutes of Health (CA42354), from Joyce Calvert, from the Cancer League of Colorado, the American Cancer Society (PDT-295), the Milheim Foundation (84-37), the Dudley Foundation, the Medical Research Council of Canada and the National Cancer Institute of Canada.

demonstrated for DT-A constructs containing immunoglobulin transcriptional regulatory elements. Another construct containing the DT-A sequence under control of the gamma-2 crystallin regulatory region was used to generate a line of transgenic mice showing specific ablation of lens fiber cells. Exquisitely tight control of tissue-specific lethal expression of DT-A in a whole organism is therefore feasible. Such targeted cell ablation should provide a valuable tool in developmental studies and might eventually be developed as a novel means of cancer therapy.

INTRODUCTION

The developmental programs for organogenesis in higher organisms involve complex interactions among differentiating cell types which are difficult to decipher by classical techniques. The ability to ablate specific cell types in a predetermined manner would provide a useful approach to elucidating these interactions. The selective destruction of specific cells is also a primary goal in cancer therapy where presently available drugs produce dose limiting toxicity to normal cell populations. A promising approach to the latter goal involves the use of a toxin, conjugated with a ligand (e.g. a polypeptide hormone or a monoclonal antibody) capable of binding selectively to a receptor on the cancer cell surface and allowing the subsequent internalization of the toxin by the cell. The protein toxins produced by various species of bacteria and plants are important candidates for such a therapeutic use. As an alternative to directing such a toxin to the cell surface, we have developed the approach of using regulated expression of an introduced gene encoding a toxin to direct selective cell suicide. Experiments demonstrating the feasibility of this idea, both in cell culture and in whole animals, are presented below.

We chose to use diphtheria toxin (DT) because of its well-characterized mechanism of action and extreme toxicity for eukaryotic cells (1). Mutants with diminished toxicity were also available (2) and, most importantly, the DT gene had been cloned and sequenced (3,4). DT is encoded in the genome of a bacteriophage capable of lysogenizing *Corynebacterium diphtheriae*. The DT gene is expressed by the lysogen (in response to Fe deprivation) as a precursor polypeptide containing a hydrophobic signal sequence which is cleaved from the secreted product. The latter undergoes a further proteolytic cleavage yielding an A-fragment (193 amino acids) and a B-fragment (342 amino acids) which remain associated via a disulfide bridge. The B-fragment is responsible for adsorption to the surface of mammalian cells and also functions to introduce the A-fragment (DT-A) efficiently into the cytoplasm after internalization of the toxin into endosomes. DT-A then enzymically inactivates elongation factor 2 by ADP-ribosylation of diphthamide, a specifically modified histidine residue found in this protein. The consequent inhibition of protein synthesis leads to cell death. Experiments on the controlled introduction of DT-A into mouse L-cells suggest that the presence of a single molecule in the cytoplasm can kill a cell (5). We sought to express the DT-A coding sequence in mammalian cells under tight genetic control so that selective suicide of specific cell types could be obtained. We describe below how tissue-specific, *cis*- acting transcriptional regulatory elements have been used to achieve this goal in cell culture and in transgenic mice.

RESULTS

DT-A Expression Plasmids

The DT-A coding sequence was adapted for expression in mammalian cells, without secretion, by making a gene fusion between the promoter region of the human metallothionein IIA gene and the mature DT-A sequence, utilizing a HhaI site

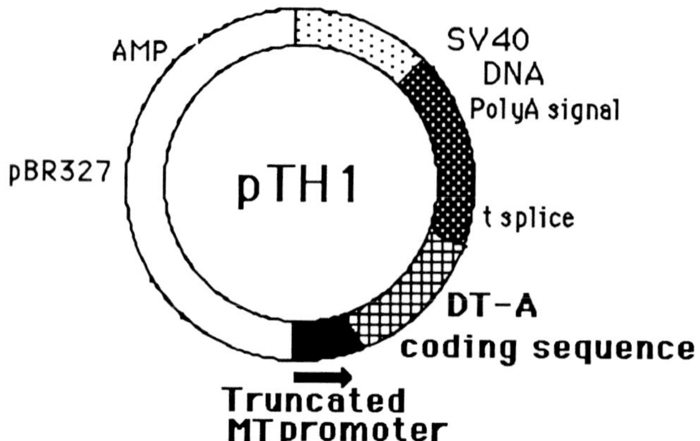

Figure 1. pTH1, the prototype DT-A expression plasmid.

conveniently located in the second codon. The resulting construct, in a pSV2 type expression plasmid, was designated pTH1 (see Fig. 1) (6) and this formed the prototype for further constructs containing various transcriptional regulatory elements. The 3' terminus of the DT-A sequence is ligated to SV40 DNA such that there is an in-frame terminator two codons downstream. The protein expressed from pTH1 and its derivatives is therefore identical to mature DT-A except for the first three and the last two amino acids:

MatureDT-A:
Gly-ala-asp-asp-val.....arg-val-arg-arg
pTH1product:
Met-asp-pro-asp-asp-val.....arg-val-arg-arg-ser-leu

For control experiments a frameshift mutation was introduced into the DT-A sequence at codon 35 by cleavage at the unique AccI site, filling-in and re-ligating. All even-numbered plasmids (pTH2, pTH4, etc.) contain this frameshift; odd numbering designates the wild-type DT-A sequence. The frameshift plasmids should express an inactive, truncated protein terminating ≈30 codons after the frameshift.

Table 1.

Frequency of G418 resistant transformants obtained after transfection of 293 cells with *neo* plasmids containing wild-type or frame-shift mutant DT-A gene. pSV2-327neo: control, no DT-A sequence. pTH1neo, pTH1eon: Wild-type DT-A plasmids differing only in orientation of *neo*. pTH2neo: DT-A frameshift control. Transfections were by the Ca phosphate method using 10 μg plasmid DNA.

Plasmid	Number of colonies per 5×10^5 cells transfected
None	0
pSV2-327neo	64
pTH1neo	0,0
pTH1eon	0
pTH2neo	79,71

To test for lethal expression, pTH1 was transfected into several types of cultured cells. Since transfection procedures usually introduce DNA into only a small fraction of the cells, indirect assays were used to assess the effect of DT-A expression on this fraction. First, a *neo* transcription unit was cloned into both pTH1 and pTH2 to provide a dominant selectable marker. We expected that transfection of cells with pTH2neo should allow selection of clones resistant to antibiotic G418 as a result of *neo* expression whereas expression of DT-A from pTH1neo should preclude such selection. As shown in Table 1, this expectation was confirmed using the human 293 cell line. The absence of G418 resistant colonies after transfection with pTH1neo strongly suggested that this construct expressed a lethal product. In similar experiments with HeLa cells, colonies were obtained after transfection with pTH1neo at a reduced but significant frequency (6). We have not investigated whether these clones had rearranged the transfected construct or had acquired resistance through some other mechanism.

We next developed a transient co-transfection assay to enable more rapid assessment of the level of DT-A expression. A reporter plasmid capable of expressing chloramphenicol acetyltransferase (CAT) was co-transfected with varying amounts of a plasmid of the pTH series. As shown in Fig.2, expression of the reporter was diminished in a dose dependent manner by pTH1 but not by the frameshift control plasmid, pTH2. This diminution in reporter expression is assumed due to a general inhibition of protein synthesis in the transfected cells by DT-A. The use of the reporter plasmid allowed us to detect this phenomenon in the relatively small

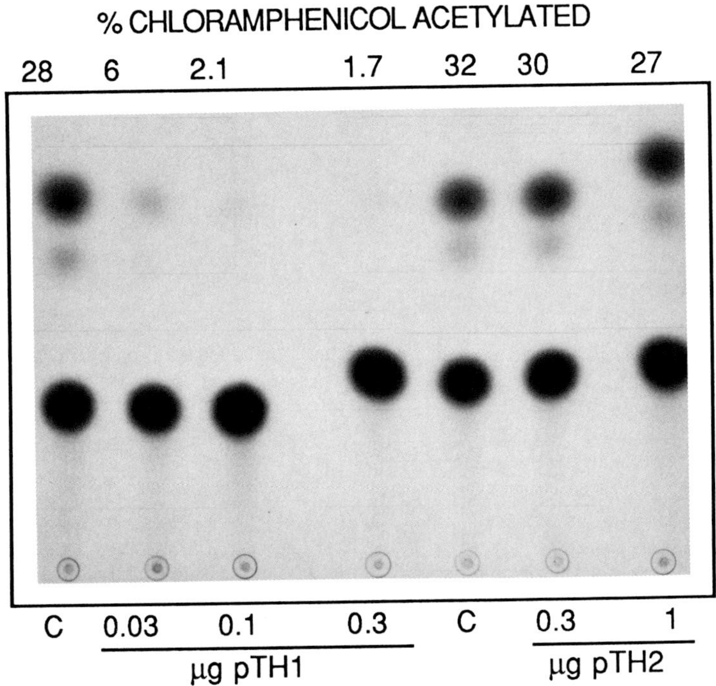

Figure 2. Assay of CAT activity in extracts of 293 cells 48 h after transfection with 5 µg pSV2CAT together with the indicated amounts of pTH1 or pTH2. Tracks C: pSV2cat only. Shown is the autoradiogram of a thin layer chromatogram separating acetylated (upper spots) from non-acetylated ^{14}C-chloramphenicol (lower spot).

percentage of cells which take up DNA in transfection experiments. The transient co-transfection assay was then used in subsequent experiments, described below, to assess toxicity resulting from expression of an attenuated mutant DT-A, and of DT-A constructs designed to show cell-type specificity.

The *tox* 176 Attenuated DT-A Mutant

Mutant *tox* 176 was isolated in 1973 among a series of corynephage ß mutants generated using MNNG mutagenesis (2). *Tox* 176, the only A-chain mutant that retained substantial toxicity, was reported to show ≈8% of the in vitro enzyme activity (ADP-ribosylation of EF2) of wild-type DT-A (2). On introduction of the toxin A chains into mouse L-cells by fusion with loaded erythrocyte ghosts it was estimated that 200 fold more *tox* 176 than wild-type A-chain was required for lethality (5). These results suggested that expression of *tox* 176 DT-A might be useful for selective, controlled lethality in situations where the wild-type DT-A might prove too toxic owing to minimal levels of leaky expression. We therefore cloned the *tox* 176 DT-A coding sequence from genomic DNA of an appropriate *C. diphtheriae* lysogen (7) and constructed a prototype expression plasmid, pTH1-176, analogous to pTH1. pTH1-176 contains a BamH1 site, absent from pTH1, providing a linked marker which we have found useful when substituting the mutant for the wild-type DT-A sequence in further expression constructs.

To characterize the mutation in *tox* 176 we determined the sequence of the cloned DNA. There was a single G to A transition in the second position of codon 128 resulting in replacement of glycine by aspartic acid. This result has been confirmed by Comanducci et al. (8) who also reported the absence of any mutation in the *tox* 176 B-chain. Fig.3 shows the position of the *tox* 176 mutation in the DT-A amino acid sequence in relation to other characterized mutations.

In Fig.4, the activity of pTH1-176 is compared with that of pTH1 in the transient co-transfection assay in 293 cells. About 30 fold

1 gly-ala-asp-asp-val-val-asp-ser-ser-lys-ser-phe-val-met-glu-
asn-phe-ser-ser-tyr-his-gly-thr-lys-pro-gly-tyr-val-asp-ser-
31 ile-gln-lys-gly-ile-gln-lys-pro-lys-ser-gly-thr-gln-gly-asn-
tyr-asp-asp-asp-trp-lys-gly-phe-tyr-ser-thr-asp-asn-lys-tyr-
61 asp-ala-ala-gly-tyr-ser-val-asp-asn-glu-asn-pro-leu-ser-gly-
lys-ala-gly-gly-val-val-lys-val-thr-tyr-pro-gly-leu-thr-lys-
91 val-leu-ala-leu-lys-val-asp-asn-ala-glu-thr-ile-lys-lys-glu-
leu-gly-leu-ser-leu-thr-glu-pro-leu-met-glu-gln-val-gly-thr-
121 glu-glu-phe-ile-lys-arg-phe-**Asp**-asp-gly-ala-ser-arg-val-val-
leu-ser-leu-pro-phe-ala-glu-gly-ser-ser-ser-val-glu-tyr-ile-
151 asn-asn-trp-glu-gln-ala-lys-ala-leu-ser-val-glu-leu-glu-ile-
asn-phe-glu-thr-arg-gly-lys-arg-gly-gln-asp-ala-met-tyr-glu-
181 tyr-met-ala-gln-ala-cys-ala-gly-asn-arg-val-arg-arg

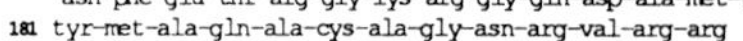

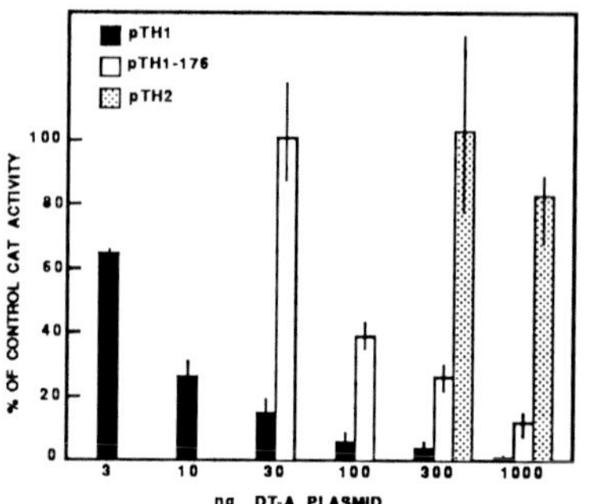

Figure 3. (Left) Deduced amino acid sequence of *tox* 176 DT-A. The amino acid substitution observed in *tox* 176 (asp for gly at position 128) is shown in bold print. The positions of amino acid substitutions in inactive DT-A mutants, *tox* 197 (glu for gly at 52)(13) and *tox* 228 (asp for gly at 79, and lys for glu at 162)(4) are underlined, as is the glu at position 148 where deletion, or substitution of asp, also result in inactivation (14,15).

Figure 4. (Right) Effect of co-transfection with DT-A expression plasmids on transient expression of CAT activity from pSV2cat in line 293 cells.

---

more of the mutant plasmid was required for similar inhibition, a result consistent with reported levels of activity of the *tox* 176 DT-A protein (2,5). The use of the wild-type or the *tox* 176 DT-A coding sequence should therefore facilitate manipulation of levels of toxic expression as appropriate for particular experimental situations.

Regulated and Cell-specific DT-A Expression in Cultured Cells

To facilitate the insertion of various promoters or enhancers, we removed the MTIIA promoter from pTH1, generating a promoterless construct designated pTH7. The ATG originally derived from the MT gene was retained for translation initiation of DT-A. Tests with pTH7 revealed substantial toxicity in some cell types despite the absence of a eukaryotic promoter, implying the occurrence of transcriptional initiation in plasmid sequences. In agreement with published results (9), we found that this problem could be overcome by inserting an oligomer of a DNA fragment containing SV40 polyadenylation signals at the boundary of the bacterial plasmid sequences. Insertion of such an "A-trimer" into pTH7, generating pTHA7, eliminated inhibition by this promoterless plasmid in the transient co-transfection assay. The A-trimer was then inserted upstream of the metallothionein promoter in pTH1 (Fig. 1), generating pTHA1. In 293 cells, DT-A expression from pTHA1 was substantially less than from pTH1 and was increased by treatment of the transfected cells with Cd^{++}. We previously had difficulty in demonstrating significant heavy metal induction of the truncated MT promoter in pTH1 and these results suggest that induction may have been obscured by a substantial level of DT-A expression from plasmid-initiated transcripts. It will clearly be important to avoid such spurious transcription when testing constructs designed for cell-type specificity; the A-trimer promises to be very useful for this purpose.

As an initial model for directing DT-A expression to specific cell types we investigated the use of immunoglobulin promoters and enhancers for targeting B-lymphoid cells. The human Ig heavy chain enhancer was first cloned as a 279 bp fragment into pTH1, generating pTH3. This plasmid showed some B-cell preference for DT-A expression (6) but there was still substantial expression in other cells. It is probable that transcripts initiated from the relatively non-specific MT promoter, as well as in plasmid sequences, contributed to this expression. To improve B-cell specificity we substituted an Ig heavy chain promoter for the MT promoter, also inserting the A trimer upstream of this promoter, generating pTHA15. Finally, an additional Ig heavy chain enhancer was placed upstream of the promoter giving plasmid pTHA17. Transient co-transfection assays were performed in B-cells and in HeLa with pTHA15 or pTHA17. Under the conditions used, 80 - 90% inhibition of co-transfected reporter expression was observed in the B-cells and the extent of inhibition was similar with both plasmids. In contrast, no inhibition was detected in HeLa cells, showing that a high degree of B-cell preferential DT-A expression had been achieved.

The use of inducible systems in which DT-A expression from a resident gene construct could be turned on by supplying an appropriate external stimulus would be of considerable value in developmental studies and potentially in cancer therapy. We have begun to explore this possibility by using transcriptional regulatory sequences from heat shock genes and from the interleukin 2 gene. So far, we have not succeeded in obtaining stable transformants inducible for cell suicide.

Ablation of Specific Tissue in Transgenic Mice

To demonstrate the potential for using DT-A expression to kill specific cell types in an intact organism, we constructed transgenic mice containing the DT-A coding sequence in the transgenome. A lens crystallin regulatory sequence was chosen since previous work had shown rigorous restriction of crystallin expression to the fiber cells of the ocular lens and ablation of these cells should produce an obvious phenotype. Furthermore, an 804 bp 5' flanking sequence from the mouse gamma-2 crystallin gene was sufficient to direct expression of a chimeric ß-galactosidase gene to lens fiber cells in transgenic mice (10). The MT promoter in pTH1 was replaced by this gamma crystallin flanking sequence and the DT-A transcription unit isolated from the resulting plasmid was used to generate transgenic mice. Four of the resulting animals displayed microphthalmia (Fig. 5A) and two of these had cataracts, but no other abnormalities were detected. One male with microphthalmia was bred and 16 of its 49 F1 progeny were microphthalmic; segregation of this phenotype showed complete concordance with the presence of the DT-A transgenome (11).

Histological examination of sectioned eyes from F1 individuals showed considerable variation ranging from almost complete lens ablation (Fig. 5C) to a morphologically normal lens of diminished

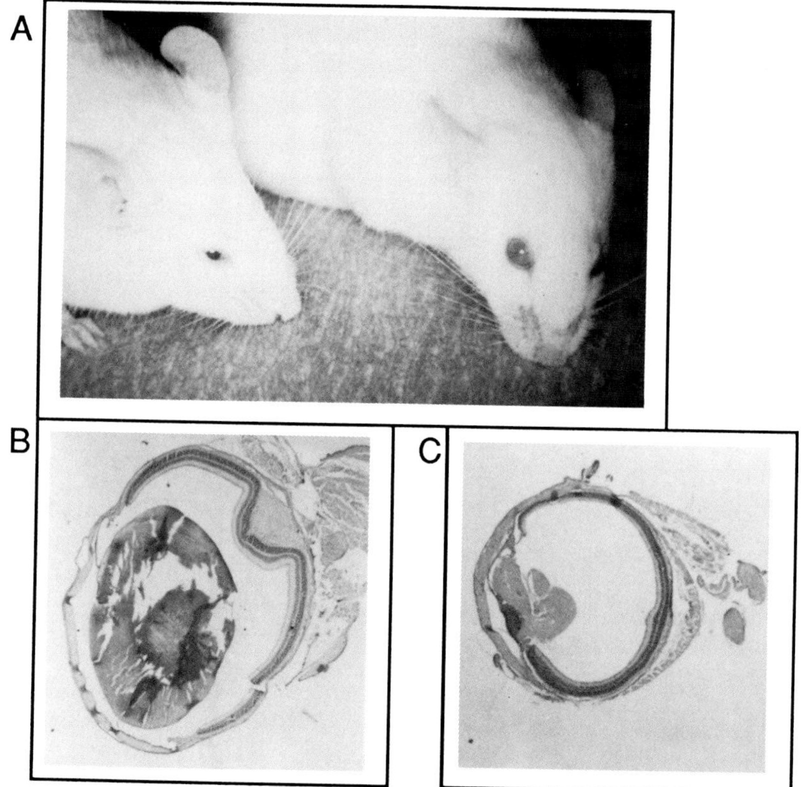

Figure 5. A. Microphthalmic appearance at age 4 months of founder mouse (left), transgenic for gamma-2 crystallin DT-A construct. (Age matched control on right.) B.Histological section (7 μm) of eye from control 3 week old mouse. (Fractures in the lens occurred during sectioning.) C. Eye section from 3 week old mouse of F1 progeny of the transgenic mouse in A., showing most severe lens ablation.

size (11). Three types of experiment were performed to elucidate the mechanism of this variable penetrance. First, DT-A transgenic mice were cross-bred with the previously derived transgenics expressing the *lac Z* gene from the same gamma-2 crystallin promoter. Histological sections from eyes of resulting progeny receiving both transgenomes showed, on staining for ß-galactosidase, that most lenses retained a few intensely staining fiber cells in the core. Secondly, the F1 DT-A transgenic mice were bred to obtain homozygotes and, in nine cases out of ten, a much more severe phenotype was observed in which the eyes were greatly diminished in size in comparison with the F1 microphthalmics. Thirdly, the *tox 176* attenuated DT-A mutant coding sequence (7), also attached to the gamma-2 crystallin promoter, was used to generate transgenic mice. The resulting transgenics showed an interesting range of phenotypes: two founder animals had severe microphthalmia while the other three founder mice displayed cataracts but not gross evidence of ablation. The latter phenotype could not simply be attributed to expression of any foreign protein in the lens fiber cells since the *lac Z* transgenics do not display cataracts. The variable penetrance in DT-A transgenic mice may result from a minority of developing lens fiber cells failing to activate the DT-A transgenome, suggestive of the involvement of a stochastic event in the normal activation of gamma-2 crystallin gene expression.

DISCUSSION

We have shown that DT-A expression can be used to induce cell suicide both in cell cultures and in whole animals. Tissue-specific transcriptional regulatory sequences can be used to target killing to specific cell types, as has been convincingly demonstrated by the ablation of predicted cell populations in transgenic mice. In transfection experiments in cell culture, specificity of expression may sometimes be compromised due to plasmid initiation of transcription but this problem can be minimized by placing an appropriate blocking sequence (A-trimer) upstream of the regulated transcription unit. This is unnecessary for cell targeting in transgenic mice generated by injection of the isolated DT-A transcription unit.

In the experiments described above, ablation of lens fiber cells was observed in mice containing a DT-A transgene under control of the gamma-2 crystallin 5' regulatory sequence. Similar ablation was obtained using the cDNA coding sequence for the A-chain of ricin, attached to an alpha-crystallin promoter (results reported by Carlisle Landel at this meeting). Presumably, a variety of toxin genes from bacterial, plant or other sources could be used in this way. Regulated DT-A expression in transgenic mice has also been used successfully to ablate the exocrine pancreas (using a rat elastase enhancer/promoter (12)), pituitary somatotropes (using rat growth hormone 5'-flanking sequence), and the adrenal medulla together with certain cells in the retina (using a regulatory sequence from the gene encoding the terminal enzyme in epinephrine synthesis; results reported by Richard Palmiter). In these examples, the transgenic mice showed normal development except for the targeted tissue. It is therefore likely that transcriptional regulatory sequences from many tissue-specific genes may be capable of directing sufficiently tight control to allow targeted DT-A lethality. This conclusion is surprising in view of the extreme toxicity of the DT-A protein (reported to be lethal at a level of one molecule per cell (5)). In anticipation of possible non-specific lethality due to failure to restrict basal expression of DT-A, we cloned the coding sequence (7) for the *tox* 176 attenuated mutant described above. Use of this sequence together with the gamma-2 crystallin promoter in transgenic mice produced a less severe phenotype (cataracts without microphthalmia) in most cases than did the wild-type DT-A construct. However, two mice transgenic for *tox* 176 displayed a very severe microphthalmic phenotype similar to that seen with the wild-type DT-A homozygotes (see above). This result, most likely due to a chromosome position effect, suggests that the *tox* 176 mutant will be useful for ablation studies in situations where high levels of expression can be expected in the target tissue. Since minimal levels of expression of the mutant in other tissues might be tolerated, *tox* 176 DT-A may allow the ablation technique to be extended to the use of regulatory elements from genes showing high level target tissue expression and more generally distributed low level expression. This consideration could be of considerable importance in attempting to ablate cancer cells by toxin gene expression (see below).

The ability to generate transgenic animals ablated for specific cell types will have numerous applications in studying interactions among cell lineages. The technique need not be restricted to mammals since DT-A is toxic in a variety of eukaryotic cells including yeast. The potential exists for exploiting translational, or other post-transcriptional regulatory mechanisms, in addition to control at the level of DT-A transcription, for achieving cell-specific lethality. It will also be interesting to explore the possibility of combining inducibility with tissue-specificity of toxin expression (for example using IL2 regulatory sequences). An analogous approach using tissue-specific expression of herpes thymidine kinase to permit tissue ablation by certain antiviral drugs was described by Borrelli et al. at this meeting. However, this ablation method is probably applicable only to dividing cell populations, in contrast to the direct expression of a toxin such as DT-A.

Aside from its application in developmental biology, regulated toxin gene expression has potential as a novel means of cancer therapy. Thus, for example, cancer cells might be killed by DT-A expressed under the control of regulatory elements from a gene encoding a protein expressed abundantly in that particular tumor. Such a gene might encode either a product ectopic to the tissue of origin (e.g. an oncofetal antigen) or a common product in the case of a non-essential tissue. Successful therapy would require a means of delivering a toxin gene construct to the cancer cells with reasonably high efficiency. This could potentially be achieved by incorporation of the gene together with the required regulatory sequences into appropriate viral vectors. We are initiating the construction of such delivery vehicles.

ACKNOWLEDGMENT

We are indebted to Jack Murphy for supplying pDT201 which was the source of the DT-A coding sequence.

REFERENCES

1. Pappenheimer, AM,Jr (1977).Diphtheria toxin. Ann Rev Biochem 46: 69.

2. Uchida, T, Pappenheimer, AM, Greany, R (1973). Diphtheria toxin and related proteins. I. Isolation and properties of mutant proteins serologically related to diphtheria toxin. J Biol Chem 248: 3838.

3. Greenfield, L, Bjorn, MJ, Horn, G, Fong, D, Buck, GA, Collier, RJ, Kaplan, DA (1983). Nucleotide sequence of the structural gene for diphtheria toxin carried by corynephage ß. Proc Natl Acad Sci USA 80: 6853.

4. Kaczorek, M, Delpeyroux, F, Chenciner, N, Streeck, RE, Murphy, JR, Tiollais, P (1983). Nucleotide sequence and expression of the diphtheria *tox* 228 gene in *Escherichia coli*. Science 221: 855.

5. Yamaizumi, M, Mekada, E, Uchida, T, Okada, Y (1978). One molecule of diphtheria toxin fragment A introduced into a cell can kill the cell. Cell 15: 245.

6. Maxwell, IH, Maxwell, F, Glode, LM (1986). Regulated expression of a diphtheria toxin A-chain gene transfected into human cells: possible strategy for inducing cancer cell suicide. Cancer Res 46: 4660.

7. Maxwell,F, Maxwell, IH, Glode, LM (1987). Cloning, sequence determination, and expression in mammalian cells of the coding sequence for the tox 176 attenuated diphtheria toxin A chain. Mol Cell Biol 7: 1576.

8. Comanducci, M, Ricci, S, Rappuoli, R, Ratti, G (1987). The nucleotide sequence of the gene coding for diphtheria toxoid CRM176. Nucl Acids Res 15:5897.

9. Kadesch, T, Berg, P (1986). Effects of the position of the simian virus 40 enhancer on expression of multiple transcription units in a

single plasmid. Mol Cell Biol 6: 2593.

10. Goring, DR, Rossant, J, Clapoff, S, Breitman, ML, Tsui, L-C (1987). In situ detection of ß-galactosidase in lenses of transgenic mice with a gamma-crystallin/*lacZ* gene. Science 235: 456.

11. Breitman, ML, Clapoff, S, Rossant, J, Tsui, L-C, Glode, LM, Maxwell, IH, Bernstein, A (1987). Genetic ablation: targeted expression of a toxin gene causes microphthalmia in transgenic mice. Science 238: 1563.

12. Palmiter, RD, Behringer, RR, Quaife, CJ, Maxwell, F, Maxwell, IH, Brinster, RL (1987). Cell lineage ablation in transgenic mice by cell-specific expression of a toxin gene. Cell 50: 435.

13. Giannini, G, Rappuoli, R, Ratti, G (1984). The amino-acid sequence of two non-toxic mutants of diphtheria toxin: CRM45 and CRM197. Nucl Acids Res 12: 4063.

14. Tweten, RK, Barbieri, JT, Collier, RJ (1985). Diphtheria toxin. Effect of substituting aspartic acid for glutamic acid 148 on ADP-ribosyltransferase activity. J Biol Chem 260: 10392.

15. Emerick, A, Greenfield, L, Gates, C (1985). Enzymatically inactive diphtheria A fragment: expression in *E. coli* and toxicity characterization DNA 4: 78.

Gene Transfer and Gene Therapy, pages 205–214

TARGETED INTRODUCTION AND EXPRESSION OF ACTIVATED GENES INTO POST-IMPLANTATION RAT EMBRYOS VIA INJECTION OF TRANSFECTED RAT EMBRYO CELLS[1]

Stephen A. Schwartz, Ross Couwenhoven, and Stanley Welch

Department of Pathology, The Chicago Medical School
North Chicago, Illinois 60064

ABSTRACT The use of transgenic animals has become a valuable tool to investigate molecular mechanisms responsible for regulation of cell growth, differentiation, and neoplasia. Retroviral vectors and micro-injection techniques have been widely used to successfully introduce foreign genes into mammalian embryos. Due to the randomness and non-specificity of these methods, however, it is difficult to target or generate transgenic animals with respect to a single organ or tissue derived from a particular germ line. We developed a technique to conveniently and reproducibly generate genetic mosaic rats containing exogenous genes predominantly in tissues of mesenchymal derivation. Primary cultures of 16-day old rat embryo cells were transfected with the pSV2neo gene in an expression vector, selected with G418, and immediately injected into 9.5-day post-implantation rat embryos. Inasmuch as the fibroblast vectors are predominantly totipotent mesenchymal presursors, rat offspring were genetic mosaics in cells and tissues essentially of mesenchymal origin. Nearly 60% of the injected embryos survived to birth, at which approximately 50% contained and expressed pSV2neo as determined by Southern and Northern blotting, respectively. The tissues differentially enriched in neo sequences included spleen, bone marrow, cardiac and skeletal muscle, and stroma from major visceral organs. When activated Ha-ras-1 sequences were similarly introduced into rat embryos, nearly all of the major, mesenchyme-derived tissues examined contained the oncogene. Although the surviving offspring were apparently healthy up to 4 months

1. This work was supported by Grant DE-07938, USPHS NIH

postpartum, 2 of 5 female mosaics developed adenocarcinomas of the breast subsequent to pregnancy and lactation. Southern blotting of tumor cell DNA revealed the presence of neo as well as the oncogene nucleotide sequences in addition to the germ-line rat Ha-ras-1 and Ha-ras-2 genes. Similar studies are underway to introduce cloned DNAs into totipotent epithelial stem cells in order to generate mosaic rats with respect to tissues exclusively of epithelial origin.

INTRODUCTION

The generation and study of transgenic animals has proved to be a powerful new tool in the assessment of gene function in developmental, physiologic, and pathologic conditions (1,2). For the first time, the regulation and expression of finite genes have been examined in the most natural model system of all, the living animal. However, the most widely used methods by which such animals have been successfully generated are undesirable to many workers for at least several important considerations. First, the use of genetically engineered retroviruses as vectors for the infection of either post-fertilization (3) or mid-gestational (4) embryos necessitates tedious and complex molecular genetic methods requiring highly skilled and experienced personnel. A second drawback is the requirement for expensive laboratory equipment and devices such as micro-manipulators, micro-injectors,.. etc. Finally, and most importantly, the use of retroviral vectors or micro-injection intranuclearly preclude the targeted, selective insertion of cloned, expressible genes into finite cells and tissues inasmuch as the integrated DNA sequences are typically distributed in a random and widespread fashion throughout the transgenic animal.

In certain types of investigation, the capability to target a finite expressible gene into particular organs or tissues is highly desirable. For instance, the logical experimental therapy of a genetic disorder involving specific tissues or organ systems would require the targeted delivery of a correcting factor exclusively to the affected cells. Another instance in which directed genetic manipulation would be advantageous involves the characterization of the molecular basis for cell- and tissue-specific activation and expression of certain genes. Members of the keratin gene family, for instance, are expressed solely by cells and tissues of epithelial derivation (5). Inasmuch as the precise intracellular functions of this complex gene family remain unclear (5), the targeted introduction of particular keratin genes into cells descended from the mesenchymal compartment of mid-gestational, experimental mammals, would be of significant interest and scientific value.

A recent report by Jaenisch (6) described the introduction of cultured neural crest cells derived from a non-albino (C57BL/6J) mouse embryo into mid-gestation albino (BALB/c) embryos as a result of intra-embryonic injection. Some of the newborn chimeric animals demonstrated irregular patterns of ocular and cutaneous pigmentation due to the donor-derived melanocytes. We expanded and modified this approach for our experimental requirements by substituting primary rat embryo cells for neural crest cells in rats. Activated genes in selectable expression vectors were first transfected into primary rat embryo cells prior to injection into mid-gestation rat embryos. In this manner, cells, tissues, and organs descended from mesenchymal lineages were shown to be non-randomly enriched in the exogenous genes. The application of this general method to study the potential physiologic, developmental, or pathologic roles of activated oncogenes in programmed embryogenesis was explored as well.

MATERIALS AND METHODS

Embryonic cell culture

Timed-pregnant Sprague-Dawley rats were obtained from the Holzmann Co. (Madison, Wis.). On the 15-16th day of gestation, whole embryos were aseptically removed from the uterine horns, minced with sterile knives, dispersed with sterile trypsin-EDTA for 1 hour at 37°C with gentle agitation, and finally redistributed into several 60mm plastic Petri dishes. All cultures were maintained in DMEM medium containing 10% FCS and appropriate antibiotics. After 18 hours of incubation at 37°C, 5%CO_2, in a humidified environment, the non-adherent cells were discarded, and the attached cells were prepared for immediate DNA transfection.

DNA Transfection

Plasmid and cloned oncogene DNAs were precipitated exactly as before (7) essentially according to the methods of Graham and van der Eb (8) and Cooper et al (9). The precipitated DNA vectors were initially applied at a ratio of

2-5 ug DNA/2-5 x 10^5 cells/60mm dish for 16-18 hours. Inasmuch as the precipitated DNAs contained the pSV2neo antibiotic resistance gene, successfully transfected cells were selected following exposure of all cultures to 400 ug/ml of Geneticin (G418, Sigma Corp). After 2-5 days, surviving cells were removed from the dishes with trypsin-EDTA, and either directly injected immediately into rat embryos, or frozen and stored in liquid nitrogen for subsequent use.

Intra-embryonic injection

Timed-pregnant Sprague-Dawley rats were anesthetized by an IP injection of sodium pentathol 9.5 days after the documentation of conception. Both uterine horns were sterilly exposed through a small midline incision, and each embryo was identified as a small swelling. Cell suspensions were previously prepared in sterile saline (500-2500 cells/ul) and kept on ice. Graduated (1-5 ul) glass micropipettes were previously heated in an open flame, drawn apart briskly, and carefully snapped when cool resulting in an extremely narrow but sharp tip. Cell suspensions were drawn into the pre-sterilized micropipettes via a tight-fitting metal plunger, and were carefully injected into the ventral third of the decidual swelling (approx. 0.1 to 0.2 ul/embryo). The uterine horns were reolacedinto the abdominal cavity, and the incision was closed with sterile metal clips.

Determination of genetic mosaics

To identify rat offspring containing foreign genes and cells, small tips were biopsied from the tails of each newborn animal. DNA was extracted by conventional methods, and a small volume of blood was sampled as well. The blood was diluted in physiologic saline, and typed with respect to blood group antigens.

DNA extraction, blotting, and hybridization

High molecular weight DNA was extracted from tail biopsies as well as from various organ samples as before (7). Samples were blotted onto nitrocellulose membranes (BA-85, Schleicher and Schuell) directly through a dot-blot device, or digested with restriction enzymes, electrophoresed through 1% agarose gels, and transferred according to Southern (10). Nick-translated DNA probes were hybridized at high stringencies as before (7) for 72 hours prior to washing of the blots and autoradiographic analysis.

RESULTS

Primary embryonic target cells were readily plated immediately following tryptic dispersion. Although the majority of adherent cells assumed a spindle-shaped, fibroblastic phenotype, it was apparent that epithelial-like, round-to-polygonal cells were adherent as well. After 24 hours of incubation, the primary cells were transfected with the pSV2neo selectable marker gene alone in order to easily isolate transfected cells as well as to determine their eventual distribution in the various organs and tissues of the mosaic offspring. Following transfection and treatment with G418, the surviving cells represented a small proportion of the original cultures. Approximately 10-20 colonies per ug of DNA of G418-resistant cells were routinely generated by this method. Most of the cells appeared fibroblastic in morphology.

In order to assure that transfected embryonic cells would be optimally recruited and maximally distributed on the basis of germ line lineage in the developing embryos, it was necessary to identify an appropriate developmental stage at which to inject intra-embryonically. According to published data (11), it was reported that at developmental stage 15, the head folds, allantois, and soon the first pair of somites appeared 9.5 days post-conception. Since none of the major organ systems had yet to develop, we decided to use this particular stage of development at which to introduce foreign embryonic cells to hopefully "seed" as many differentiative pathways as possible.

After the first few litters of injected embryos were born, it became apparent that identification of genetic mosaics from normal littermates in a rapid and accurate manner would be necessary. The reliance upon the use of tail biopsies necessitated the the relatively time-consuming, tedious, and expensive process of DNA purification, blotting, hybridization, and autoradiography. On a large scale involving many litters of animals, this process seemed undesirable. We therefore decided to alter our strategy based upon 2 assumptions. First, because most of the descendents of the initial G418-resistant cells were in mesenchymal tissues, it seemed likely that bone marrow stem cells would contain the neo gene as well as explanted rat embryo cell derivatives. Second, we learned that whereas nearly all strains of albino rats were Type A blood group, the majority of wild and non-albino rats were exclusively Type AB (11). Therefore, our protocol was altered in the following manner: primary embryonic cells from a non-albino rat strain (Zucker) would be employed as targets for DNA transfection prior to injection into embryos of an albino (Sprague-Dawley) strain as before. Screening of the blood from newborns would expedite and markedly simplify the identification process of chimerics.

In order to characterize all the component factors of our system, primary cultures of 16-day old Zucker embryos were first transfected with the pSV2neo marker, and G418-resistant cells were immediately injected into 9.5-day post-implantation albino embryos. To ascertain the baseline level of trauma induced by the surgical procedures alone, 6 litters consisting of 78 mid-gestation embryos were sham injected with 0.1 ul of sterile saline. It was apparent that the manipulative procedures alone were responsible for a nearly 40% mortality rate (Table 1). Overall, however,

TABLE 1
INJECTION OF 9.5-DAY OLD POST-IMPLANTATION RAT EMBRYOS

Injected with	#Injected	#Surviving	#Mosaics
Sterile saline	78	45 (58%)	----
Primary embryo cells (albino) +neo	45	28 (62%)	21 (75%)
Primary embryo cells (Zucker)+ neo	68	44 (65%)	32(73%)

it appeared that nearly 2/3 of the injected embryos survived the procedure and that nearly 3/4 of the newborn offspring were genetic mosaics with respect to the neo gene.

Periodically, DNA was isolated from a variety of tissues one month postpartum. After dot-blotting and hybridization against the neo probe, intensities of the autoradiographic signal were subjectively scored from 0 to 4+ (Table 2).

TABLE 2
TISSUE DISTRIBUTION OF PSV2NEO IN MOSAIC RATS

Organ	Relative Intensity
Spleen, bone marrow	++++
Heart, skeletal muscle, kidney	+++
Skin, lung, liver	++

It was apparent that tissues primarily derived from mesenchyme were differentially enriched in pSV2neo-containing cells when equimolar amounts of DNA were blotted and hybridized. The positive signal from visceral organs of mixed derivation was unclear, although it was assumed that stromal cells accounted for the less intense hybridization, or that some epithelial cells in the primary cultures were transfected as well.

As our first experimental application of this model system, we introduced an activated Ha-ras-1 oncogene (EJ) in the same neo expression and selection vector. The new vector (pHO6T1) was introduced into Zucker embryonic cells and injected into 42 Sprague-Dawley albino embryos (Table 3).

TABLE 3
INTRODUCTION OF ACTIVATED HA-RAS-1 ONCOGENE

Injected with	# Animals	#Surviving	#Mosaics
Primary rat embryo cells (Zucker) + pHO6T1	42	28 (67%)	20 (71%)
Malignant rat tumor cells expressing activated Ha-Ras	28	6 (21%)	---

Approximately 67% of the injected embryos were born, of which 20 proved to contain the foreign oncogene. After 4 months of observation, these particular animals have remained disease-free and appear phenotypically normal. On the other hand, an established line of malignant rat tumor cells expressing high levels of the Ha-ras-1 oncogene failed to support embryogenesis when injected into mid-gestation rat embryos in that a nearly 80% mortality rate resulted. Autopsy of the mothers receiving the tumor cell injections intra-embronically failed to reveal any active tumorigenesis or uterine lesions post-partum.

Inadvertently, one litter of 6 month old rats mosaic for the Ha-ras-1 oncogene were not separated according to sex after reaching sexual maturity. Two of the 5 female litter-mates became pregnant and nursed normal litters. Shortly after weaning, both mothers developed rapidly-growing adeno-carcinomas of the breast. The lesions were solitary and had failed to metastasize at the time of autopsy. Studies are in progress to intentionally breed such genetic mosaic females on a larger scale in order to reproduce this interesting and potentially exciting phenomenon.

DISCUSSION

With increasing interest and investigation, improved methods of gene transfer will most likely be focused away from the research laboratory and onto clinical applications. In order to accomplish this overall objective, however, it will be necessary to refine and target gene transfer to appropriate tissues and organs employing simple, reproducible, and reliable methods. In that regard, our efforts have been directed to the establishment of non-random targeting of cloned DNAs into finite tissues and organs. Our rationale for such a model system was to investigate the action of dominant-acting, activated oncogenes on embryogenesis, and subsequently on postnatal growth and development as well.

The preliminary choice of total primary embryo cells as targets assured that the great majority of successfully transfected cells were of mesenchymal origin. Hence, the concentration of cellular descendents in mature tissues of mesenchymal derivation was anticipated. On the other hand, in order to direct cloned and expressible genes into different specialized tissues, especially the major visceral organs, embryonic target cells of epithelial derivation will be necessary. Studies are currently underway to pre-select pure populations of primary epithelial stem cells from the majority of spindle-shaped cells. Ultimately, we hope to target mature tissues from each of the major germ line lineages with activated, expressible genes.

The choice of the pSV2neo vector as a model for tissue-specific targeting of transfecting genes was based upon 2 criteria. First, it readily permitted the selection of the successfully transfected from the untransfected cells. Second, it served as a molecular marker to determine the final localization of the cellular descendents of the original embryo cells. Although the G418-resistant, embryonic cells expressed the neo immediately post-transfection and postnatally in direct proportion to the concentration of transfected cells according to Northern blotting (data not shown), it is unclear whether and how long expression persists in the absence of G418 stimulation in the various tissues of the mosaic rats. Work is in progress to address that question. Certainly, if remedial gene therapy is to be considered as an important application of tissue-directed gene transfer, questions of regulation of gene expression must be resolved.

With respect to the initial experimental use of the EJ ras oncogene, the results are unusual in that they are consistent and inconsistent with some previously published work. For example, Sinn et al (12) recently reported the introduction of an activated Ha-ras oncogene in transgenic mice was generally compatible with normal embryogenesis. Moreover, breast tumors were observed in some of the transgenic female offspring. However, the incidence of spontaneous carcinogenesis was significantly elevated synergistically when such animals were mated with transgenic mice carrying a c-myc oncogene. The offspring hybrid mice experienced a rapid and elevated level of spontaneous tumor formation. What was curious in our preliminary observations with the pH06T1 ras gene mosaics was the fact that breast carcinomas were generated immediately post-pregnancy and lactation. The tumor cells were derived from epithelial cells; the exogenous Ha-ras was presumably concentrated in cells of mesenchymal derivation. Indeed, we anticipated the appearance of hematologic and lymphoid neoplasms in the mosaic rats. One likely explanation would be that some of the low proportion of epithelial stem cells in the primary embryo cultures became transfected, and subsequently amplified in the epithelial-derived organs. This notion is likely due to the presence of an additional Ha-ras band in addition to the 2 germ-line Ha-ras fragments in tumor DNA (data not shown). However, on the basis of only 2 animals, more definitive conclusions must await greater samplings from mosaic females currently being bred in our laboratories. Most certainly, the presence of an expressed Ha-ras oncogene in a frank rat carcinoma cell terminated embryogenesis, suggesting that the cell carrying the activated oncogene is a highly important factor in the successful generation of mosaic offspring.

In conclusion, increasingly refined methods for gene transfer, including the generation of transgenic animals, will be available in the near future. One important goal of such investigation is the targeted introduction of expressible genes into defined cells and tissues. The experimental approach we have undertaken is based upon the pre-selection of subpopulations of totipotent epithelial or mesenchymal stem cells as recipients by transfection of exogenous DNAs, with subsequent re-introduction back into appropriate embryos. Such capabilities will eventually allow the characterization of cell-cell and tissue-tissue interactions during mammalian development which are crucial for normal growth, development and morphogenesis, as well as for abnormal interactions associated with disease processes such as cancer. Perhaps in combination with molecular modifications to expressible genes such as the inclusion of tissue-specific promoters or enhancers, it will become a routine procedure to design and engineer the successful gene transfer in the laboratory as well as in the clinic.

ACKNOWLEDGEMENTS

The authors wish to thank Ms. Christine Magura and Joel Hass for their important contributions to this study.

REFERENCES

1. Palmiter, RD, Brinster, RL, Hammer, RE, Trumbauer, ME, Rosenfeld, MG, Birnberg, NC, and Evans, RM (1982) Nature 300, 611.
2. Palmiter, RD, and Brinster, RL (1985) Cell 41, 343.
3. Soriano, P, Cone, RD, Mulligan, RC, and Jaenisch, R (1986) Science 234, 1409.
4. Jaenisch, R (1980) Cell 19, 181.
5. Steinert, PM, and Parry, DAD (1985) Ann Rev Cell Biol 1, 41.
6. Jaenisch, R (1985) Nature 318, 181.
7. Schwartz, SA, Shuler, CF, and Freebeck, P (1988) Cancer Res, In press
8. Graham, FL, and van der Eb, AJ (1973) Virol 52, 456.
9. Cooper, CS, Blair, DG, Oskarsson, MK, Tainsky, MA, Eader, LA, and Vande Woude, GF (1984) Cancer Res 44, 69.
10. Southern, EM (1975) J Mol Biol 98, 503.
11. Farris, EJ, and Griffith, J "The Rat In Lab. Investigation", Hafner Publishing Co, New York, NY, 1963.
12. Sinn, E, Muller, W, Pattengale, P, Tepler, I, Wallace, R, and Leder, P (1987) Cell 49, 465.

Gene Transfer and Gene Therapy, pages 215–223

COMPLEMENTATION OF EXCISION-REPAIR DEFICIENCY IN A HUMAN CELL LINE: ADVANTAGE IN THE USE OF A cDNA CLONE LIBRARY FOR GENE-TRANSFER

T. Teitz, T. Naiman, D. Eli, M. Bakhanashvili and D. Canaani

Department of Biochemistry, Tel Aviv University, Tel Aviv 69978, Israel

ABSTRACT Two methods of gene transfer were compared in ability to complement the UV sensitivity of a xeroderma pigmentosum group C (XP-C) established cell line. In the first, a human cDNA clone library constructed in a mammalian expression vector, and itself incorporated in a neo containing lambda phage vector, was introduced into the cells as a calcium phosphate precipitate. Following selection to G418 resistance, transformants were selected for UV resistance. Twenty-one cell clones were obtained with UV-resistance levels typical of normal human fibroblasts. Upon further propagation in the absence of selection for G418 resistance, about half of the primary transformants remained UV-resistant. Secondary UV-resistant transformants were generated by transfection with a partial digest of total chromosomal DNA from one of these stable primary transformants. The other primary transformants lost UV resistance rapidly upon subculturing, but several retained UV resistance under G418 selection pressure.

In contrast, transfection of the XP-C cell line with genomic DNA covalently linked to neo, resulted in only one transformant having intermediate UV-resistance levels. Upon subculturing, this phenotype was lost rapidly and could not be stabilized by reinstitution of selection for G418 resistance. Hence, stable correction of the excision-repair deficiency in an XP-C cell line, was achieved more efficiently by transfer of an expressable cDNA clone library than transfection with genomic DNA.

INTRODUCTION

Various mammalian genes, including several oncogenes, were isolated following transfection of foreign DNA into mutant cultured cells selected for genetic complementation or altered phenotype. Originally, genomic DNA was used for transfection (1), but the advent of cDNA clone libraries constructed in mammalian expression vectors offers an alternative approach (2,3). We have compared these two approaches during attempts to stably complement the excision-repair deficiency of a previously established xeroderma pigmentosum group C cell line (4).

Xeroderma pigmentosum (XP) is an autosomal recessive human disease manifested as an extreme sensitivity to UV light resulting in a very high incidence of skin cancer and, in many patients, neurological abnormalities. Cells from XP patients are defective in the excision repair of pyrimidine dimers and are extremely sensitive to killing by UV radiation. So far, nine complementation groups (A-I) defective in excision repair have been defined (5). The molecular basis for the different XP mutations has not yet been characterized - neither genes nor gene products have been identified. Stable corrections of the UV sensitivity of simian virus 40 (SV40) - transformed XP-A cell lines by transfection with normal human DNA (6), or following fusion with X-ray-irradiated Chinese hamster ovary (CHO) cells (7) or microcells containing a single human chromosome (8) were reported. A human gene (ERCC-1) that complements group-2 excision-deficient CHO mutants had been molecularly cloned (9,10), but the relationship of this gene to human DNA-repair genes defective in XP is not clear.

RESULTS

Complementation of the UV Sensitivity of an XP-C Cell Line

In an attempt to complement the UV sensitivity of the XP-C established cell line GM2096-SV3(4), we have used two methods of gene transfer for transfection. In the first, a human cDNA library constructed from SV40-transformed human fibroblasts (GM637) in the pcD mammalian expression vector (2), was introduced into the XP-C recipient cell line. The pcD-GM637 human cDNA library is in turn incorporated in the bacteriophage lambda vector NMT (3). λNMT contains, 0.25 kb from the pcD-GM637 library insertion site, a transcription unit of neo, a dominant selectable marker which confers in mammalian cells resistance to the toxic drug G418. Transfection was accomplished with lambda phage particles under conditions similar to those specified before (3). As a second method, total genomic DNA from Hela cells (with normal UV resistance) was partially cut by the restriction endonuclease MboI to yield fragments of an average size of 50 kb. These were ligated at 1:2 molar ratio to BamHI cleaved mpSV2-neo DNA. The ligation product was introduced into the GM2096-SV3 cell line as a calcium phosphate precipitate. About $2x10^7$ cells were transfected by each method, followed by selection for G418 resistance. The efficiency for transformation to G418 resistance was $2.5x10^{-4}$ for the cDNA library and $5.6x10^{-4}$ for the genomic DNA transfections. The surviving colonies were trypsinized, replated and subjected to several consecutive (1-2 days apart) UV irradiations. The UV dose employed ($4J/m^2$) is lethal to the GM2096-SV3 recipient cells, but causes little or no damage to normal human fibroblasts. The surviving colonies were isolated and grown to mass culture. The UV sensitivity of the isolated transformants was assayed by UV-survival measurements as indicated in (11).

Twenty-one UV^R clones were recovered using the cDNA clone library, a partial list of which is shown in Table 1, whereas only one UV^R transformant (D/3) resulted from the genomic DNA transfection. Inspection of the initial levels of UV resistance shows [Table 1 and (11)] that most clones from the former group gained UV-resistance levels similar to those of normal human fibroblasts transformed by SV40(GM637). The resistance levels obtained at UV dose of $2J/m^2$ exceeded that of the parental cells (GM2096-SV3) by three orders of magnitude, while being close to that of GM637. However, the initial level of UV-resistance of cell clone D/3, generated by genomic DNA transfection, was only intermediate between those of normal SV40-transformed fibroblasts (GM637) and the UV sensitive XP-C cell line GM2096-SV3.

Stability of UV-Resistant Phenotype

Following the initial identification of UV^R transformants, these cell clones were tested periodically for retention of the UV-resistant phenotype, while subcultured in the absence of selection for G418 resistance. As shown in Table 1 and (11) most of the

TABLE 1

UV RESISTANCE OF XP-C PRIMARY TRANSFORMANTS

Clone	Passage no. since clone's establishment	Fraction of cells surviving UV dose 2J/m^2	3J/m^2
GM637 (Normal)	-	1.0	0.44
GM2096-SV3(XP-C)	-	0.0007	<0.0003
GM2096-SV3/cD1[+]			
17-20/1	5	0.94	0.88
	55	0.60	0.40
17-20/3	3	0.93	0.93
	50	0.83	0.55
17-20/6	2	0.80	ND*
	32	0.93	0.80
17-20/8	6	0.30	0.24
	16	0.14	0.07
	25	<0.0003	<0.0003
5/1	4	0.45	0.34
	6	0.20	0.15
	21	0.04	0.01
5/1-G418^R	12	0.50	0.38
	28	0.26	0.20
	50	0.20	0.15
5/5	3	0.57	0.57
	10	0.05	0.04
5/5-G418^R	8	0.37	ND*
	13	0.25	0.27
	69	0.30	0.20
5/10	3	0.75	0.62
	10	0.33	0.20
	18	0.08	0.07
5/10-G418^R	6	0.37	0.33
	14	0.52	0.34
	69	0.50	0.35
GM2096-SV3/HgD[#]			
D/3	2	0.25	0.01
	4	0.12	0.02
	12	0.06	0.01

*ND, not done; +cDl, cDNA library; #HgD, Hela genomic DNA

transformants initially derived from dishes 17-20 (except for clones 17-20/8 and 17-20/9) retained the UV-resistant phenotype during subsequent passaging. However, all five UV^R transformants isolated from dish 5, and the single transformant resulting from genomic DNA-transfer (D/3) showed decreased UV resistance upon subsequent subculturing. Particularly interesting are the cases of transformants 5/1, 5/5 and 5/10, in which the instability of UV resistance was accompanied by loss of G418 resistance. We therefore reinstituted G418 selection in these cell lines at the earliest passage available. As shown in Table 1, selection for neo expression, resulting in the three cell clones 5/1-$G418^R$ (G418 resistant), 5/5-$G418^R$ and 5/10-$G418^R$, led also to stabilization of the UV-resistant phenotype. A similar attempt in clone D/3 to select a subpopulation of cells resistant to both G418 and UV-irradiation did not succeed, as upon subculturing there was a dissociation of the two phenotypes, rendering D/3 cells G418-sensitive. Such a rapid loss of G418 and UV resistances can not be attributed to reversion, but rather to the frequent instability of transduced foreign DNA, as indicated by Axel and co-workers (1). Moreover, the stabilization of UV resistance by G418 selection, reinforced a linkage between the resistances to UV and G418 in these three cell transformants, implying that UV resistance was acquired by gene transfer and not by reversion.

UV Resistant Secondary Transformants

It could be argued that those transformants which had a phenotype of stable UV resistance, arose by spontaneous or UV-induced reversion. Although, control experiments yielded no UV^R cell clones, we set out to test this argument by trying to obtain secondary UV^R transformants without the use of UV irradiation for selection. Total genomic DNA of transformant 17-20/6, which had a low number of neo copies in its nuclei (11), was partially digested with MboI restriction endonuclease to fragments averaging 50 kb, and tranferred into GM2096-SV3 cells. Fifteen colonies selected for G418 resistance (due to expression of endogenous neo genes) were isolated and grown for further characterization. Of these, thirteen were shown by UV-survival measurements to be as UV-sensitive as the recipient cell line GM2096-SV3. However, two independent transformants (derived of different trans-

TABLE 2
UV RESISTANCE OF XP-C SECONDARY TRANSFORMANTS

	Fraction of cells surviving UV dose	
Clone	$1.5J/m^2$	$2.0J/m^2$
GM637	1.0	1.0
GM2096-SV3 (XP-C)	0.004	0.0008
GM2096-SV3/Secondaries		
17 - 20/6 - 2/24	0.15	0.04
17 - 20/6 - 3/9	0.12	0.03

fected plates) acquired partial but significant UV-resistance levels, 30-50 fold higher than the parental cell line GM2096-SV3 (Table 2). Since no UV selection was employed in the generation of the secondary transformants, and the proportion of UV^R clones among $G418^R$ ones (two out of fifteen) was relatively high, this result indicates that these UV^R clones arose due to DNA-mediated gene transfer and not by spontaneous or UV-induced reversions. Even though transfection was carried out with fragments of approximately 50 kb, further experiments are needed to determine whether _neo_ is physically linked to the DNA element that provides UV resistance.

DISCUSSION

Transduction of an immortalized XP-C cell line with either a human cDNA library incorporated in a mammalian expression vector, or genomic DNA derived from Hela cells led to complementation of the UV sensitivity of the recipient cells. Two lines of evidence strongly suggest that at least the cDNA library transformants acquired UV resistance by DNA-mediated gene transfer and not by reversion: (i) the acquisition of UV^R secondary transformants, derived by transfection of DNA from a stable UV^R primary transformant and selected on the basis of resistance to G418 only; (ii) the linkage of G418 and

UV resistances in the group of unstable UV^R clones.

It is possible that not all 21 UV^R cell clones were independent isolates, because we trypsinized and replated the transformed colonies to allow effective UV irradiation. While increasing the chances of complementation, this protocol may amplify the initial number of independent transformants. Furthermore, the distinct hybridization patterns observed with the different clones (11), do not necessarily signify independent transformation events, since the transfected DNA may rearrange. Indeed, we have noticed (data not shown) DNA rearrangements of the transfected λNMT-pcD-GM637, at least during propagation of transformants in the absence of G418 selection. This property of the system may be partly responsible for the relatively low number of transfected DNA molecules per cell. Therefore, we can only set a minimal value of two independent transformants based on the number of initially transfected cell dishes from which UV^R transformants originated, indicating a minimal efficiency of transformation to UV resistance of 10^{-7}. No matter what the actual efficiency of XP-C cells-transformation is to UV resistance with the cDNA clone library, it is clear that this method is superior, in this particular system, to gene transfer by genomic DNA. One possible explanation is that the repair gene defective in XP-C cells is very large, and therefore unlikely to be transduced efficiently by a preparation of genomic DNA fragments averaging 50 kb in length. Still, there are other possible explanations. Unlike transfection with genomic DNA fragments, in the cDNA-library case the elements that drive transcription and posttranscriptional processing are of nonhuman origin, being derived from SV40. Yet, apparently, they are recognized sufficiently well so that the expression of one or a few integrated cDNA copies per cell (11) provides a wild-type level of UV resistance.

The complementing cDNA library was constructed with mRNAs from the SV40-transformed human fibroblast line GM637, itself not UV-irradiated or chemically treated. This indicates that, at least in SV40-transformed human fibroblasts, there is a constitutive level of mRNA that can complement the particular DNA-repair defect. In transformant 5/10-$G418^R$, the resistances to G418 and UV are linked. This transformant has also a single integration site for neo sequences, confined to the high molecular weight DNA of the cell (11).

Recently, we constructed in lambda EMBL3 vector a genomic library from transformant 5/10-G418^R. Using a probe derived from one of the neo containing phages in this library, to screen the expressable cDNA library, we recovered several clones that confer increased UV resistance upon the XP-C recipient cell line.

ACKNOWLEDGEMENTS

We thank Hiroto Okayama for providing the λNMT-pcD-GM637 recombinant phage library, Sofia S. Avissar and Sara Bar for excellent assistance in various stages of this project.

REFERENCES

1. Pellicer A, Robins D, Wold B, Sweet R, Jackson J, Lowy I, Roberts JM, Sim GK, Silverstein S, Axel R (1980). Altering genotype and phenotype by DNA-mediated gene transfer. Science 209:1414.
2. Okayama H, Berg P (1983). A cDNA cloning vector that permits expression of cDNA inserts in mammalian cells. Mol Cell Biol 3:280.
3. Okayama H, Berg P (1985). Bacteriophage lambda vector for transducing a cDNA clone library into mammalian cells. Mol Cell Biol 5:1136.
4. Canaani D, Naiman T, Teitz T, Berg P (1986). Immortalization of xeroderma pigmentosum cells by simian virus 40 DNA having a defective origin of DNA replication. Somat Cell Mol Genet 12:13.
5. Friedberg EC (1985). "DNA Repair." New York: Freeman, p. 506.
6. Takano T, Noda M, Tamura T (1982). Transfection of cells from a xeroderma pigmentosom patient with normal human DNA confers UV resistance. Nature 296:269.
7. Karentz D, Cleaver JE (1986). Repair-deficient xeroderma pigmentosum cells made UV light resistant by fusion with X-ray-inactivated Chinese hamster cells. Mol Cell Biol 6:3428.

8. Schultz, RA, Saxon PJ, Glover TW, Friedberg EC (1987). Microcell-mediated transfer of a single human chromosome complements xeroderma pigmentosum group A fibroblasts. Proc Natl Acad Sci USA 84:4176.
9. Westerveld A, Hoeijmakers JHJ, van Duin M, de Wit J, Odijk H, Pastink A, Bootsma D (1984). Molecular cloning of a human DNA repair gene. Nature 310:425.
10. Rubin JS, Prideaux VR, Huntington FW, Dulhanty AM, Whitmore GF, Bernstein A (1985). Molecular cloning and chromosomal localization of DNA sequences associated with a human DNA repair gene. Mol Cell Biol 5:398.
11. Teitz T, Naiman T, Avissar SA, Bar S, Okayama H, Canaani D (1987). Complementation of the UV-sensitive phenotype of a xeroderma pigmentosum human cell line by transfection with a cDNA clone library. Proc Natl Acad Sci USA 84:8801.

Gene Transfer and Gene Therapy, pages 225–234

DOMINANT NEGATIVE MUTATION IN GALACTOSYLTRANSFERASE CREATED BY OVER-EXPRESSION OF A TRUNCATED cDNA[1]

Vincent J. Kidd, Helen Fillmore[2], Paula Gregory and Bruce Bunnell

Departments of Cell Biology and Anatomy and Microbiology, University of Alabama at Birmingham, Birmingham, Alabama 35294

ABSTRACT. Using a truncated cDNA clone for the glycosyltransferase, β1-4 galactosyltransferase (GalTase), and a vector containing a DHFR minigene we have produced a CHO cell line that produces less than 20% of the normal endogenous GalTase gene products. Analysis of GalTase enzyme activity from cells containing increasing amounts of the truncated GalTase mRNA shows proportionately decreasing amounts of endogenous Golgi (40 kD) and cell-surface (80 kD) enzyme activities. Likewise, Western blot analysis reveals a similar proportional decrease in both GalTase proteins. This mutation is, however, lethal at high levels of exogenous truncated GalTase mRNA, presumably due to the involvement of cell-surface GalTase in cellular growth. The GalTase mutant cells also demonstrate alterations in glycoconjugate biosynthesis that appear to alter the micro-structure of the Golgi.

INTRODUCTION

Eukaryotic genes are now being isolated at a rapid rate, but their biological function is not always obvious. This is particularly true for cellular oncogene homologues where their function can only be inferred from the predicted protein sequnce and/or homology with another gene product. In addition, genes

---

[1]This work was supported by the NIH (DK 36973) and March of Dimes/Birth Defects Fd. (#5-577).
[2] Present address: Dept. of Anatomy, Univ. of Tennessee, Memphis, Tennessee.

corresponding to well characterized proteins have been shown to have functions and properties not associated with the previously isolated protein (1). One way to address the problem of biological function for an isolated gene product is to inactivate the protein or corresponding gene locus in a cell and examine the change in cellular phenotype. Elegant methodologies have been devised to accomplish this task, but these techniques are not applicable to all systems (2-5). Herskowitz has suggested an additional approach that involves the utilization of truncated or highly mutated cDNAs which are over-expressed in a cell, inactivating the normal endogenous gene product and thereby creating a dominant negative mutation (6). We have used this approach to create a dominant negative mutation in the β1-4 galactosyltransferase gene loci in eukaryotic cells. This report describes our approach and the changes in cellular phenotype that accompany the mutation.

RESULTS

Isolation of β1-4 galactosyltransferase cDNA clones and construction of expression plasmids. We have previously reported the cloning and characterization of human GalTase cDNA clones (7). Northern blot analysis of various human tissues and cell lines has revealed two homologous, but distinct, GalTase mRNAs that are 1.7 kb (GTI) and 4.5 kb (GTII) in size. GTI has recently been shown to encode a 56 kD protein with abundant β1-4 galactosyltransferase activity when expressed in COS cells (Kidd et al., submitted). The larger GTII transcript is highly homologous at the nucleotide sequence level but contains additional sequence and an open reading frame that would encode a much larger GalTase protein. This data corresponds to reports detailing two human GalTase isoenzymes, 56 kD and 78 kD in size, that are highly homologous yet quite distinct in size, amino acid and carbohydrate content (8,9).

Figure 1 shows a schematic representation of the isolated GalTase cDNA clones. Several cDNA clones of various sizes have been isolated containing sequences corresponding to the larger GTII transcript (Fig. 1, panel A, 1, 2 and 3). Another cDNA isolate corresponding to an expressable full-length version of the GTI transcript has also been isolated (Fig. 1, panel A, 4). The GalTase cDNA clone shown in Figure 1 (panel A, 1) was chosen for the purposes of potentially creating a dominant negative mutation. This truncated GalTase cDNA was cloned into the eukaryotic expression vector p91023 (B) (10). The vector was chosen for its strong regulated promoter and the presence of a dihydrofolate reductase (dhfr) minigene, allowing selection and amplification

A.

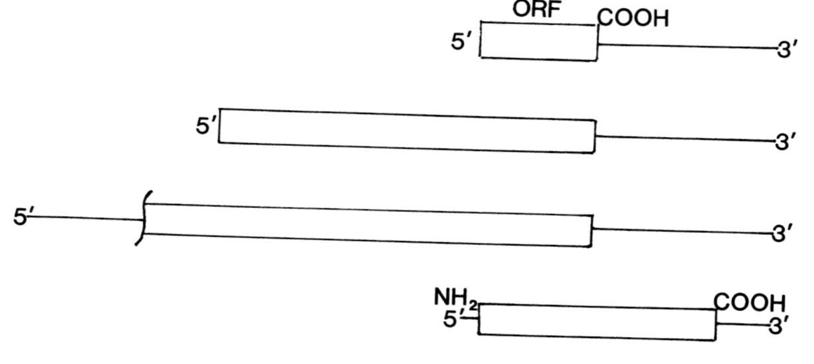

B.

FIGURE 1.(A) Schematic representation of isolated GalTase cDNA clones. ORF, open reading frame; NH_2, amino terminus; COOH, carboxyl terminus. The orientation is shown by 5' and 3' designations. (B) GalTase mutant construct.

of the construct in certain cell lines. A schematic representation of a portion of the final construct is shown in Figure 1 (panel B). The truncated GalTase expression plasmid was then introduced into a chinese hamster ovary (CHO) cell line lacking normal dhfr enzyme production (11) by calcium phosphate precipitation. Simultaneously, the p91023 (B) vector DNA alone was also transfected into the CHO (dhfr$^-$) cells. The cells containing the dhfr minigene construct were selected and the exogenous DNA sequences amplified by slowly increasing the concentration of methotrexate (MTX) in the cell media (12).

Demonstration of a dose-dependent decrease in endogenous β1-4 galactosyltransferase in cells containing the truncated cDNA. As the transfected cells were selected and amplified by MTX we concomitantly isolated DNA, RNA and cellular protein for analysis. Both the cellular DNA and RNA demonstrated dose-dependent increases in the GalTase construct copy number with increasing amounts of MTX (data not shown). The cellular protein, however, demonstrated a dramatic decrease in specific GalTase proteins and enzyme activity. We assayed the cellular protein by Western blot analysis using a monospecific polyclonal bovine GalTase antibody (13) and the enzyme activity by incorporation of [^{14}C]-galactose onto macromolecular acceptors (14). To quantitate the decrease of immunoreactive GalTase proteins we ran parallel denaturing polyacrylamide gels containing equivalent amounts of intracellular proteins. One was stained with coomassie blue (Fig. 2, panel A) while the other was Western blotted with the GalTase antibody (Fig. 2, panel B). Normally, CHO cells have two immunoreactive bands, similar to the human GalTase isoenzymes, that are approximately 40 kD and 80 kD in size. Western blot analysis demonstrated normal amounts of the two proteins in CHO cells containing amplified vector sequences (Fig. 2, panel B, V), but decreasing amounts of the same two protein species in CHO cells containing the truncated GalTase cDNA construct (Fig. 2, panel B, lanes 1, 10, 20). The decrease in the intracellular GalTase proteins appeared to be dose-dependent, directly associated with the increasing copy number of the truncated GalTase construct in these cells.

Analysis of the specific β1-4 galactosyltransferase enzyme activity in these same cells verified the results seen by Western blot analysis. As can be seen in Figure 2 (panel C), GalTase enzyme activity was not affected by the presence of vector alone while it decreased in a dose-dependent fashion in cells containing increasing copy-numbers of the truncated GalTase construct. Therefore, two separate and independent analysis demonstrated that the presence of increasing amounts of a truncated GalTase cDNA and its resulting mRNAs resulted in a dose-dependent loss of endogenous cellular GalTase activity.

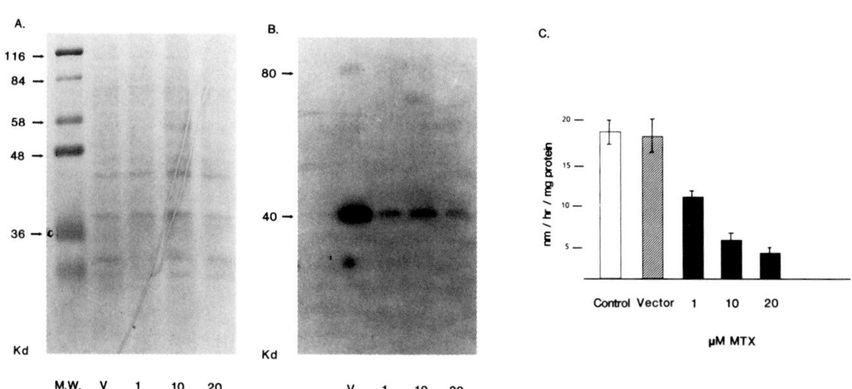

FIGURE 2. Quantitation of decreasing β1-4 galactosyltransferase proteins and enzyme activity in dominant mutant cell lines. Fifty micrograms of total intracellular protein for each sample was loaded on two 8% polyacrylamide gels for (A) coomassie blue staining and (B) Western blot analysis using a monospecific bovine galactosyltransferase antibody (13). The samples represent the following: V, vector transfected cells; 1, GalTase mutant (GT^-)cells in 1 µM MTX; 10, GT^- cells in 10 µM MTX; 20, GT^- cells in 20 µM MTX. The specific β1-4 galactosyltransferase enzyme activity was also assayed for these protein samples (C). Assays were performed as previously described (15). The results represent nm/hr/mg protein of β1-4 galactosyltransferase enzyme activity and represent the results of two separate experiments performed in triplicate. Standard deviations are indicated for each samples. The samples are identical to those analyzed by Western blotting with the addition of untreated parental CHO cells.

Alterations in the normal growth characteristics of the GalTase mutants. Changes in normal cell phenotype are carefully noted in mutant cells to ascertain possible functions of the mutated gene product. We and others have noted that specific increases in cell-surface associated GalTase enzyme are associated with β-agonist induced hypertrophy of rat parotid glands (14) and transformed cell growth (15). These studies have also shown that specific modifiers of the GalTase enzyme can block this abnormal growth dramatically (14,15). Therefore, we chose to closely monitor cell growth in these GalTase mutants to determine whether or not a decrease in normal GalTase proteins would also have an affect on cell growth. A dramatic difference was seen in the mitotic index of the mutants and vector or CHO control cells. The GalTase mutants, at 20 μM MTX, had a mitotic index that was 75% less than either control cell line (data not shown). We then looked at incorporation of ^{3}H-thymidine (Td) into DNA as a measure of cell growth using in situ analysis on intact cells as well as trichloroacetic acid (TCA) precipitation and quantitation of cellular DNA (14). In situ analysis of ^{3}H-Td incorporation showed that while approximately 50-60% of the control CHO cells were actively synthesizing DNA (Fig. 3, panel A), greater than 90% of the GalTase mutant cells were in the same stage of the cell cycle (Fig. 3, panel 3). Quantitation of ^{3}H-Td incorporation by TCA precipitation was consistent with the in situ analysis, demonstrating a 3 fold increase in DNA synthesis in the GalTase mutant cells versus control or vector treated cells (Fig. 3, panel C). Hoescht staining and chromosome analysis of the GT^{-} cells also support these results. Many of the GalTase mutant cells are polyploid and multi-nucleated (data not shown). Further analysis of these alterations will provide more definitive information concerning the changes in normal cell growth that accompany changes in the GalTase proteins.

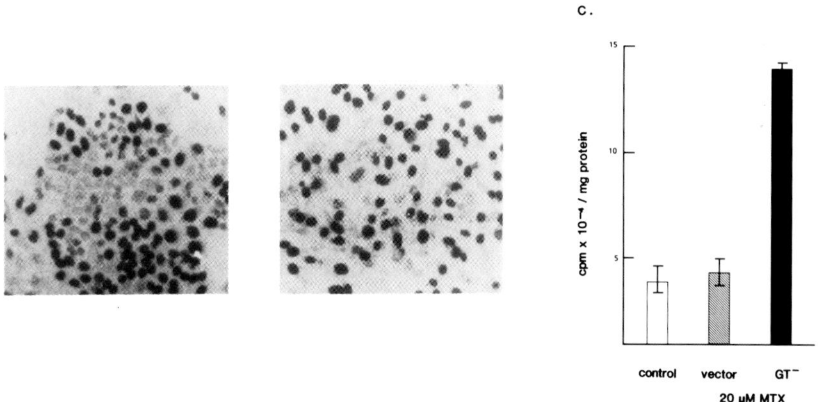

FIGURE 3. DNA replication in GalTase mutant and control cell lines. Autoradiography of [^{3}H[Td incorporated into (A) CHO (dhfr^{-}) parental cells and (B) GT^{-} cells in 20 μM MTX. Cells were analyzed during exponential cell growth. (C) Quantitation of [^{3}H]Td incroporated into the nuclei of the indicated cell lines by TCA precipitation (14).

Alterations in glycoconjugate biosynthesis inthe GalTase mutant cells. Initial analysis of the GalTase mutant cells revealed changes in the physical appearance of the cells in addition to the changes in cellular growth. Upon initial examination greater than95% of the mutant cells (GT^-, 20 µM MTX) had gross changes in cellular morphology, including the presence of numerous vacuoles scattered throughout the cells. One gross measure of changes in glycoconjugate biosynthesis is the sensitivity to specific plant lectins (16). Preliminary analysis of the GalTase mutant cells and the CHO parental cells has shown that the GalTase mutants have become more resistant to lectins from C. ensiformis (CON A) but not T. vulgaris (WGA) (data not shown). We are now in the process of quantitating these alterations as well as examing the sensitivity of resistance of these cells to additional lectins. These observations are not totally unexpected due to the normal function of the Golgi form of GalTase. However, the presence of the large vacuoles was somewhat surprising and suggested possible changes in the micro-structure of certain cellular organelles in the mutants.

A. B.

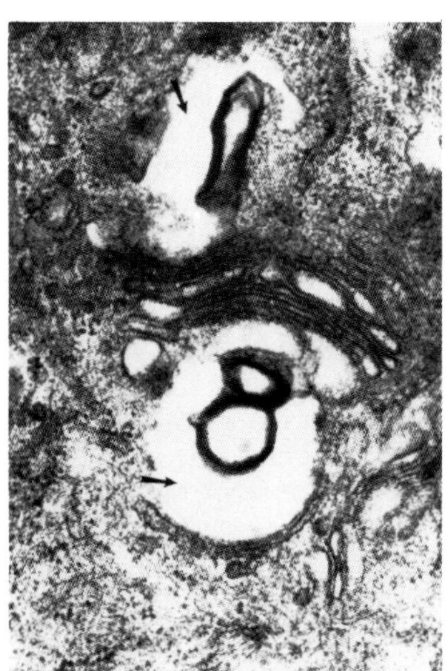

FIGURE 4. Electron micrographs of Golgi from (A) vector transfected CHO cells and (B) GalTase mutant cells. Cells were prepared for microscopy as described by Brinkley and Chang (17). The monolayers were viewed on a Hitachi H-7000 TEM at magnification X 40,000. The arrows indicate the Golgi regions in (A) and the large vessicles associated with the trans Golgi in (B).

To examine this latter possibility we decided to compare the GalTase mutants with the cells transfected with vector sequences only using electron microscopy. A striking difference was immediately observed between the control cells and the GalTase mutant cells; namely, the presence of large vessicles attached to the trans Golgi of the mutant cells and its complete absence in the control cells (Fig. 4). The control cells have no obvious microstructural changes in the vicinity of the Golgi or elsewhere (Fig. 4A). However, greater than 95% of 200 mutant cells examined had identical large vessicles attached to the trans Golgi membrane and containing electron dense material of unknown origin (Fig. 4B). These micro-structural alterations are obviously manifested by changes in GalTase enzyme expression, but their possible correlation with changes in glycoconjugate biosynthesis is not known.

DISCUSSION

In this communication we have demonstrated the successful use of amplification of a truncated cDNA construct in creating a dominant negative mutation for known gene products in a eukaryotic cell line. As originally envisioned by Herskowitz (6) this technique was based on over-expression of a truncated or highly mutated protein molecule generated by its corresponding cDNA. Even though we have the carboxyl terminal portion in our construct, it was not designed to encode a translatable mRNA sequence. Our data indicates that this methodology has much broader uses than even those described by Herskowitz. While we cannot conclusively rule out the possible production of a highly unstable protein product in our mutant cells, it seems more likely that regulation of endogenous GalTase protein production is occurring transcriptionally. The high degree of DNA sequence conservation in the 3' untranslated regions of the two human cDNAs, the DNA sequence conservation between species (7) and the tightly regulated expression of these gene products lends credence to the possibility of transcriptional regulation. We are now in the process of determining whether or not this is true using more definitive molecular analysis.

One can envision the use of the methodology for examining genes that are tightly regulated by specific trans-activating factors, such as the heat shock proteins or cellular oncogene homologues that are developmentally regulated. In these instances, flanking DNA sequences that normally bind these activating factors could be amplified in a particular cell line using a construct containing, for example, a dhfr minigene. This amplified flanking sequence would then serve as a sponge for this activating factor, trapping the protein and preventing normal activation of its targeted gene.

We have also demonstrated the ability of this methodology to provide new insights into functions of known gene products. Previous reports have documented the effects of increased cell-surface GalTase on growth (14,15) and now we have shown that dramatic changes in growth are also associated with decreased GalTase protein levels. Additionally, we have been successful in creating a cell line with a specific glycosyltransferase defect that is manifested by structural changes within the cell. These changes may be due to altered secretion of glycoconjugates, much the same as the results seen in certain cell lines that have been treated with tunicamycin (18). Further investigation will, hopefully, answer the questions raised by these mutants as well as demonstrate the efficacy of this approach for additional gene products.

ACKNOWLEDGEMENTS

We would like to thank Dr. R. Kaufman, Genetics Institute, for providing the plasmid p91023 (B) and Dr. B. Shur for providing the polyclonal bovine GalTase antibody. We would also like to thank Mr. Bill Mollon for help with the electron microscopy, Dr. J. Lahti for helpful discussions and Ms. Cynthia Webster for preparing this manscript.

REFERENCES

1. Morgan, D.O., Edman, J.C., Standring, D.N., Fried, V.A., Smith, M.C., Roth, R.A., Rutter, W.J. (1987). Nature 329: 301.
2. Smithies, O., Gregg, R.G., Boggs, S.S., Korakwski, M.A., Kucherlapati, R.S. (1985). Nature 317: 320.
3. Izant, J.G., Weintraub, H. (1984). Cell 36: 1007.
4. Meeks-Wagner, D., Hartwell, L.H. (1986). Cell 44: 43.
5. Blose, S.H., Meltzer, D.I., Feramisco, J.R. (1984). J. Cell Biol. 98: 847.
6. Herskowitz, I. (1987). Nature 329: 219.
7. Humphreys-Beher, M.G., Bunnell, B., VanTuinen, P., Ledbetter, D., Kidd, V.J. (1986). Proc. Natl. Acad. Sci., U.S.A. 83: 8918.
8. Podolsky, D.K., Isselbacher, K.J. (1979). J. Biol. Chem. 254: 1807.
9. Podolsky, D.K., Isselbacher, K.J. (1984). Proc. Natl. Acad. Sci., U.S.A. 81: 2529.

10. Wong, G.G., Witek, J.S., Temple, P.A., Wilkens, K.M., Leary, A.C., Luxenberg, D.P., Jones, S.S., Brown, E.L., Kay, R.M., Orr, E.C., Shoemaker, C., Golde, D.W., Kaufman, R.J., Hewick, R.M., Wang, E.A., Clark, S. (1985). Science 228: 810.
11. Urlaub, G., Chasin, L.A. (1980). Proc. Natl. Acad. Sci., U.S.A. 77: 4216.
12. Kaufman, R.J., Wasley, L.C., Furie, B.C., Furie, B., Shoemaker, C.B. (1986). J. Biol. Chem. 261: 9622.
13. Lopez, L.C., Shur, B.C. (1987). J. Cell Biol. 105: 1663.
14. Humphreys-Beher, M.G., Schneyer, C.A., Kidd, V.J., Marchase, R.B. (1987). J. Biol. Chem. 262: 11706.
15. Klohs, W.D., Wilson, J.R., Weiser, M.M., Frankfurt, O., Bernacki, R.J. (1984). J. Cell Physiol. 119: 23.
16. Stanley, P. (1983). Methods. Enzymol. 96: 157.
17. Brinkley, B.R., Chang, J.F. (1973). In: Tissue Culture Methods and Applications. Academic Press. p. 438.
18. Olden, K., Parent, J.B., White, S.L. (1982). Biochimica et Biophysica Acta. 650: 209.

Gene Transfer and Gene Therapy, pages 235–242

TRANSGENIC SHIVERER MUTANT MICE: EXPRESSION OF A MYELIN BASIC PROTEIN GENE[1]

Carol Readhead, Brian Popko, Naoki Takahashi[2]
H. David Shine[3] Raul Saavedra, Richard L. Sidman[3]
and Leroy Hood

Division of Biology, 147-75,
California Institute of Technology
Pasadena, California 91125

ABSTRACT Mice homozygous for the autosomal recessive mutation shiverer (shi) lack myelin basic protein (MBP). The central nervous system of these mutant mice is hypomyelinated and myelin, where it exists, is uncompacted. Shiverer mice exhibit a distinct behavioral pattern including tremors, convulsions and early death. We have previously demonstrated that shiverer mice have a partial deletion of the gene encoding MBP. Now we are able to demonstrate that the abnormal myelination and shivering phenotype are due to a single gene defect. This was achieved by introduction of the wild-type MBP gene into the germline of shiverer mice by microinjection into fertilized eggs (1). Shiverer mice homozygous for the transgene have MBP mRNA and protein levels that are approximately 25% of normal. The expression of the MBP transgene is tissue specific and developmentally regulated. Moreover, the four different forms of MBP produced by alternative patterns of RNA splicing are present. These mice have compacted myelin with

---

[1]This work was supported by National Institutes of Health grants NS 22223 and NS 20820 to H.D.S. and R.L.S., NIH grant NS 14069 to L.H., Multiple Sclerosis Society grant MS RG 1683-A-1 to L.H. and a Seaver Foundation grant to L.H.
[2]Present address: Department of Medicine, Tokyo University, Faculty of Medicine, Hongo, Bunkyo-ku, Tokyo, Japan
[3]Present address: Harvard Medical School, Boston, MA 02115

major dense lines though the myelin sheaths are thinner than normal. Transgenic shiverer mice homozygous for the MBP transgene no longer shiver nor die prematurely.

INTRODUCTION

Shiverer (shi) is an autosomal recessive mutation first observed by Biddle et al. (1973) in Swiss Vancouver (SWV) mice (2). Shiverer mice display a generalized, coarse action tremor beginning at about the end of the second postnatal week, and becoming more prominent later. Tonic seizures (convulsions) commence at a few months and increase in severity, eventually leading to a shortened lifespan of about 90-150 days.

Myelin is markedly deficient in the shiverer CNS (3,4), but the sheaths are only slightly thinner than normal in the peripheral nervous system (5,6). The major dense line, recognized in the CNS myelin by transmission electron microscopy, is characteristically missing in these mice (7). MBP, which is associated with the major dense line is also absent (8-12). Little, if any, MBP mRNA is present (13), and the MBP structural gene is deleted from the second intron to 2 kb 3' of the last exon (14-16). The shiverer mutation has been mapped to chromosome 18 (14) and so has the mouse MBP gene (14), consistent with the hypothesis that the deleted MBP gene causes the shiverer phenotype.

The introduction of the normal MBP gene into the germ line of the shiverer mouse by microinjection into fertilized eggs allowed us to test the hypothesis that the shivering phenotype and abnormal myelination of the CNS are due to a simple deletion in the structural gene for MBP.

RESULTS

The normal mouse MBP gene from the cosmid clone 138 (14) was injected into the pronuclei of shiverer and C57BL/6J fertilized eggs. Two transgenic mice were obtained; one a shiverer that did not express the transgene and the other a C57BL/6J mouse. Since we were unable to distinguish the expression of the MBP transgene from that of the wild-type gene, the mouse was backcrossed to homozygous shiverer mice so that the MBP transgene could be transferred onto the shiverer germline in the absence of the wild-type gene. Progeny that lacked the normal MBP gene but had the MBP transgene (shi/shi; MBP/-) were identified by Southern blot analysis. These mice were behaviorally like homozygous shiverer mice except that they lived about 1-4 months longer than their littermates. The mice hemizygous for the MBP transgene (shi/shi; MBP/-) were mated inter se to produce mice that were homozygous for the transgene (shi/shi; MBP/MBP). These mice are behaviorally normal and have lived so far for 18 months.

The MBP transgene is tissue-specifically expressed. The tissues examined were brain, liver, spleen, kidney, lung and heart. Northern blot analysis showed that only RNA from the brains of transgenic mice contained a 2.1 kb transcript complementary to the MBP cDNA. Roach et al. (13) had shown previously that shiverer brain RNA contained nor detectable MBP transcripts.

Myelination of the CNS in mice occurs almost entirely in the first month of postnatal development (18). Likewise, the level of expression of the normal MBP gene is high on postnatal day 18 and then drops to a lower, steady-state level at about 4 weeks (19,20). The MBP transgene has a pattern of expression that is identical to that in normal mice.

The level of expression of the MBP transgene was estimated by comparing the intensity of hybridization of the MBP transcript with that of normal mice on Northern blots. The level of expression of the transgene from homozygous transgenic mice (shi/shi; MBP/MBP) was estimated to be 25% of normal. The amount of MBP in the transgenic shiverer brain was measured using a radioimmunoassay for MBP and was found to be 20% of normal. This is in keeping with the relative amount of MBP message found in the transgenic brain.

Purified MBP from he brains of normal and transgenic mice were analyzed by SDS-polyacrylamide gel analysis. Three of the four forms of MBP, the 18.5, 17 and 14 kd form were detected in normal and transgenic mice. The 21.5 kd form was not detected in either wild-type or transgenic mice probably due to its low abundance. The relative expression of the four forms of MBP in the transgenic mice was indistinguishable from that of normal mice. Thus, the genetic elements necessary for normal differential RNA splicing and quantitative control of the expression of the four forms of MBP are present in the microinjected cosmid MBP clone.

Myelination in the optic nerves of wild-type of transgenic mice was examined by electron microscopy. Wild-type mice have myelin sheaths around all the optic nerve fibers with the larger fibers demonstrating thicker sheaths. The myelin is compacted with prominent major dense lines and interperiod lines. Very little myelin is observed in optic axons of shiverer mice and when present, the myelin fails to compact, is wrapped loosely around the axons and does not contain major dense lines. Myelination in the optic nerves of shiverer mice hemizygous for the transgene (shi/shi; MBP/-) occurred around a few of the larger axons; in these cases the myelin wrapping was limited although predominant dense lines could be seen. The optic axons of shiverer mice homozygous for the transgene (shi/shi; MBP/MBP) clearly contained myelin around most of the axons and the myelin was well compacted with prominent major dense lines and interperiod lines.

Shiverer mice homozygous for the transgene (shi/shi; MBP/MBP) have been drastically improved in several aspects of their phenotype: they no longer shiver, no convulsions have been noted, and these mice appear normal and healthy at 18 months.

DISCUSSION

These transgenic shiverer mice provide support for the hypothesis that the phenotypic changes in the shiverer mice are the result of a simple deletion of the MBP gene. The complex array of phenotypic traits associated with this mutation - shivering, tonic seizures, early death, failure of myelin compaction and failure to express MBP RNA or protein - can largely be reversed by the integration and expression of the MBP gene in shiverer mice.

The cosmid clone used for these experiments contains seven exons of the MBP gene as well as 4 kb or 5' flanking sequence and 1 kb of 3' flanking sequence. The MBP transgene is expressed in a tissue-specific manner and it follows the normal temporal patterns of expression. In addition, the different forms of MBP are generated. Accordingly, all the sequences necessary for these controls must be included in the clone. Mice homozygous for the MBP transgene express MBP RNA and protein at approximately 20-25% of normal levels. This reduced level of expression could arise because enhancer sequences are not included in the clone or because the gene was integrated at an unfavorable integration site. Enhancer sequences have been found as far as 6.1 kb 5' of the chicken lyzosome gene (21). Altered expression levels due to the position of chromosomal integration have been noted for other transgenes (22,23).

The mutation myelin deficient, mld, is allelic to shiverer (10). The phenotype exhibited by mld homozygous mutants is very similar to that of shiverer mice. The CNS of these mice is poorly myelinated although with increasing age the amount of compacted myelin increases but remains less than that of normal mice (24). In mld mice, the amount of MBP protein increases with age although it never reaches normal concentrations (24), whereas, in shiverer mice it remains undetectable throughout the life of the animal (25). The MBP gene in mld mice has undergone a duplication and an inversion (26,27), and the developmental expression of the gene(s) is abnormal. The MBP transgene has been introduced into mld mice by crossing and mld mice that are homozygous for the transgene no longer shiver and appear to have a normal life expectancy (26). By breeding, it is possible to vary the combinations of the MBP transgene and the mld and shi MBP genes. Mice are generated that have varying degrees of myelination and they are being used to examine the effect of increasing levels of myelin on the physiology and development of the central nervous system.

REFERENCES

1. Readhead C, Popko B, Takahashi N, Shine HD, Saavedra R A, Sidman RL, Hood LE (1987). Expression of a myelin basic protein gene in transgenic shiverer mice: Correction of the dysmyelinating phenotype. Cell 48:703.
2. Biddle F, March E, Miller JR. (1973). Research News. Mouse News Lett 48:24.
3. Bird TD, Farrell DF, Sumo SM. (1978). Brain lipid composition of the shiverer mouse: Genetic defect in myelin development. J Neurochem 31:387.
4. Rosenbluth J (1980a). Central myelin in the mouse mutant shiverer. J Comp Neurol 194:639.
5. Rosenbluth J (1980b). Peripheral myelin in the mouse mutant shiverer. J Comp Neurol 193:729.
6. Peterson AC, Bray GM (1984) Hypomyelination in the peripheral nervous system mice and in shiverer → normal chimera. J Comp Neurol 227:348.
7. Privat A, Jacque C, Bourre JM, Depouey P, Baumann N (1979) Absence of major dense line in the myelin of the mutant mouse Shiverer. Neurosci Lett 12:107.
8. Depouey P, Jacque C, Bourre JM, Cesselin F, Privat A, Bauman N (1979). Immunochemical studies of myelin basic protein in shiverer mouse devoid of major dense line of myelin. Neurosci Lett 12:113.
9. Kirschner DA, Ganser AL (1980). Compact myelin exists in the absence of basic protein in the shiverer mutant mouse. Nature (London) 283:207.
10. Bourre JM, Jacque C, Delasalle A, Nguyen-Legros J, Dumont O, Lachapelle F, Raoul M, Alverez C, Baumann N (1980). Density profile and basic protein measurements in the myelin range of particulate material from normal developing mouse brain and from neurological mutants (jimpy, quaking, trembler, shiverer and its mld (allele) obtained by zonal centrifugation. J Neurochem 35:458.
11. Mikoshiba K, Kohsaka S, Takamatsu K, Tsukada Y (1981). Neurochemical and morphological studies on the myelin of peripheral nervous system from shiverer mutant mice: Absence of basic proteins common to central nervous system. Brain Res 204:455.

12. Barbarese E, Nielson ML, Carson LH (1983). The effect of the shiverer mutation on myelin basic protein expression in homozygous and heterozygous mouse brain. J Neurochem 40:1680.
13. Roach A, Boylan K, Horvath S, Prusiner SB, Hood L (1983). Characterization of cloned cDNA representing rat myelin basic protein: Absence of expression in brain at shiverer mutant mice. Cell 34:799.
14. Roach A, Takahashi N, Pravtcheva D, Ruddle F, Hood L (1985). Chromosomal mapping of mouse myelin basic protein gene and structure and transcription of the partially deleted gene in shiverer mutant mice. Cell 42:149.
15. Takahashi N, Roach A, Teplow DB, Prusiner SB, Hood L (1985). Cloning and characterization of the myelin basic protein gene from mouse: One gene can encode both 14 Kd and 18.5 Kd MBPs by the alternate use of exons. Cell 42:139.
16. Molineaux SM, Engh H, de Ferra F, Hudson L, Lazarini RA (1986). Recombination within the myelin basic protein gene created the dysmyelinating shiverer mouse mutation. Proc Natl Acad Sci. USA 83:7542.
17. Sidman RL, Conover CS, Carson JH (1985). Shiverer gene maps near the distal end of chromosome 18 in the house mouse. Cytogenet Cell Genet 39:241.
18. Raine CS (1984). "Morphology of Myelin and Myelination." New York: Morell, p 1.
19. Zeller NK, Hunkeler MJ, Campagnonei AT, Sprague J, Lazzarini RA (1984). The characterization of mouse myelin basic protein specific messenger RNAs using a myelin basic protein cDNA clone. Proc Natl Acad Sci USA 81:18.
20. Carson JH, Nielson ML, Barbarese E (1983). Developmental regulation of myelin basic protein expression in mouse brain. Dev Biol 96:485.
21. Thiesen M, Stief A, Sippel AE (1986). The lyzosyme enhancer: Cell-specific activation of the chicken lysozyme gene by a far-upstream DNA element. EMBO J 5:719.
22. Palmiter RD, Chen HY, Brinster RL (1982). Differential regulation of metallothionein. Thymidine kinase fusion genes in transgenic mice and their offspring. Cell 29:701.

23. Chada K, Magram J, Raphael K, Radice G, Lacy E, Costantini F (1985). Specific expression of a foreign β-globin gene in erythroid cells of transgenic mice. Nature 314:377.
24. Matthieu J-M, Omlin FX, Ginalski-Winkelmann H, Cooper BJ (1984) Myelination in the CNS of mld mutant mice: Comparison between composition and structure. Dev Brain Res 13:149.
25. Jacque C, Delassalle A, Raoul M, Baumann N (1983). Myelin basic protein deposition in the optic and sciatic nerves of dysmyelinating mutants quaking, jimpy, trembler, mld and shiverer during development. J Neurochem 41:1335.
26. Popko B, Puckett C, Lai E, Shine HD, Readhead C, Takahashi N, Hunt S, Sidman RL and Hood L (1987) Myelin deficient mice: Expression of myelin basic protein and generation of mice with varying levels of myelin. Cell 48:713.
27. Akowitz AA, Barbarese E, Scheld K, Carson JH (1987) Structure and expression of myelin basic protein gene sequences in mld mutant mouse: Reiteration and rearrangement of the MBP gene. Genetics 116:447.

Gene Transfer and Gene Therapy, pages 243–253

GENETIC ANALYSIS OF HOX 3.1, A MOUSE HOMEO-GENE[1]

Hervé Le Mouellic, Yvan Lallemand, Patrice Blanchet, Danielle Boullier and Philippe Brûlet

Unité de génétique cellulaire, Institut Pasteur, 25 rue du docteur Roux, 75724 Paris Cedex 15, France.

ABSTRACT Our study of a mouse homeo-gene, Hox 3.1, has led to the characterization of the coding sequence and the distribution of transcripts in mouse embryos from 7.5 to14.5 days p.c. by in situ hybridization experiments, suggesting a regionalizing role during embryogenesis. Transcription is first detectable in the whole posterior part of the 8.5 day embryo, without tissue specificity. This distribution follows gradual location and tissue restrictions. Thus, transcription is concentrated in the median part of the embryo, in the somitic mesoderm and the neural tube by day 10.5 p.c., and only in the neural tube by day 12.5 p.c.
Gene transfer into the mouse egg appeared convenient for the genetical study of Hox 3.1. To obtain the equivalent of a dominant mutation, we choosed to place the Hox 3.1 gene under the influence of a strong and ubiquitous promoter in transgenic mice.

INTRODUCTION

Vertebrates homeo-genes have been isolated on the basis of sequence homologies with genes involved in Drosophila development (1). The conserved sequence is a183 bp coding region called the homeobox. Homeo-genes from vertebrates and Drosophila also share the particularity to be expressed during development in a spatially and

[1]This work was supported by grants from the Centre National de la Recherche Scientifique UA 1148, the INSERM CR n°851004, the Fondation pour la Recherche Médicale, the ARC CR n°6609 and the DRET CR n°86.165.

temporally restricted way (2,3).

cDNA molecules from a previously identified locus Hox 3.1, localized on chromosome 15 (4,5), were isolated and their nucleotide sequence obtained (6). They were used as probe for transcriptional analysis. In particular the Hox 3.1 transcripts distribution was established by in situ hybridization on frozen sections of mouse embryos at successive stages. Our results suggest a probable involvement of the Hox 3.1 gene in the regionalization process of the embryo after gastrulation.

To establish Hox 3.1 role during mouse embryonic development, experiments at various levels are needed. At the molecular level, the DNA binding and the regulatory function of the homeo protein has to be established. Therefore specific antibodies are needed. At a more biological level, phenotypical modifications are sought with a constitutive expression of Hox 3.1 during embryogenesis in transgenic animals. Our first results are presented.

RESULTS

The Hox 3.1 Locus

The homeobox locus, isolated in our laboratory was identified as the Hox 3.1 gene by restriction mapping and nucleotide analysis (4,5). A cDNA containing the complete open reading frame was cloned and sequenced (6).To establish hybridization conditions, various DNA fragments were initially used. Under slightly reduced stringency, two homologous loci are detected with various probes (Fig. 1) covering the 5' coding region (cPS), the intron (gPS) or the 3' coding region (HA) (Fig . 2). However under high stringency conditions a single locus is detected with the probes. Thereafter the Pst I - Sal I cDNA fragment (cPS) was mostly used in the transcriptional studies.

Northern blots of total RNA extracted from embryos aged from 9.5 to 12.5 days post coïtum (p.c.) reveal a broad band of transcripts with an apparent 3.2 kb size using cPS probes as DNA or anti-sense RNA probe generated from the same cDNA fragment.

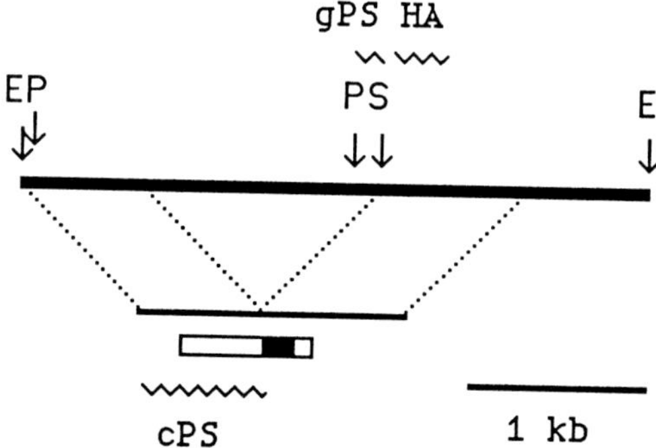

FIGURE 1. Partial restriction map of the Hox 3.1 locus (upper line) ; E: EcoRI; P:PstI; S: SalI. The exons of the cDNA (lower line) are positioned with the dotted lines. The ORF is indicated with the homeobox in black. The probes used are represented by weavy lines.

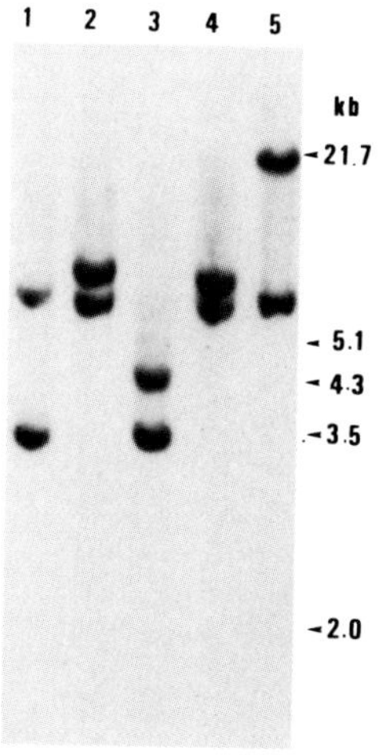

FIGURE 2. Southern blot of genomic DNA digested with EcoRI (1), PvuII (2), PstI (3), HindIII (4) and BamHI (5), and hybridized to the HA RNA probe.

In Situ Hybridization.

Cryostat sections of embryos aged from 7.5 to 14.5 days p.c. were hybridized under stringent conditions with a single-stranded, ^{35}S labelled RNA probe (cPS) as previously described (6). The results reported were obtained with the antisense probe as no labelling was ever observed using the sense probe.

At day 7.5 p.c. of gestation, no labelling could be detected on any part of the embryo. At day 8.5 p.c., labelling is present on all tissues of the caudal end of the embryo (mesenchyme, neural tube, blood vessels and gut). One day later labelling can be observed on the same areas but is more inequally distributed. Indeed the neural tube and hind gut of the 9.5 day p.c. embryo are more strongly labelled than the surrounding tissues (Fig. 3). In contrast at 10.5 day p.c., transciption was detected in a more anterior region as well as in a caudal part of the embryo. Figure 4 shows labelling on the anterior section of the neural tube and the dorsal region of somites in the thoracic part of the embryo.

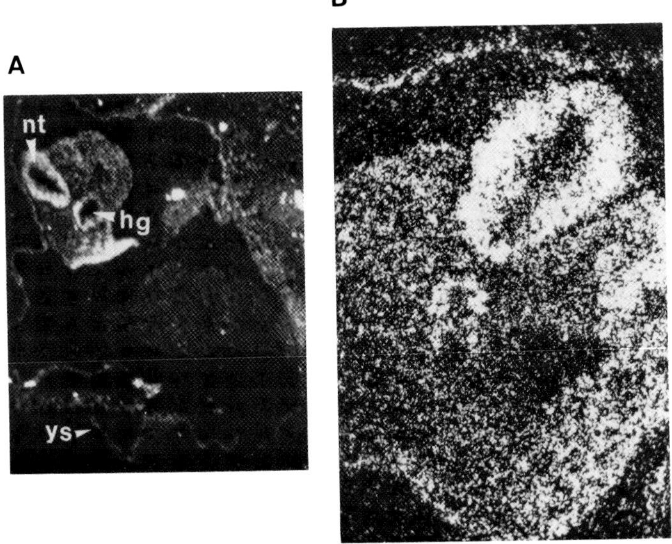

FIGURE 3. Localization of Hox 3.1 transcripts in 9.5 day p.c. embryos by dark-field illumination. (A) Section of a whole embryo. Because of the embryo folding, the posterior part is seen transversally and the anterior part tangentially. Only the caudal part is labelled. (B) Higher magnification of a transverse section of the caudal part. All tissues are labelled. An important signal is seen in the neural tube. hg: hind gut; nt: neural tube; ys: yolk sac.

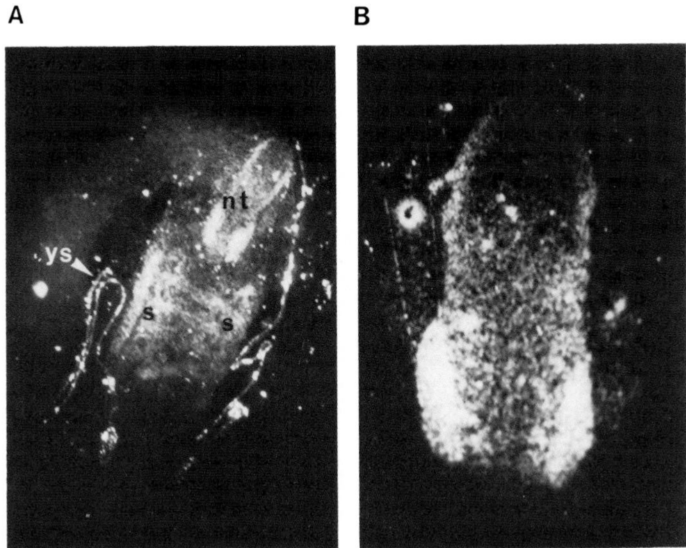

FIGURE 4. Localization of Hox 3.1 transcripts in 10.5 day p.c. embryos by dark-field illumination. (A) Transverse section of a whole embryo. Labelling is localized in the abdominal part of the neural tube and in somites. (B) Higher magnification of the neural tube seen in (A), showing a particularly intense labelling on the ventral horns. nt: neural tube; s: somites; ys: yolk sac.

Notice that the signal is strongest in the ventral horns of the neural tube and mostly absent in its dorsal part, whereas at 9.5 day p.c. all neural tube cells along a dorsal ventral axis are equally labelled .From 11.5 to 14.5 day p.c. (Fig. 5), the distribution of RNAs is stable and displayed a highly restricted tissue specificity since only a small region of the neural tube, posterior to the third cervical vertebra is labelled. Its ventral cells are more labelled than its dorsal ones.

Fusion Proteins

In order to obtain immunogens, the complete coding sequence of Hox 3.1, or the 5' coding region, without the homeobox were fused to the protein A or the

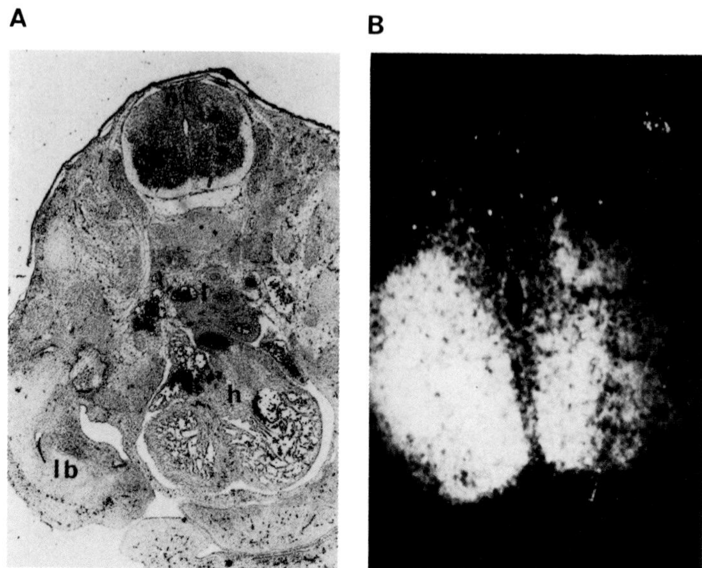

FIGURE 5. Localization of Hox 3.1 transcripts in 12.5 day p.c. embryos. (A) Transverse section of a whole embryo in the heart region by phase contrast. (B) Higher magnification, by dark-field illumination, of the neural tube seen in (A). Hox 3.1 transcripts are localized in the ventral part of the neural tube, at the bottom of the photograph.
h: heart; lb: limb bud; l: lung; nt: neural tube.

ß-galactosidase genes. Figure 6 displays the mobilities of the two proteins fused to protein A . No antisera were obtained so far.

Transgenic Mice

To overproduce Hox 3.1 gene product during embryogenesis we have fused the Hox 3.1 cDNA to the Rous Sarcoma Virus LTR. This construct (Fig. 7) was microinjected, without plasmid sequences, at 3 ug/ml into one of the pronuclei of fertilized F_2 eggs (C57BL/6 x CBA). About 200 microinjected eggs were reimplanted into F_1 (C57BL/6 x DBA/2) pseudo-pregnant females yielding 46 embryos, 8 of which died shortly after birth. The DNAs of the 38 surviving animals were analysed by Southern blotting.

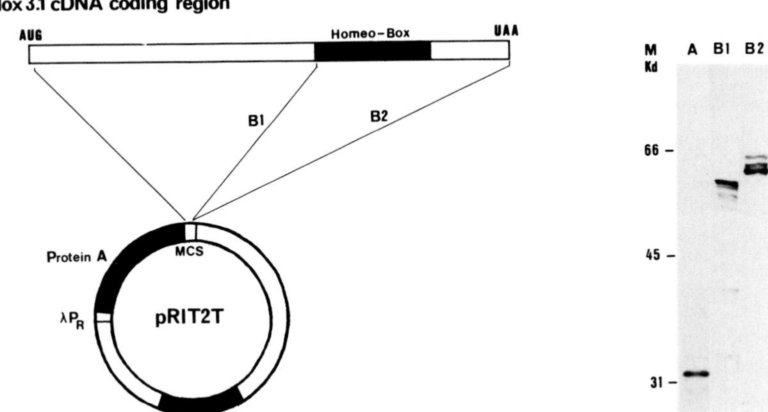

FIGURE 6. (left) Schematic representation of the cloning of the 5' part (B1) or the entire coding region (B2) of Hox 3.1 into the multiple cloning site (MCS) of pRIT2T. This provide molecules for the synthesis of fusion proteins under the control of the right promoter of phage lambda (P_R). (right) Western blots showing the corresponding fusion proteins (B1,B2) and the protein A alone (A) after migration on SDS gel.

Transgenics animals could be identified by the presence of a 2.2 kb cDNA related band in the Eco RI digested DNAs. Because of the presence of an intron the endogeneous Hox 3.1 Eco RI fragment migrates at 3.5 kb as revealed by cPS probe (Fig. 7).

Among the 38 born animals only two males, LY 11 and LY 14, were transgenic. To analyse the pattern of the transgene expression , this two males were mated with C57BL/6 females. The transmission rate of the transgene was different in both cases. LY 11 didn't transmit whereas 100% transmission is obtained with LY 14. The offsprings are being tested for their expression pattern. However none of the transgenic animals displayed a particular phenotype.

DISCUSSION

A chronological study of an homeo-gene Hox 3.1 transcription pattern during mouse embryogenesis is

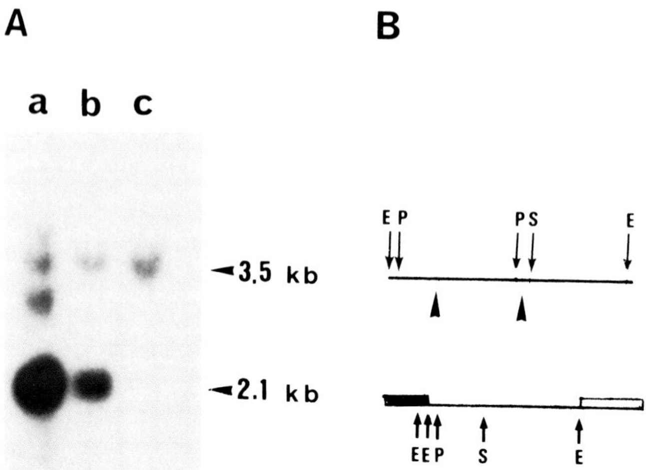

FIGURE 7. Southern analysis of transgenic animals. (A) Southern blots of EcoRI digested tail DNAs from Ly14 (a), Ly11(b) and a control mouse (c), hybridized under high stringency with the cPS probe . (B) Partial restriction map of the Hox 3.1 locus (upper line). The known intron is delimited by the two arrowheads. Schematic representation of the microinjected molecule (lower line). The Hox 3.1 cDNA transcription is driven by the RSV LTR (black box). The polyA addition and transcription termination sequences are from SV40 (white box). E: EcoRI; P: PstI; S: SalI.

---

presented. This pattern, as determined by in situ hybridization with 7.5 to 14.5 day p.c. embryos undergoes gradual modifications during the course of embryogenesis. When expression is first detected after gastrulation in the 8.5 day p.c. embryo it occurs in all tissues of the posterior region, from whichever embryonic layer they originate. By day 10.5 p.c. the posterior transcription region becomes more weakly labelled. The signal is concentrated in the median part of the embryo, in the dorsal region of somites and, more strongly, in the ventral horns of the neural tube. Day 10.5 p.c. of the mouse embryogenesis appears to be a transition step in the transcription of the Hox 3.1 locus which is reorganized along an anterior-posterior axis as well as a dorso-ventral axis. From 11.5 day p.c. to the newborn animal, expression will be in ventral cells of a highly localized area of the neural tube, posterior to the third cervical vertebra. Such a transcription pattern is suggesting a regionalizing role during embryogenesis. Some features of the Hox 3.1 transcription pattern are also shared with homeogenes in other organisms, in particular with the Xeb 1 gene in the Xenopus embryo (7). There is at first co-transcription in neuro-ectoderm and lateral mesoderm tissues, limited to the caudal region of the neurula. This labelling in the caudal neural tube decreases in the tailbud embryo while the Xeb 1 transcripts accumulate in an anterior zone of the neural tube, pearticularly in its ventral region.

This similarity of transcription patterns for two homeo-genes in two different organisms, mouse and xenopous as well as the existence of several conserved peptides in their ORFs, as already noticed (6), strongly suggest that the homeo-proteins mediate conserved mechanisms of genetic regulation during embryogenesis. Those genes could provide some kind of positional information.

To study the possible post-transcriptional regulation of Hox 3.1, the use of specific antibodies will be helpful in the analysis of proteins distribution during embryogenesis.

In a first approach to finding Hox 3.1 role during embryogenesis, phenotypical modifications are sought by overproducing Hox 3.1 protein in transgenic animals.In our first constructs we used the RSV-LTR as a strong promoter, in most cells, to drive the transcription of our cDNA. From expression studies we expected the strongest levels of transcripts in tissues of mesodermal origins, and especially in heart, abdominal and leg muscles (8). However no phenotypical modifications in our limited number of transgenic animals was yet found. Others promoters are now being used. Wrong expression of such a gene could be lethal and so transgenic animals would be dying in utero. To avoid this problem we are currently shifting our approach. We introduce our constructs in embryonic stem cells which, after a selection step, are injected into blastocysts to produce chimaeric animals. However even if we obtain phenotypical modifications with a constitutive overexpression of Hox 3.1, the interpretation of the function of such a regulatory gene will probably not be easy. One expect the interplay of genetic regulatory mechanisms during mouse embryogenesis to be complex.

In fact to establish the function of such a gene during

embryogenesis, one would ideally need to stop the activity of the gene or to introduce defined mutations in the gene. We are currently pursuing such an approach by trying to inactivate the gene by homologous recombination in embryonic stem cells after in vitro selection for the desired genotype. The cells are reintroduced into blastocysts and colonize the various embryonic tissues, in particular the germ line. This way we would obtain F1 animals carrying the mutation which could be rescued, even if homozygote lethal, in heterozygote animals.

ACKNOWLEDGMENTS

The in situ experiments were realized with Hubert Condamine. We would like to thank Pr. F. Jacob for his support and Mrs M. Maury for her excellent technical assistance.

REFERENCES

1. McGinnis W, Garber RL, Wirz J, Kuroiwa A, Gehring WJ (1984). A homologous protein-coding sequence in Drosophila homeotic genes and its conservation in other metazoans. Cell 37:403.
2. Snow MHL (1986). New data from mammalian homeobox-containing genes. Nature 324:618.
3. Schofield PN (1987). Patterns, puzzles and paradigms: The riddle of the homeobox. Trends Neurosci 10:3.
4. Awgulewitsch A, Utset MF, Hart CP, McGinnis W, Ruddle FH (1986). Spatial restriction in expression of a mouse homoeo box locus within the central nervous system. Nature 320:328.
5. Breier G, Bucan M, Franche U, Colberg-Poley AM, Gruss P (1986). Sequential expresssion of murine homeo box gene during F9 EC cell differenciation. EMBO J 5:2209.
6. Le Mouellic H, Condamine H, Brûlet P (1988). Pattern of transcription of the homeo gene Hox 3.1 in the mouse embryo. Genes Dev in press.
7. Carrasco AE, Malacinski GM (1987). Localization of Xenopus homeo-box gene transcripts during embryogenesis and in the adult nervous system. Dev Biol 121:69.
8. Overbeek PA, Lai SP, Van Quill KR, Westphal H (1986). Tissue-specific expression in transgenic mice of a

fused gene containing RSV Terminal Sequences. Science 231: 1574.

Gene Transfer and Gene Therapy, pages 255–268

DEVELOPMENTAL AND TISSUE SPECIFIC REGULATION OF ADENOSINE DEAMINASE IN MICE

Jeffrey M. Chinsky[1,2], V. Ramamurthy[1], Thomas B. Knudsen[3], Howard R. Higley[4], William C. Fanslow[5], John J. Trentin[4] and Rodney E. Kellems[1,2]

[1]Baylor College of Medicine, Department of Biochemistry, Houston, TX 77030. [2]Baylor College of Medicine, Institute for Molecular Genetics, Houston, TX 77030. [3]East Tennessee State University, Department of Anatomy, Johnson City, TN. [4]Baylor College of Medicine, Department of Experimental Biology, Houston, TX. [5]Department of Biochemistry, Rice University.

ABSTRACT The expression of adenosine deaminase (ADA) in murine tissues at different times during development were examined. There is an abrupt 10-20 fold increase in enzyme activity in fetal tissues occurring approximately 2-3 days post-implantation. The majority of ADA activity is ultimately localized to the placenta at the maternal-fetal inferface. At birth, the highest enzyme levels are observed in the thymus. Following birth, the tissues of the upper alimentary canal demonstrate a developmental increase in enzyme activity. By adulthood, these tissues demonstrate the highest levels of ADA activity in the animal, the activity localizing to the keratinized squamous epithelial layer of the mucosa. ADA may account for 10-20% of the total soluble protein in these tissue layers.

INTRODUCTION

An extensive amount of both genetic and pharmacologic evidence indicates that adenosine deaminase (ADA), an enzyme of purine metabolism, plays an essential role in the development of the functional mammalian immune system (for review see 1,2). Consistent with this evidence are numerous published reports indicating that the highest level of this enzyme is found in the thymus, where the specific activity of ADA is 10-100 fold that of many other tissues (3). Within the thymus, the level of ADA is subject to developmental regulation during thymocyte maturation (4). However, there is little insight into the exact physiologic role for ADA in mammals. In order to pursue this question, we have focused our studies on the mouse in an ongoing attempt to understand the molecular basis of tissue specific expression and developmental regulation of ADA. We present here evidence that during both fetal and postnatal development, there are specific organ systems which demonstrate much higher levels of ADA expression than that found in the thymus. In addition, the high levels of ADA observed in these tissues appear to be subject to pronounced developmental regulation.

RESULTS

Murine Tissue Distribution of ADA Activity. The level of ADA activity has been determined for a variety of tissues and cell populations derived from adult mice using the spectrophotometric assay of Agarwal and Parks (7) (see TABLE I). Murine tissues can be classified as either having high, intermediate, or low levels of ADA specific activity. The highest levels of ADA activity are observed in tissues of the upper alimentary tract (tongue, esophagus and stomach) and placenta. The murine stomach contains two visually demarcated regions, the proximal forestomach and distal glandular stomach portions, which are histologically distinct. There is an approximately 20 fold difference in

TABLE I

TISSUE DISTRIBUTION OF ADA ACTIVITY IN MICE

Tissue	Strain: Balb/c[a]	Strain: ICR[b]
	Specific Activity[c]	
Gastrointestinal		
Tongue	580 ± 40.0(3)	2619 ± 226.0(3)
Esophagus	770 ±140.0(3)	1300 ± 200.0(5)
Whole Stomach	190 ± 45.0(4)	
Forestomach	740 ± 80.0(5)	3300 ± 1040.0(5)
Glandular Stomach	37 ± 4.0(4)	200 ± 41.0(5)
Intestine	20 ± 2.0(4)	
Hematopoietic		
Thymus	62 ± 3.2(7)	180 ± 59.0(5)
T-cells (splenic)	55 ± 5.2(7)	
Bone Marrow	21 ± 6.2(4)	
Macrophages	19 ± 1.3(4)	
Spleen	19 ± 1.1(10)	
B-cells (splenic)	12 ± 3.4(4)	
Other		
Placenta	190 ± 24.0(6)	
Skin	27.0 ± 13.0(2)	
Embryo	12.5 ± 10.0(4)	
Liver	5.2 ± 0.2(4)	6.7 ± 1.6(5)
Lung	1.5 ± 0.1(2)	
Kidney	1.4 ± 0.4(2)	
Brain	0.9 ± 0.2(4)	
Striated Muscle	0.4 ± 0.3(2)	

[a]4-6 month old
[b]8 week old
[c]expressed as nm/min/mg. Shown is the mean and standard error for the number of determinations in parentheses.

TABLE II

ENRICHMENT OF ADA ACTIVITY IN UGI MUCOSAL LAYERS[a,b]

	Forestomach	Esophagus	Tongue
Whole Tissue	4,509± 151(2)	2,495±131(2)	4,007± 346(3)
Muscular Layer	226± 78(4)	215± 1(2)	1,753± 138(3)
Mucosal Layer	21,517±1,507(6)	8,295±352(2)	53,090±1,973(3)

[a]ADA activity expressed as nmol/min/mg
[b]Values given are the mean of the test individuals ± standard error with (n) being the sample size.

ADA activity between these two regions, with the forestomach levels being similar to that observed in tongue and esophagus. Intermediate levels of enzyme activity are observed in thymus, spleen, bone marrow and cells of hematopoietic origin, as well as in the skin, glandular stomach and intestine. Lower levels are found in other tissues, including liver, muscle, brain, kidney, and lung. Since there is a distinct tissue distribution of enzyme activity in adult murine tissues, we were interested in determining if developmental changes in ADA expression occurred both during prenatal and postnatal growth. The changes that occur in the thymus have been previously described (3,4,5) and only a slight variation (2-4 fold decrease) in ADA activity occurs during the first few weeks of life, related to T-cell maturation and migration. In contrast, pronounced changes occur in all of the tissues that we have identified with high levels of ADA activity.

ADA Activity During Prenatal Development. As shown in Figure 1A, there appears to be a developmentally controlled increase in ADA activity beginning after day 7 of gestation (2 days postimplantation). During the 48 hour period from day 7 to day 9 of gestation, there is an increase in ADA specific activity of more than 50 fold in the embryo-placental (EP) unit, the attained levels being more than 20 fold that of surrounding uterine tissue. By day 11, the embryo can be easily separated from the placenta by microdissection techniques. The vast majority of the ADA activity is associated with the placenta (Figure 1B), where it continues to be more than 10 fold higher than in the embryo throughout gestation. Similar results are obtained when ADA specific RNA transcripts are analyzed. Accumulation of ADA RNA transcripts increases at least 10 fold in the fetal EP unit from days 7 through 10 and these high RNA levels eventually localize to the placental tissue compared to the embryo, where little ADA RNA is detectable (data not shown). There is thus a

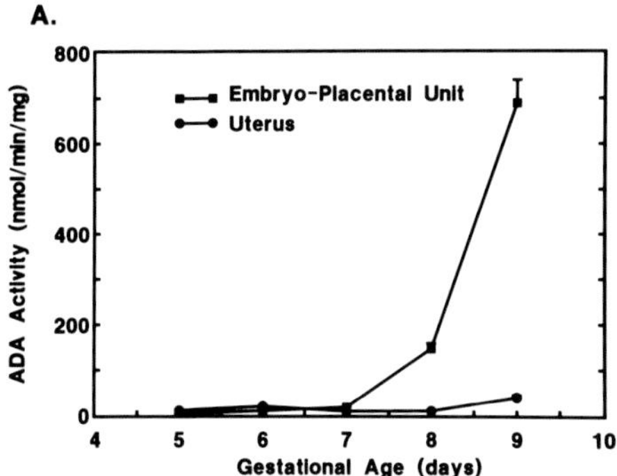

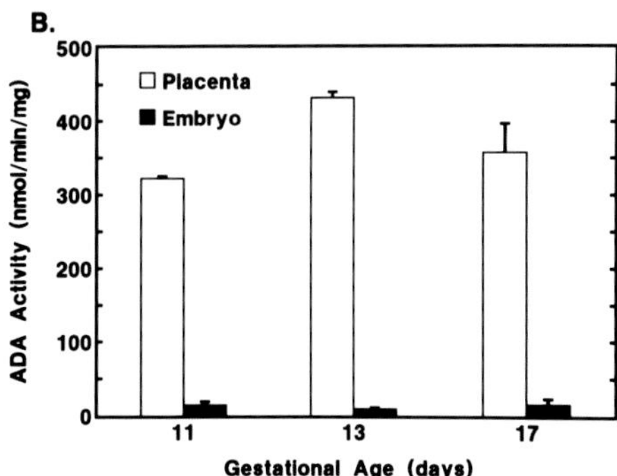

FIGURE 1. Prenatal developmental expression of ADA. Each point represents the mean of the measured ADA activity (7) on at least 4 individual samples. The gestational age is recorded as the number of days following the occurrence of a vaginal plug (day 0). The embryo-placental unit represents the enclosed sac at the implantation site, which is easily removed from surrounding uterine tissue. After day 11, the placental portion of the E-P unit is separable from the embryo body, with loss of yolk sac and amniotic fluid contents.

striking increase in ADA expression during prenatal development which results in high levels of ADA in murine placenta.

Histologic analysis by histochemical staining for ADA as well as by immunofluorescent localization confirms that the highest ADA levels are found at the maternal-embryo interface (see Figure 2). Both the fetal trophoblast layer and maternal decidua are associated with high levels of ADA, as demonstrated by histologic localization techniques (see Figure 2). In addition, it can be observed that there is ADA localization to the fetal yolk sac parietal endoderm (data not shown). High levels of ADA enzyme activity can be demonstrated in both maternal decidua and fetal trophoblast enriched tissues following microdissection of both murine and rat placentae (data not shown). *In situ* hybridization studies are being performed to detect ADA specific transcripts in the different cell types of the murine placenta (specifically fetal trophoblast versus maternally derived decidual cells). In this way, sites of actual high level synthesis of ADA, rather than simply areas of accumulation of exogenously synthesized ADA, will be identified. Nonetheless, the high level of ADA detected in rodent placenta tissues suggests that ADA has some physiologic role in the interaction between maternal and fetal tissues that occurs during gestation.

<u>ADA Activity During Postnatal Development of Upper Gastrointestinal Tissues</u>. At birth, the highest level of ADA activity is observed in the thymus, compared to spleen, liver, and upper gastrointestinal tissues. During the first two weeks of life, the specific activity of ADA in the thymus decreases two fold, and the relatively low levels observed in liver and spleen remain constant. In contrast, the level in whole stomach increases 5 fold (data not shown and (6)). Mice are born without a differentiated forestomach, which first becomes visually evident by the end of the first week and then enlarges in proportion to the glandular

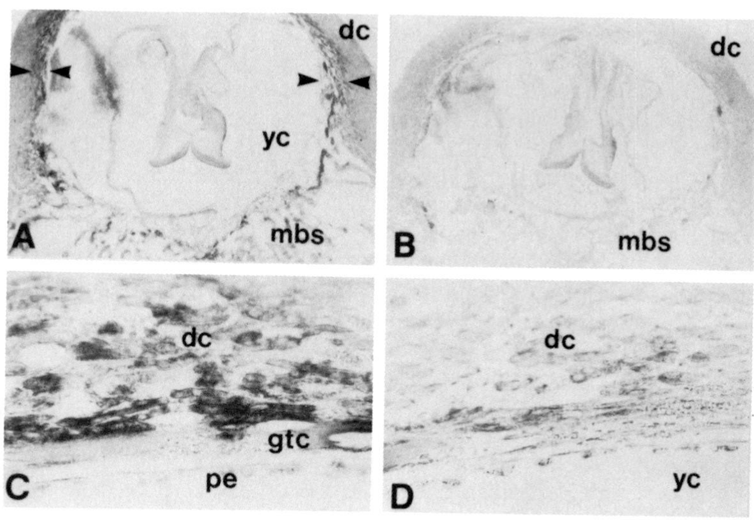

FIGURE 2. Localization of ADA activity to the maternal fetal interface. Frozen sections were prepared from 8.5 day old whole embryos (together with outer placenta and mesometrial wall) and histochemically stained for ADA (8) using adenosine as substrate and nucleoside phosphorylase, xanthine oxidase, nitroblue tetrazolium, and phenazine methosulfate as reagents. The blue formazan reaction product was photographed through a 590 nm filter. Magnification - A,B = 29x, C,D = 290x. B and D are the 2′-dCF (a potent inhibitor of ADA) control treated sections contiguous to A and C. The broad arrows in A indicate the band of intense reactivity associated with the giant trophoblast cell (gtc) layer and inner decidua capsularis (dc). mbs-maternal blodd space; yc-yolk sac cavity; pe-parietal endoderm. The staining of parietal endoderm (pe) was lost in this photographic reproduction of the slide.

stomach. The developmental changes in ADA specific activity were therefore examined in these two portions of the stomach as well as the tongue and esophagus. As shown in Figure 3, there is a pronounced increase in stomach ADA during the first few weeks of life, the majority of which appears to occur in the forestomach. The level of ADA in the tongue is also very low at birth and undergoes a large increase with age. The esophagus is also very low in ADA activity at birth but undergoes a similar developmental increase in ADA activity, occurring after 1 to 2 weeks of age. Thus, all three of these upper gastrointestinal tissues appear to be subject to developmental regulation of ADA expression during the first several weeks of postnatal life.

ADA Activity Resides in the Mucosal Layer of UGI Tissues. The upper gastrointestinal tissues studied above have a similar histologic architecture, containing an outer muscular layer surrounding the luminal mucosal layer. This mucosal layer is composed of stratified squamous epithelial cells which produce a thick keratinized layer lining the lumen of the upper alimentary canal. Separation of this mucosal layer from the muscle layer by EDTA treatment and microdissection results in ADA activity localizing to the mucosal layer (see Table II). Coincident with this enrichment for ADA activity, high levels of ADA specific RNA transcripts (Figure 4) and ADA protein (Figure 5) can be localized to this tissue layer. The specific activity of ADA in the mucosal layers is approximately 100-fold greater than that present in the muscle and connective tissue layer. In addition, there is an approximate 5-20 fold enrichment for ADA activity in the mucosal layer compared to the unfractionated tissue (Table II). Electrophoretic analysis of soluble protein extracts from these tissues indicate that ADA is an abundant protein in forestomach, esophagus and the tongue (see Figure 5). In fact, it appears to account for up to 5-20% of the soluble protein present in

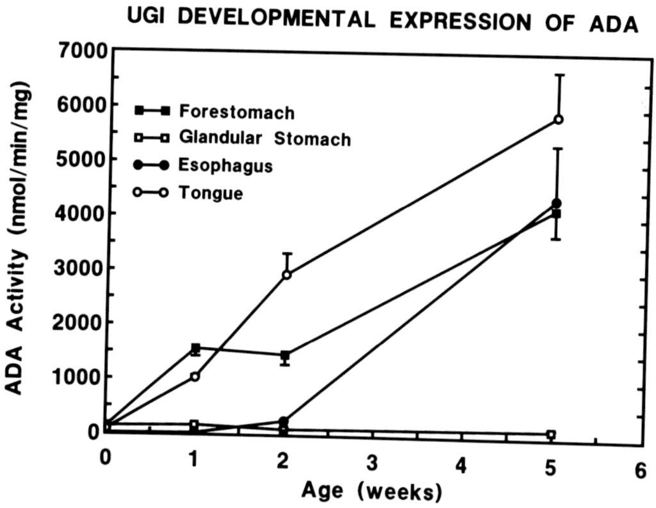

UGI DEVELOPMENTAL EXPRESSION OF ADA
ADA Activity (nmol/min/mg)
7000
6000
5000
4000
3000
2000
1000
0
Forestomach
Glandular Stomach
Esophagus
Tongue
0
1
2
3
4
5
6
Age (weeks)

FIGURE 3. ADA activity in tissues of the upper alimentary canal. Each point represents the mean of 4 individual samples assayed (7) at the indicated ages.

Forestomach
Muscle
Mucosa
1.7 kb
Specific Activity
nm/mg/min
4,357
290
18,912

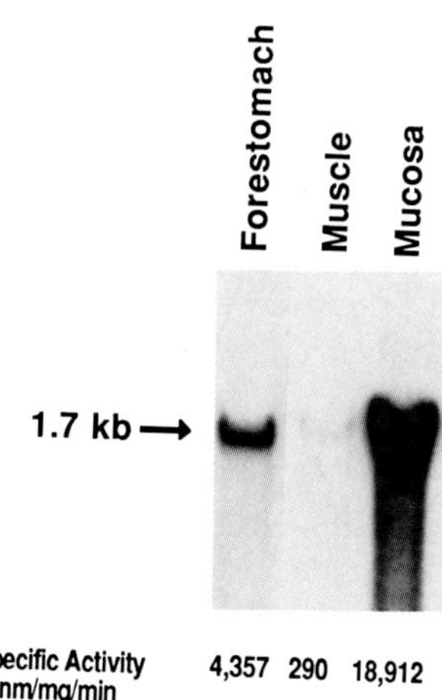

FIGURE 4. Northern blot hybridization analysis for ADA transcript (1.7 kb) in whole forestomach and separated muscle and mucosa layers.

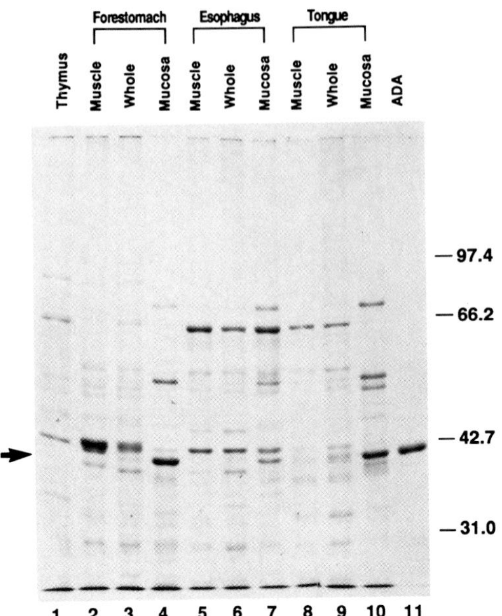

FIGURE 5. 10% PAGE-SDS analysis of total soluble protein obtained from whole tissues and fractionated muscle and mucosa layers. 10 *u*g of each tissue extract indicated was electrophoresed and subsequently stained with coomassie blue. The arrow indicates the ADA protein, confirmed by Western immunoblot analysis. Lane 11 contains 2 *u*g of purified ADA protein. Molecular weight standards are indicated.

---

the mucosal layers of these tissues. The protein band is confirmed to be ADA by comigration with purified ADA and Western blotting analysis for immunoreactive protein (data not shown). In addition, histologic analysis by immunofluorescence and histochemical staining for ADA demonstrates the striking compartmentalization of ADA in these tissues to the keratinized squamous cell mucosal layer (see

Figure 6). In contrast, we have also identified high levels of ADA activity in the nonkeratinized mucosal epithelial cells lining the intestinal villi (data not shown). ADA levels appear to be very low in the proliferating undifferentiated stem cells of the intestinal crypts and increase substantially as the immature cells leave the crypts and migrate to the tips of the villi.

DISCUSSION

The phenotype of ADA deficiency in humans and the previously held belief that the mammalian thymus had the highest levels of enzyme focused attention on the essential role of the enzyme in immune function. However, we have identified even higher relative levels of tissue specific expression in multiple seemingly unrelated non-hematopoeitic tissues. This suggests that there are other functional roles for ADA. The dramatic turn on of ADA activity in fetal tissues suggests that the enzyme provides some function during prenatal development. It remains to be determined specifically which cells of the placenta synthesize the high levels of ADA noted to be present. In this way, it can be determined whether the enzyme's function during fetal development occurs solely at the site of synthesis or in other regions of fetal tissue which have been noted to accumulate high levels of ADA. The finding of high levels in fetal parietal endoderm may in some way relate to the noted high levels detected in murine upper gastrointestinal tissues.

The high levels of ADA observed in the keratinized squamous epithelium of the forestomach, esophagus and tongue, as well as the non-keratinized villus epithelium of the small intestine, suggest a functional role in the physiology of the alimentary canal. This role may be directly related to the enzyme's involvement in purine metabolism, since we have recently established that an additional purine metabolic enzyme, xanthine oxidase (XO), is

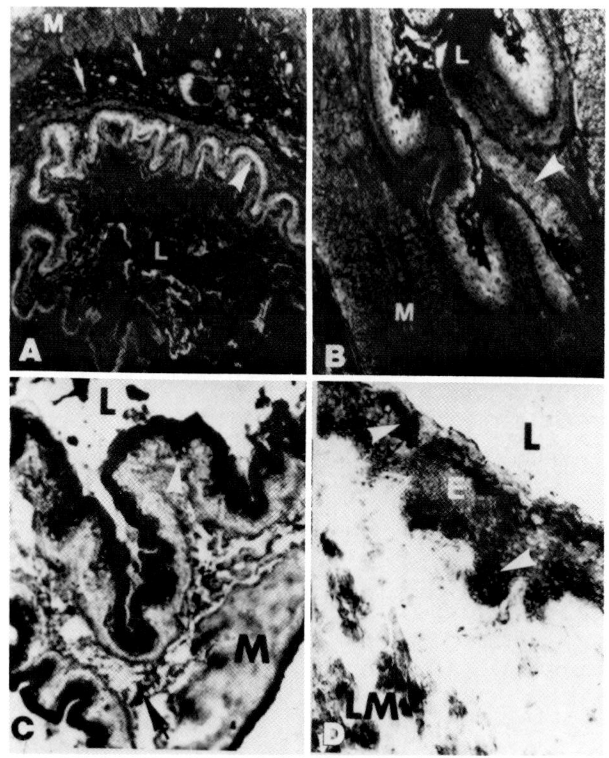

FIGURE 6. Localization of ADA to the keratinized squamous epithelial layer of upper gastrointestinal tissues. Immunofluorescent staining using ADA specific antibody (indicated by the broad white arrows in panels A (forestomach) and B (esophagus)). Histochemical staining (as per Figure 2) is indicated by the white arrows in panels C (forestomach) and D (tongue). L-lumen, M-muscle layer, LM-lingual muscle, E-mucosal epithelial layer.

---

expressed at high levels in regions of the alimentary canal similar to ADA (data not shown and (6)). It will be of interest to know if high levels of a third enzyme, purine nucleoside phosphorylase, colocalize with ADA and XO and if levels of all three enzymes are coordinately regulated during development in the gastrointestinal tissues. This might then suggest a physiological role of this pathway to prevent a cytotoxic accumulation of adenine nucleosides or to produce uric acid for some yet unrecognized purpose.

It does not appear that the high levels of ADA identified are solely in tissues involved in rapid cell turnover. Not all cells with this characteristic (such as skin) demonstrate the very high levels of ADA observed in the mucosal layer of the tissues of the upper alimentary canal. Similarly, there are several tissues which do not contain a keratinized squamous epithelium which nonetheless demonstrate high levels of ADA (placenta, intestinal villous epithelium, thymus). A unifying hypothesis for the role of ADA in murine physiology in diverse organs may involve the relationship of cells which appear to be involved in either programmed cell death or terminal differentiation from a stem cell population.

ACKNOWLEDGMENTS

This research was supported by Public Health Service grants HD21452, AI25255, CA14030 and CA35492; Robert A. Welch Foundation grants Q-893 and C-1041; and March of Dimes grant 6-393. R.E.K. was supported by a Research Career Development Award K04 CA00828; J.M.C. was supported by NIH Postdoctoral Fellowship F32 GM12059; and J.J.T. was supported by a Research Career Development Award K06 CA14219. T.B.K. gratefully aknowledges the support of the Research Development Committee, East Tennessee State University.

REFERENCES

1. Kellems RE, Yeung C-Y and Ingolia DE (1985). Adenosine deaminase deficiency and severe combined immunodeficiencies. Trends in Genetics 1:278-283.

2. Martin DW and Gelfand WW (1981). Biochemistry of diseases of immunodevelopment. Ann Rev Biochem 50:845-877.
3. Hirshhorn R, Martiniuk R and Rosen FS (1978). Adenosine deaminase activity in normal tissues and tissues from a child with severe combined immunodeficiency and adenosine deaminase deficiency. Clin Immun and Immunopath 9:287-292.
4. Kizaki H, Habu S, Ohsalea F and Sakurada T (1983). Purine nucleoside metabolizing enzyme activities in mouse thymocytes at different stages of differentiation and maturation. Cellular Immun 82:343-351.
5. Barton R, Martiniule F, Hirschhorn R and Goldschneider I (1986). Inverse relationship between adenosine deaminase and purine phosphorylase in rat lymphocyte populations. Cell Immun 49:208-214.
6. Lee PC (1973). Developmental changes of adenosine deaminase, xanthine oxidase, and uricase in mouse tissues. Dev Biol 31:227-233.
7. Agarwal RP and Parks RE (1964). Adenosine deaminase from human erythrocytes. Methods Enzymol 51:502-507.
8. Norstrand IF (1985). Histochemical distribution of adenosine deaminase in the human neuraxis. Neurochem Path 3:73-82.

Gene Transfer and Gene Therapy, pages 269–281

LARGE DNA TECHNOLOGY AND POSSIBLE APPLICATIONS IN GENE TRANSFER[1]

Charles R. Cantor and Cassandra L. Smith

Department of Genetics and Development and Departments of Microbiology and Psychiatry, College of Physicians and Surgeons, Columbia University, New York, NY 10032

ABSTRACT Methods now exist that allow routine manipulation of DNAs ranging in size from 50 kb to more than 5 Mb. These methods can provide specific large DNA fragments as input for DNA transfection procedures, and they should be especially useful in monitoring the results of such procedures.

INTRODUCTION

Separation techniques are one of the key steps in facilitating studies on nucleic acids. Size fractionation of single stranded DNA and RNA, with one nucleotide resolution, forms the basis of all current nucleic acid sequencing technologies. Somewhat lower resolution size separations are the routine methods used to monitor virtually all experiments involving recombinant DNA techniques or using the results of such techniques for probing biological questions. The upper limit of conventional recombinant DNA methods is around 50 kb, and this limit is largely determined by the difficulty of fractionation of DNA by ordinary electrophoretic methods, although it also reflects limitations of common bacteriophage packing and DNA transfection procedures.

---

[1]This work was supported by grants from the NIH, GM 14825, the NCI, CA 39782, the DOE, DE-F602-87ER-GD582, the Hereditary Disease Foundation, and LKB Produkter, AB.

Recently, methods have been developed that allow convenient manipulation of DNA molecules one hundred times larger than those used in conventional studies. These methods are leading to marked changes in the way chromosomes and even whole genomes are studied and manipulated. They have considerable implications for two aspects of gene transfer: preparing DNA samples destined for transfer, and analyzing the results of gene transfer. In this article we shall summarize the state of the art of large DNA technology and indicate how this technology is likely to be used in the near future for enhanced gene transfer methodologies.

LARGE DNA TECHNOLOGY

A series of new methods has been developed to handle DNA molecules in the 100 kb to 10 Mb size range. These include methods for preparing unbroken chromosomal DNAs (1,2), methods for specific cleavage of these DNAs into large fragments (3,4), techniques for separating these fragments by size (5), and determining their size (6), and methods for cloning such molecules (7). In addition, approaches used for constructing restriction maps of DNA on the ordinary size scale have been modified and optimized for handling special constraints seen with large DNAs, especially those of mammalian origin (8). These methods have already led to complete restriction maps of several simple genomes including *E. coli* (9) and *S. pombe* (10) and maps of extensive regions of the human genome such as the major histocompatibility complex (11), and the neighborhoods of the genes involved in Duchenne Muscular Dystrophy (12-14), Cystic Fibrosis (15), and Huntington's Disease.

Preparation of Chromosome-sized DNA Molecules

DNAs larger than a few hundred kb are difficult to handle by ordinary solution biochemical techniques because of their extreme sensitivity to shear breakage and to nucleases. The former problem can be circumvented completely by preparing DNA *in situ* in agarose, either as blocks (2,4) or as microbeads (16). Live cells ranging from bacteria to mammals are suspended in liquid low gelling agarose, and, after solidification, successive enzymatic, detergent and salt treatments are applied to the samples until all cellular constituents are removed except DNA. These procedures are convenient; they require almost no equipment, are applicable to hundreds of samples simultaneously, and appear to provide a quantitative recovery of unbroken total cellular DNA. The agarose procedures work because the gel pore sizes are large enough to allow free diffusion of proteins and detergents, but they are too small to allow significant diffusion of DNA. The DNA is protected against nucleases by the presence of proteinase K and high concentrations of EDTA. Together these agents appear to repress all common DNases so that samples stored in this way are stable for years at 4°C and months at room temperature.

Specific Fragmentation of Chromosome-sized DNA Molecules

Most restriction nucleases have recognition sites that are far too frequent in DNA to generate hundred kb or Mb fragments. However, for almost every DNA source, it seems possible to find one or more restriction enzymes that yield relatively large fragments (2,3). Two enzymes, Not I and Sfi I have 8 base recognition sequences and on average should yield 64 kb pieces from random sequence DNA. Fortunately, in many genomes, these sites are much rarer than statistical expectations, and so much larger fragments are produced. For mammalian DNA samples, the sequence CpG occurs at only 20% of the expected frequency and most of the CpG's are methylated (17). Restriction nucleases that recognize sequences containing one or more CpG's and are inhibited by methylation are thus excellent candidates for yielding large fragments. The enzyme Mlu I, which cleaves at ACGCGT, has particularly rare cleavage sites. In most cases, the methylation pattern of individual sites is all or none. This is particularly useful because, if partial methylation occurs, enzymatic digestion will inevitably yield a complex set of DNA fragments.

The mammalian recognition sites cleaved by most CpG-specific enzymes preferentially lie in unmethylated G-C rich regions called HTF islands (18). These are typically upstream from housekeeping genes, and they often contain clusters of cleavage sites for enzymes that have very few cleavages elsewhere. Thus, restriction maps generated by using such enzymes are, in part, also maps of the location of widely expressed genes. However, because of the site clustering, many different enzymes often yield the same size large DNA fragment, to within the resolution of existing methods. Thus it is difficult to use the overlap of different enzyme digests to construct maps in the way ordinarily done with smaller DNAs.

Limitations in the number and specificities of available enzymes that yield large DNA fragments are major hurdles in current large DNA technology. Two approaches under development promise to relieve current limitations. In principle, combinations of site-specific DNA methylation with subsequent cleavage by enzymes like *Dpn I*, that require methylation, can yield a large variety of specific DNA cleavages (19,20). Such methods work well in model systems, and there is no fundamental reason why they should not eventually become generally applicable to large DNA. An alternative is cleavage of DNA by reactive chemicals attached to oligonucleotides. Homopurine-homopyrimidine stretches are particularly attractive targets for such approaches because one can form a sequence-specific triple helix prior to cleavage and thus avoid having to partially denature a segment of the DNA (21-23). The yield of current double strand chemical cleavage is still low, but it is likely to be improved as reagents are optimized.

Fractionation of Large DNA Molecules

Pulsed field gel electrophoresis (PFG) provides high resolution size separations of DNA molecules ranging from 20 kb to more than 5 Mb. In PFG, separation is achieved by forcing DNA to undergo repeated reorientations in an agarose gel. Larger molecules reorient slower than smaller ones, so their overall migration rate is also slower. A large number of variations on the original experimental apparatus have been described (24-27). In general, variants that produce migration in straight lanes appear to have reduced size resolution when compared with those where molecules travel on curved paths. This forces a compromise between ease of data reading, number of samples analyzed on a single gel, and the ability to fractionate finely spaced bands. Under the best circumstances, the resolution of PFG is sufficient to detect megabase pieces that differ in size by less than 1%.

In PFG, DNA is concentrated in very fine bands. Such large molecules will have insignificant band broadening from diffusion. In addition, electrical field gradients can be used to provide band sharpening (1,26). With simple genomes the bands can be detected by direct ethidium staining. Even in genomes as complex as mammals, a discrete pattern of banding can be seen in stained DNA. However most analysis requires blotting and detection by hybridization with DNA probes. Ordinary procedures are used, except that the DNA must be fragmented prior to Southern transfer. When sufficient care is taken, the fine bands in the original PFG separation allow very sensitive detection of the resulting blots.

Construction of Macro-restriction Maps

In ordinary DNA electrophoresis, and virtually all other size fractionation methods, resolution inevitably gets worse as the molecular size increases. Thus, restriction mapping strategies are usually based on dividing the sample into quite small fragments to facilitate quantitative measurements. In PFG, under any given conditions, the resolution is constant for a ten-fold range of DNA size and then actually improves for somewhat larger molecules for another half an order of magnitude before a zone is reached where all resolution is lost (6). This encourages alternative mapping strategies where the sample fragments are kept as large as possible, consistent with the fineness of the overall data analysis desired. Thus the Smith-Birnstiel method (28) or other partial digestion techniques are particularly well suited for constructing macro-restriction maps, and a number of detailed examples of such approaches have been described.

An alternative that seems particularly powerful, in principle, is the use of linking probes (3,4). These are ordinary size cloned DNA fragments that contain the cutting sites recognized by the same enzyme used to generate large genomic DNA fragments. Such probes will detect two adjacent DNA bands, and thus a complete set of such probes contains sufficient information to link up an entire restriction map. A number of different approaches to selecting or screening for linking probes within a library have been outlined. Jumping probes containing just the ends of large DNA fragments can also be prepared and are useful for various rapid chromosome walking strategies (3).

Cloning Large DNA Pieces

A number of potential strategies for cloning large DNA fragments exist. These range from mammalian hybrid cells which can harbor intact heterologous mammalian chromosomes or fragments of these chromosomes, to various potential bacterial techniques for introducing and maintaining large DNA pieces or reconstructing large DNA domains by homologous recombination (29). However the most successful general strategy described to date involves yeast artificial chromosomes (YACs). Artificial linear *S. cerevisiae* chromosomes can be prepared by ligating two ends (each containing a telomere and a selectable marker, at least one containing a replication origin, and one containing a centromere) to large linear DNA, and then transforming yeast cells with these constructs (7). The longer the artificial chromosome the more stably it appears to be maintained. This fact, plus the dual selection, provides useful, albeit low, transformation frequencies. The YAC methodology is still in its infancy but its potential power is considerable, and improvements in methodology are sure to be forthcoming.

A second yeast approach employs chromosomes from the fission yeast *S. pombe* which can be transferred from yeast to mammalian cells by cell fusion methods (30). This should lead the way to shuttle vectors allowing significant segments of DNA to be moved between mammals and yeast. The major limitations at present are the large size of the *S. pombe* centromere and its apparent lack of function in a mouse cell. Neither of these obstacles seems insurmountable.

APPLICATIONS OF LARGE DNA TECHNOLOGY

Large DNA methods allow megabases of DNA to be viewed at once. This provides the equivalent of an aerial view of large segments of a genome or even whole small chromosomes compared to the ground level perspective of ordinary DNA approaches. The power of the large DNA approach is realized whenever a phenomenon encompasses megabases of genomic material.

Examples of Prior Large DNA Applications

The genome overview provided by large DNA methods is particularly evident with simple organisms, where the entire genome can be viewed at once by ethidium staining. Any genomic rearrangement involving 5 kb or more of DNA is likely to be detected directly by PFG analysis. For example, bacteriophage λ has no sites recognized by the restriction nuclease Not I. λ insertions into the *E. coli* chromosome increase the size of the Not I fragment at the site of the insertion by approximately 50 kb. This is directly visualized on ethidium stained gels by a shift in fragment mobility, which can be confirmed by hybridization with radiolabeled λ DNA (9). In an analogous way, chromosomal insertions of F+ DNA or Tn5 DNA can be mapped because these species contain Not I cleavage sites, and thus DNA insertion introduces extra cleavage sites into the genome (31). More complex rearrangements of the *E. coli* genome are also easily studied by PFG (9).

PFG methods are equally useful for monitoring genomic rearrangements in more complex organisms. For example, recombination of psoralen-damaged plasmids into a yeast chromosome containing homologous DNA often produces inserts consisting of tandem duplications. The shift in chromosome size allows the number of tandem inserts to be determined, directly (32).

In trypanosomes, chromosomal translocations often accompany switches in the particular surface antigen gene being expressed. These can be detected as shifts in the sizes of the chromosomes involved and confirmed by hybridization with DNA probes specific for the particular surface antigen gene (33). Even rearrangments of mammalian DNA are easily studied by PFG. As one example, in lymphocyte differentiation, DNA splicing events occur which rearrange the germ line distribution of immunoglobulin genes to make a single expressed antibody chain. Some events involve deletions of up to a megabase or more of DNA, such as the splice between a variable chain coding sequence and a D region sequence. These show up as a loss of several large DNA bands from the region when viewed with probes that cross-hybridize with most variable chain sequences and thus detect the entire DNA region (34).

The genome overview provided by large DNA methods is particularly useful for comparing closely related micro-organisms. For example, Not I or Sfi I digests reveal patterns of relatedness among various *E. coli* strains. Similar digests can be used to examine other bacteria where less well developed genetic analysis is possible. For organisms with many small chromosomes, such as most yeasts and parasitic protozoa, PFG yields a molecular karyotype which allows a cytogenetic analysis of their genome organization, evolution, and geographic variations.

In organisms with genomes as complex as mammals, not all the genome can be viewed in fine detail even by the highest resolution PFG analysis available. However, it is possible to focus attention on small regions of the genome by using hybrid cell lines containing just a single human chromosome in a rodent cell. By hybridizing a restriction digest of such a cell line with total human DNA, or the highly repeated human Alu sequence, one detects just the human DNA fragments. This allows direct comparison of different isolates of the same chromosome and direct detection of polymorphisms arising either from insertions, deletions or other rearrangements or from restriction site differences. The current need to work with hybrid cell lines limits the generality of this approach. Eventually, by using pools of chromosome-specific DNA probes, the same results will be obtainable by direct observation of any type of cell.

Potential of Large DNA Methods in Gene Transfer Studies

A number of the PFG applications described above should find direct parallels in studies of gene transfer. For example, in some DNA transfection procedures, large tandem integrations are produced. The relative number of inserts in different clonal cell lines will be revealed by the size of the macro-restriction band on which the inserted DNA now lies (Fig. 1). Comparisons of fragment size and hybridization intensity should yield the absolute size of the insert.

FIGURE 1. Mapping DNA insertions by PFG. The thick solid line shows a tandem set of insertions, containing one site each for a particular restriction nuclease. Cleavage outside the region at a, monitored by any probe inside the region will allow the size of the insertion to be estimated by detecting band c. Cleavage at a, and partial cleavage internally, monitored by a flanking probe, b, will reveal fragments, d, that provide a fine map of the region.

If sequences flanking the insertion site can be isolated, a much more detailed picture can be obtained. These can serve as probes for a Smith-Birnstiel analysis of the organization of the insert by partial cleavage with a restriction nuclease that cleaves just once per inserted monomer. This will reveal the regularity or continuity of the tandem array. The flanking probes can also be used to screen the nearby region for rearrangements. In some cases the process of DNA insertion may lead to or be accompanied by other rearrangements in the genome, and it is important to understand where and why these have occurred.

In many applications DNA transfer is used to produce localized mutagenesis by gene disruption. Where an interesting phenotype results, the next step is often to try to isolate the region where the insertion occurred. The disrupted gene may be too large to capture on an ordinary plasmid, bacteriophage or cosmid, or it may be in the midst of a family of related genes. The new large DNA cloning methods should be especially useful in studying such cases.

Large DNA methods are likely to play an increasingly important role in gene transfer studies over the next few years. Unless the size of functional centromeres in S. pombe (35), and probably most higher organisms, can be reduced substantially, large DNA methods will be required for most chromosome engineering since the size of the minimum stable chromosome is likely to be more than 200 kb.

Gene transfer of DNA regions hundreds of kb in size is more likely to preserve the cis-acting regulatory sequences needed for proper tissue and temporal control of expression (36). Genes as large as 2 Mb have now been identified, and large DNA methods will clearly be needed if such genes are to be transferred intact from one cell to another. Homologous recombination is required for site-directed gene transfer which is the best way of guaranteeing the correction of a deleterious dominant allele or the creation of a null mutation by eliminating a normal allele completely. The frequency of homologous recombination increases with the length of the homologous DNA region (37), so that DNA stretches hundreds of kb or longer may recombine very well. Thus we will need to develop large DNA gene transfer methods to fill the gap between ordinary gene transfer and chromosome mediated gene transfer. Attempts to transfer DNA immobilized in agarose directly into cells would seem to be one possible route worth exploring.

REFERENCES

1. Schwartz, DC and CR Cantor (1984). Separation of yeast chromosome-sized DNAs by pulsed field gradient gel electrophoresis. Cell 37:67-75.
2. Smith, CL and CR Cantor (1987). Purification, specific fragmentation, and separation of large DNA molecules. In Wu R (ed):"Methods in Enzymology: Recombinant DNA," San Diego: Academic Press 155:449-467.
3. Smith, CL, Lawrence, SK, Gillespie, GA, Cantor, CR, Weissman, SM, Collins, FS (1987). Strategies for mapping and cloning macro-regions of mammalian genomes. In Gottesman, M (ed): "Methods in Enzymology: Recombinant DNA," San Diego: Academic Press, Vol. 151:461-489.
4. Smith, CL, Warburton, PE, Gaal, A, and CR Cantor (1986). Analysis of genome organization and rearrangements by pulsed field gradient gel electrophoresis. Genetic Engineering 8:45-70.
5. Schwartz, DC, Saffran, W, Welsh, J, Haas, R, Goldenberg, M and CR Cantor (1983). New techniques for purifiying large DNAs and studying their properties and packaging. Cold Spring Harbor Symposia on Quantitative Biology XLVII:189-195.

6. Mathew, MK, Smith, CL and CR Cantor (1988). High resolution separation and accurate size determination in pulsed field gel electrophoresis of DNA. I. DNA size standards and the effect of agarose and temperature. Submitted.
7. Burke, DT, Carle, GF, Olson, MV (1987). Cloning of Large Segments of Exogenus DNA into Yeast by Means of Artificial Chromosome Vectors. Science 236:806-811.
8. Cantor, CR, Smith, CL, and Argarana, C (1987). Strategies for finishing physical maps of macro-DNA regions. In: "Symposium on integration and control of metabolic processes: pure and applied aspects," Cambridge, England: ICSU Press, p.327-341.
9. Smith, CL, Econome, JG, Schutt, A, Klco, S, Cantor, CR (1987). A physical map of the Escherichia coli K12 genome. Science 236:1448-1454.
10. Fan, JB, Smith, CL, Chikashige, Y, Niwa, O, Yanagida, M, and CR Cantor (1988). Construction of a Not I restriction map of the fission yeast Schizosaccharomyces pombe genome. MS in preparation.
11. Lawrance, SK, Smith, CL, Srivistava, R, Cantor, CR, Weissman, S (1987). Megabase-scale mapping of the HLA gene complex by pulsed field gel electrophoresis. Science 235:1389-92.
12. Kenwrick, S, Patterson, M, Speer, A, Fishbeck, K and K Davies (1987). Molecular analysis of the Duchenne muscular dystrophy region using pulsed field gel electrophoresis. Cell 48:351-357.
13. Burmeister, M and H Lehrach (1986). Long-range restriction map around the Duchenne musuclar dystrophy gene. Nature 324:582-585.
14. van Ommen, GJB, Verkerk, JMH, Hofker, MH, Monaco, AP, Kunkel, LM, Ray, P, Worton, R, Wieringa, B, Bakker, E and PL Pearson (1986). A physical map of 4 million bp around the Duchenne Muscular Dystrophy gene on the human X-chromosome. Cell 47:499-504.
15. Drumm, ML, Smith, Cl, Dean, M, Cole, JL, Iannuzzi, MC, and Collins, FS (1988). Physical mapping of the cystic fibrosis region by pulsed field gel electrophoresis. MS submitted.
16. Cooke, PR (1984). A general method for preparing intact nuclear DNA. EMBO J 3:1837-1842.
17. Gardiner-Garden, M and M Frommer (1987). CpG islands in vertebrate genomes. J Mol Biol 196:261-282.

18. Brown, WRA and AP Bird (1986). Long-range restriction site mapping of mammalian genomic DNA. Nature 322:477-481.
19. McClelland, M, Kessler, L and M Bittner (1984). Site-specific cleavage of DNA at 8- and 10-base-pair seqeunces. Proc Natl Acad Sci USA 81:983-987
20. McClelland, MN, and CR Cantor (1985). Purification of Mbo II methylase (GAAG^{m}A) from Moraxella bovis: site specific cleavage of DNA at nine and ten base pair sequences. Nucleic Acids Research 13:7171-7182.
21. Moser, HE and PB Dervan (1987). Sequence specific cleavage of double strand DNA by triple strand formation. Science 238:645-652.
22. Le Doan, T, Perrouault, L, Praseuth, D, Habhoub, N, Decout, J-L, Thuong, NT, Lhomme, J, Hélène, C (1987). Sequence-specific recognition, photocrosslinking and cleavage of the DNA double helix by an oligo-[α]-thymidylate covalently linked to an azidoproflavine derivative. Nucleic Acids Research 15:7749-7760.
23. Lyamichev, VI, Mirkin, SM, Frank-Kamenetskii, MD, and CR Cantor (1988). A stable complex formation between homopyrimidine oligomers and the homologous regions of duplex DNAs. Nucleic Acids Research, in press.
24. Carle, GF, Frank M, and Olsen, MV (1986). Electrophoretic separations of large DNA molecules by periodic inversion of the electric field. Science 232:65-68.
25. Chu, G, Vollrath, D, Davis, RW (1986). Separation of large DNA molecules by contour-clamped homogeneous electric fields. Science 234:1582-1585.
26. Cantor, CR, Warburton, PE, Smith, CL, and A Gaal (1986). Voltage ramp pulsed field gel electrophoresis separation of large DNA molecules. In Dunn, M (ed): "Electrophoresis '86," VCH Publishers, p.161.
27. Southern, EM, Anand, R, Brown, WRA and DS Fletcher (1987). A model for the separation of large DNA molecules by crossed field gel electrophoresis. Nucleic Acids Research 15:5925-5943.
28. Smith, HO and ML Birnstiel (1976). A simple method for DNA restriction site mapping. Nucleic Acids Research 3:2387-2399.
29. McCormick, M, Gottesman,ME, Gaitanaris, GA, Howard, BH (1987). Cosmid vector systems for genomic DNA cloning. In Gottesman, MM (ed): "Methods in Enzymology: Isolation and Detection of Mutant Genes," San Diego: Academic Press, p. 397.

30. Allshire, RC, Cranston, G, Gosden, JR, Maule, JC, Hastie, ND, Fantes, PA (1987). A fission yeast chromosome can replicate autonomously in mouse cells. Cell 50:391-403.
31. Smith, CL, and RD Kolodner (1988). Physical mapping of E. coli chromosomal Tn5 and F insertions by PFG electrophoresis. Genetics, in press.
32. Saffran, WA and CR Cantor (1988). Stimulation of plasmid integration by DNA damage. MS in preparation.
33. Van der Ploeg, LHT, Schwartz, DC, Cantor, CR and P Borst (1984). Antigenic variation in trypanosoma brucei analysed by electrophoretic separation of chromosome-sized DNA molecules. Cell 37:77-84.
34. Berman, JE, Mellis, SJ, Pollock, R, Smith, CL, Suh, H, Heinke, B, Kowal, C, Surti, U, Chess, L, Cantor, CR, Alt, FW (1988). Content and organization of the human IG V_H locus: Definition of three new V_H families and linkage to the IG C_H locus. EMBO J, in press.
35. Nakaseko, Y, Adachi, Y, Funahashi, S, Niwa, O, Yanagida, M (1986). Chromosome walking shows a highly homologous repetitive sequence present in all the centromere regions of fission yeast. EMBO J 5:1011-1021.
36. Grosveld, F, van Assendelft, GB, Greaves, DR, Kollias, G (1987). Position-independent high-level expression of the human β-globin gene in transgenic mice. Cell 51:975-985.
37. Thomas, KR, and MR Capecchi (1987). Site-directed mutagenesis by gene targeting in mouse embryo-dervied stem cells. Cell 51:503-512.

Gene Transfer and Gene Therapy, pages 283–292

MAPPING AROUND THE CYSTIC FIBROSIS LOCUS[1]

W.Bautsch (1), D.Grothues (1), G.Maass (1), J.Hundrieser (2), B.Kroll (3), E.Frömter (3), N.Ponelies (4), K.O.Greulich (4), A.Claass (5), and B.Tümmler (1)

Cystic Fibrosis Research Group, Zentrum Biochemie (1) and Zentrum Humangenetik (2), Medizinische Hochschule Hannover, D-3000 Hannover 61, Physiologisches Institut, Universität Frankfurt (3), Physikalisch-Chemisches Institut, Universität Heidelberg (4), Universitäts-kinderklinik Kiel (5), West Germany

ABSTRACT For the construction of a physical genome map around the cystic fibrosis (CF) locus single gene markers were isolated from metaphase 1p:7q2-qter chromosomes by laser-microdissection and subsequent microcloning. Probes were localized by hybridization to somatic cell hybrids and by long-range restriction mapping. Pedigree studies in 75 German CF families revealed three recombinants between the closely linked marker loci D7S8 and MET. No significant correlation between the clinical status of the CF patient and any of the MET and D7S8 haplotypes was observed suggesting that at least in the German population genetic microheterogeneity within the CF locus is unlikely. A case report illustrates the usefulness of RFLP typing for the diagnosis of CF in patients with borderline sweat tests. - The basic defect in CF leads to decreased chloride secretion and elevated sodium absorption of airway epithelia. mRNA from N and CF epithelial cells was injected into Xenopus laevis oocytes, and an amiloride-sensitive sodium channel typical for the apical epithelial membrane was found to be expressed.

---

[1]This work was supported by grants from the Deutsche Forschungsgemeinschaft, the Bundesministerium für Forschung und Technologie and the Deutsche Förderungsgesellschaft zur Mucoviscidoseforschung.

INTRODUCTION

Cystic fibrosis (CF) is the most common severe genetic disorder in the Caucasian population (1-3). This autosomal recessive disease affects all exocrine glands of the body. The major clinical features are the progressive lung disease, the exocrine pancreatic insufficiency, and hepatobiliary complications. The growing clinical experience in the treatment of CF has greatly improved prognosis. The majority of CF patients is now reaching adulthood.

The last few years brought considerable progress toward understanding the underlying basic defect in CF, mainly because of advances in two different areas: Electrophysiological studies showed decreased chloride permeability of CF exocrine epithelia (3-10). Secondly, the techniques of reverse genetics were successfully applied to localize the CF gene to a small part of chromosome 7 (11-14). Linkage studies in CF families with restriction fragment length polymorphisms (RFLPs) (11-13) and the subsequent hybridization of probes to somatic cell hybrids (11-15) allowed to assign the CF gene to the band 7q22-31. However, these methods of genetic analysis give a map resolution only to within several million base pairs. Hence, as a further step forward to the identification of the CF gene the physical genetic analysis of this chromosomal region was initiated using pulsed-field gradient electrophoresis and rare-cutter restriction mapping (14,16,17). In addition, further polymorphic markers were isolated that display strong linkage disequilibrium with the CF locus (14). Candidates for the CF gene closely linked to these RFLPs are currently being investigated, mainly by R.Williamson's group in London.

This paper reports on long-range restriction mapping from the region of the CF locus and on the generation of novel markers by laser-microdissection and microcloning. In addition results of a linkage study in CF families are presented, and the implications of RFLP typing for the postnatal diagnosis of CF are discussed. Finally, we outline our current approaches for bioassays on gene expression.

RESULTS AND DISCUSSION

RFLP Linkage Study in German Cystic Fibrosis Families

The loci MET (12) and D7S8 (11) display significant linkage disequilibrium with CF. We investigated the question if any of the MET and D7S8 haplotypes on CF chromosomes are associated with the severity of the clinical course. DNA typing was performed with the probes met-H, met-D (12) and pJ3.11 (11) in 75 German families with two and more children. The CF patients were classified into three groups of weakly, moderately, and severely affected patients according to the course of clinical (18) and X-ray scores (19), relative underweight, and lung function parameters. No significant correlation between the clinical course and any haplotype could be evaluated. This result suggests that at least in the German population genetic microheterogeneity within the CF locus is unlikely. Hence, the clinical heterogeneity of CF is rather a consequence of the different genetic and social background of patients than of different mutations within the CF gene itself.

Pedigree analysis revealed three recombinants between D7S8 and MET. Interestingly, a considerably larger number of gene carriers inherited the CF allele from his mother than from his father.

Diagnosis of Cystic Fibrosis by DNA Typing in Subjects with Borderline Sweat Tests

The quantitative pilocarpine iontophoresis sweat test (20) represents the most accurate physiological assay to diagnose cystic fibrosis. However, about 2% of tested individuals consistently display sweat chloride concentrations in the borderline range of 50 - 70 mEq/L. In this case the uncertainty of the diagnosis of CF can be overcome by DNA typing (Fig. 1). If in a family one child is known to be affected by CF, the status of all other siblings can be determined from the RFLP inheritance pattern provided that the family is informative for the used polymorphic marker. The validity of the linkage analysis is only limited by the recombination frequency between the marker loci and the CF locus. If borderline sweat test values are found in a person with no documented case of CF in his family, RFLP typing of the family will allow the exclusion of CF if the index case and at least one other sibling share the same pair of informative haplotypes. E.g., 25% of the typing analyses with two children will give an unequivocal result. For the remaining cases at least risk assessments are feasible by using RFLP markers that display strong linkage disequilibrium with the CF locus(14). In conclusion, in all subjects with borderline sweat test values cystic fibrosis should be definitely diagnosed or excluded only after the DNA typing of the family with RFLP markers has been performed.

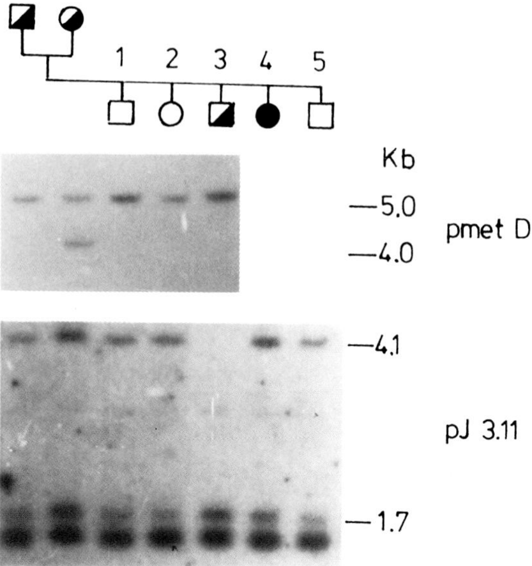

FIGURE 1. Segregation of the partially informative MET-D and fully informative D7S8 loci with CF in a family with borderline sweat electrolytes. Repetitive sweat tests revealed normal values for siblings 1, 2, 5 and borderline chloride concentrations in both male 3 (light lung symptoms) and female 4 (marked pulmonary disease). The three siblings 1, 2, 5 are typed homozygous N, girl 4 homozygous CF. The carrier male 3 inherited the paternal CF chromosome.

Laser - Microdissection and Microcloning

Cystic fibrosis still belongs to the human genetic diseases where neither the gene nor its gene product are known with certainty. Since this information is lacking, the unequivocal identification of the CF gene requires a complete physical map of the gene locus and its environment. We isolated single gene markers by microdissection (21) of the long arm of chromosome 7 and subsequent microcloning.

A laser microbeam apparatus (22) was employed for microdissection of metaphase chromosomes from a patient with a 1p:7q2-qter translocation. The apparatus used an excimer laser as the primary source of light which pumped a dye laser for tunability and improvement of beam quality. The 20 nsec pulses of the laser were directed into an inverted microscope and focused to power densities of more than 10^{14} W/cm² which

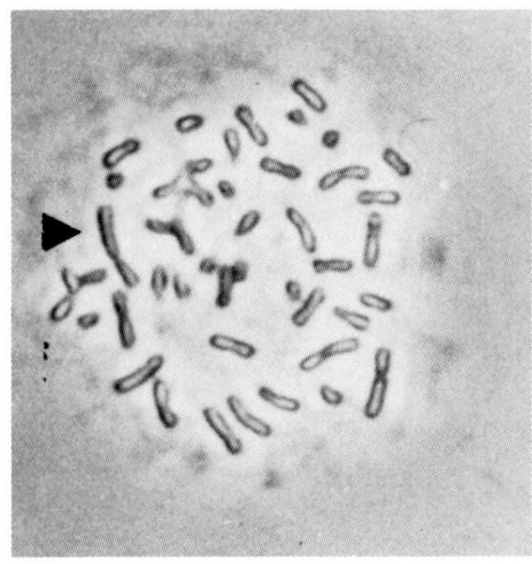
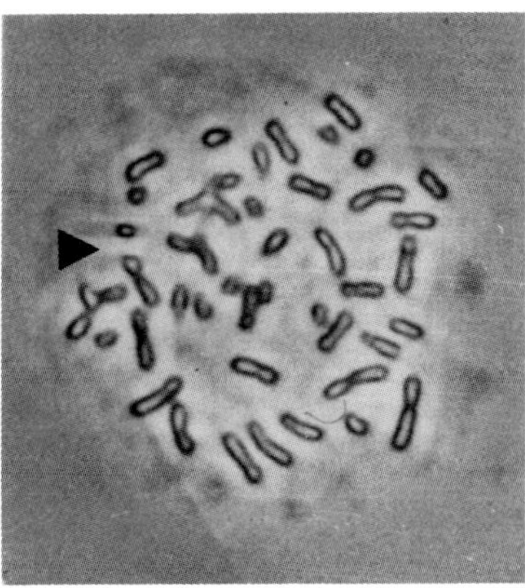

FIGURE 2. Metaphase chromosomes from a patient with a 1p:7q2-qter translocation prior (left) and after (right) laser-microdissection.

are sufficient to cut biological material. Single chromosome slices were prepared by combined use of the diffraction rings and the central disc of the laser microbeam (Figure 2). The slices were taken up by a micromanipulator and microcloned into the EcoRI site of λNM641 (23) in a few nl drop under oil.

The yield was about one λ clone per EcoRI digested chromosome slice. 24 out of 76 analyzed clones contained inserts from 1.5 to 5 kb in size, in the other clones the inserts were smaller than 0.5 kb. The significant proportion of small inserts suggests that the acidic pH during preparation of chromosomes is critical for the integrity of DNA. After the exclusion of clones containing repetitive DNA the remaining λ clones with large inserts were assayed for their chromosomal localization by hybridization to a panel of cell lines each containing various portions of chromosome 7 (15,24). Clones from the bands 7q2-3 are currently being subjected to long-range restriction mapping (see below) and to isolation of cosmids from a genome walking library. These cosmids will be assayed for presence of transcripts, RFLPs, and rare-cutter restriction sites.

Long - Range Restriction Mapping

Genomic DNA was digested with rare-cutting restriction endonucleases, separated by field inversion gel electrophoresis (FIGE) (25), and probed with various RFLP markers, microdissection, jumping, and walking clones from the region of the

CF locus. As an example Figure 3 shows the FIGE analysis for the probe met-H. The restriction map roughly corresponds with the data reported by Collins et al. (16) and Michiels et al. (17), however, an extra MluI site was detected, and the BSSH II and Sfi I fragments were 15-20 % smaller than described previously (16).

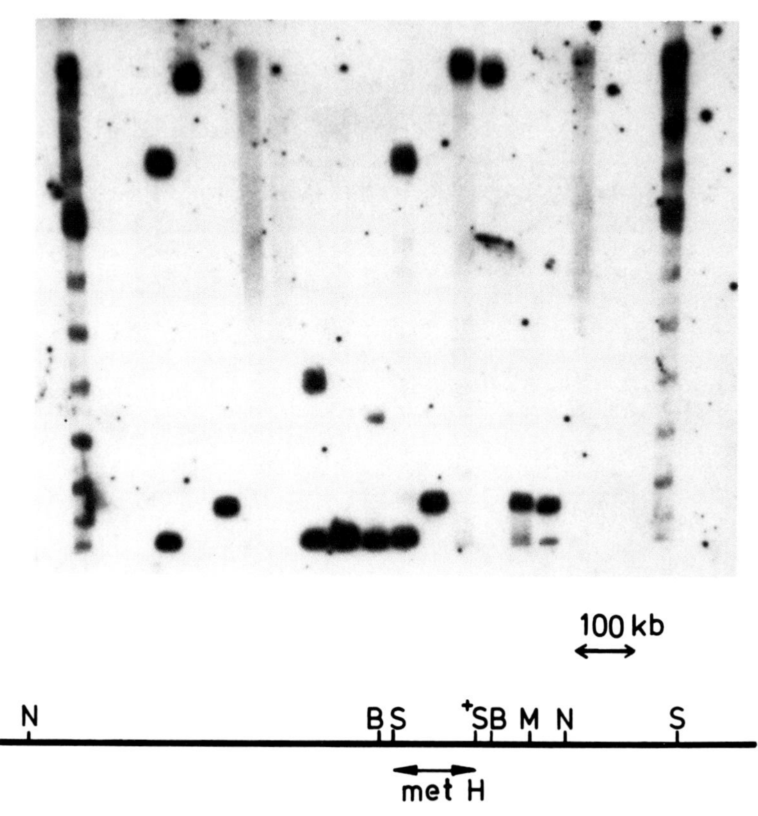

FIGURE 3. FIGE analysis of met-H. From left to right: λ oligomers starting with the trimer at the bottom (lane 1), DNA of human leucocytes digested with SfiI (lane 4), NotI (lane 5), BSSHII (lane 6), MluI (lane 7), SfiI + NotI (lane 9), Sfi I + BSSHII (lane 10), SfiI + MluI (lane 11), SfiI (lane 12), NotI + BSSHII (lane 13), NotI + MluI (lane 14), NotI (lane 15), BSSHII + MluI (lane 16), BSSHII (lane 17), MluI (lane 18), λ oligomers (lane 21). In the restriction map B is BSSHII; M, MluI; N, NotI; S, SfiI. The asterisk indicates an SfiI polymorphism.

Assays on Gene Expression and Gene Complementation

The basic defect in CF leads to abnormal fluid and ion transport of exocrine glands. The CF epithelia from airways and sweat gland are impermeable for chloride and fail to secrete chloride when stimulated with isoprotenerol (4-8), In the CF cell the catalytic subunit of the cAMP-dependent protein kinase fails to activate the Cl^--channels of the apical membrane (9). Na^+-absorption is elevated in CF respiratory epithelia (26).

We focus our interest on nasal epithelial cells which have been demonstrated to express the defect of Cl^--impermeability (5,26). The apical membrane contains amiloride-sensitive Na^+-channels and 40-50 pS Cl^--channels (E.Frömter, unpublished).

Cell culture. Epithelial cells from N and CF nasal polyps were grown in hormone-supplemented Ham's F12 medium with and without 1% fetal calf serum (27). Collagen, cationic polymers, and feeder layers of irradiated fibroblasts were found to be suitable supports. The primary cultures consisted to more than 95% of cytokeratin-positive cells indicating that indeed epithelial cells were cultured and that the outgrowth of fibroblasts was efficiently suppressed. mRNA was prepared from N and CF cells for Northern blot analysis, construction of representative cDNA libraries in the vector λgt10, and for *in vivo* translation assays.

Gene expression in Xenopus laevis oocytes. The microinjection of mRNA or cDNA clones from N and CF exocrine epithelial cells in Xenopus oocytes offers the opportunity to measure the function of ion channels by electrophysiological means provided that the injected genetic information is properly translated and assembled into a functioning gene product. In principle the amphibian egg represents an ideal test tube for assays on gene complementation and candidate genes.

We investigated the expression of human Na^+- and Cl^--channels in Xenopus oocytes after microinjection of mRNA from nasal epithelial cells. Previous microelectrode measurements and patch-clamp experiments had revealed that the oocyte contains endogeneous sodium and chloride channels. However, in contrast to the human channels the endogeneous Na^+-channel is insensitive to amiloride, and the Cl^--channel has a low conductivity of 3-5 pS and is regulated by Ca^{2+}, but not by cAMP (28).

Figure 4 demonstrates the expression of a human amiloride-sensitive sodium channel in Xenopus oocytes. For cloning the corresponding gene(s), size-selected mRNA preparations, and, later on, size-selected cDNA libraries will be tested

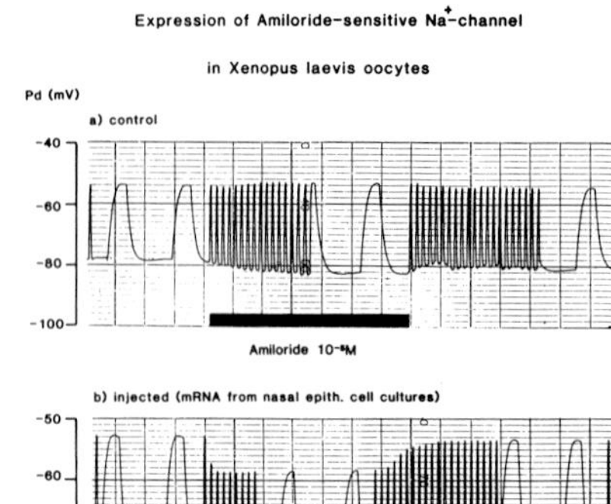

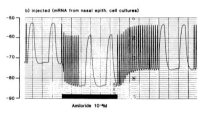

FIGURE 4. Original microelectrode recordings of membrane potential three days after injection of 50 ng mRNA from human nasal epithelial cells (below) or buffer (above) into Xenopus laevis oocytes. Oocytes were incubated in modified Barth's solution. 0.1 Hz pulses of 5 nA were applied for 3 sec.

for the expression of the channel in oocytes. Since according to our current knowledge the Cl^--channel and its membrane-associated regulatory proteins are the probable site of the CF defect (9), this approach of electrophysiological testing of size-selected cDNA libraries in the oocyte should allow to clone the normal counterpart of the CF gene. However, if for any reason the cAMP-dependent protein is not expressed in the oocyte in sufficient amounts, transfection and complementation assays of CF candidate genes will have to be performed in epithelial cells. Immortalized N and CF cell lines of defined electrophysiological properties have meanwhile become available.

ACKNOWLEDGMENTS

We cordially thank CF patients, their families, and our clinical colleagues for continuous support and encouragement. The experimental help by S.Monajembashi, Heidelberg; S.Clauss, Kiel; J.Disser, S.Heintz, Frankfurt; A.Aschendorf, S.Bautsch, S.Bremer, C.Bürger, T.Darnedde, K.Fryburg, T.Heuer, V.Neubauer, M.Wehsling, M.Wilke, B.Wulff, all Hannover; is gratefully acknowledged. K.Miller, Hannover, prepared metaphase chromosomes for microdissection. S.Jakubiczka, J.Schmidtke, Göttingen; F.Michiels, Gent; and G. vande Woude, NCI; kindly donated RFLP probes, walking cosmids, and jumping clones. K.H.Grzeschik, Marburg, generously supplied DNA from cell hybrids and deletion cell lines. We thank J.Bijman, B.J.Scholte, Rotterdam, and G.Rechkemmer, Hannover, for valuable discussions and support in cell culturing and electrophysiological measurements. D.Grothues is a fellow of the 'Mucoviscidose-Hilfe'.

REFERENCES

1. Wood RE, Boat TF, Doershuk CF (1976). State of the art: Cystic fibrosis. Amer Rev Respir Dis 113: 833.
2. Talamo RC, Rosenstein BJ, Berninger RW (1983). Cystic fibrosis. In Stanbury JB, Wyngaarden JB, Frederickson DS, Goldstein JL, Brown MS (eds): "The metabolic basis of inherited disease," New York: McGraw Hill, p. 1889.
3. Welsh MJ, Fick RB (1987). Cystic fibrosis. J Clin Invest 80: 1523.
4. Quinton PM (1983). Chloride impermeability in cystic fibrosis. Nature 301: 421.
5. Knowles M, Gatzy J, Boucher R (1983). Relative ion permeability of normal and cystic fibrosis nasal epithelium. J Clin Invest 71: 1410.
6. Frizzell RA, Rechkemmer G, Shoemaker RL (1986). Altered regulation of airway epithelial cell chloride channels in cystic fibrosis. Science 233: 558.
7. Welsh MJ, Liedtke CM (1986). Chloride and potassium channels in cystic fibrosis airway epithelia. Nature 322: 467.
8. Cotton CU, Stutts MJ, Knowles MR, Gatzy JT, Boucher RC (1987). Abnormal apical cell membrane in cystic fibrosis respiratory epithelium. J Clin Invest 79: 80.
9. Schoumacher RA, Shoemaker RL, Halm DR, Tallant EA, Wallace RW, Frizzell RA (1987). Phosphorylation fails to activate chloride channels from cystic fibrosis airway cells. Nature 330: 752.
10. Frizzell RA (1987). Cystic fibrosis: a disease of ion channels? TINS 10: 190.
11. Wainwright BJ, Scambler PJ, Schmidtke J, Watson EA, Law HY, Farrall M, Cooke HJ, Eiberg H, Williamson R (1985). Localization of cystic fibrosis locus to human chromosome 7cen-q22. Nature 318: 384.
12. White R, Woodward S, Leppert M, O'Connell P, Hoff M, Herbst J, Lalouel JM, Dean M, Vande Woude G (1985). A closely linked genetic marker to cystic fibrosis. Nature 318: 382.
13. Knowlton RG, Cohen-Haguenauer O, Nguyen VC, Frezal J, Brown V, Barker D, Bramann JC, Schumm JW, Tsui LC, Buchwald M, Donis-Keller H (1985). A polymorphic DNA marker linked to cystic fibrosis is located on chromosome 7. Nature 318: 380.
14. Estivill X, Farrall M, Scambler PJ, Bell GM, Hawley KMF, Lench NJ, Bates GP, Kruyer HC, Frederick PA, Stanier P, Watson EK, Williamson R, Wainwright BJ (1987). A candidate for the cystic fibrosis locus isolated by selection for methylation-free islands. Nature 326: 840.

15. Zengerling S, Olek K, Tsui LC, Grzeschik KH, Riordan JR, Buchwald M (1987). Mapping of DNA markers linked to the cystic fibrosis locus on the long arm of chromosome 7. Am J Hum Genet 40: 228.
16. Collins FS, Drumm ML, Cole JL, Lockwood WK, Vande Woude GF, Iannuzzi MC (1987). Construction of a general chromosome jumping library, with application to cystic fibrosis. Science 235: 1046.
17. Michiels F, Burmeister M, Lehrach H (1987). Derivation of clones close to met by preparative field inversion gel electrophoresis. Science 236: 1035.
18. Kraemer RH, Tschäppeler H, Rüdeberg A, Stoll E, Rossi E (1979). Verlauf und quantitative Erfassung des pulmonalen Befalls bei der zystischen Fibrose. Schweiz med Wochenschr 109: 39.
19. Chrispin AR, Norman AP (1974). The systematic evaluation of the chest radiograph in cystic fibrosis. Pediatr Radiol 2: 101.
20. Gibson LE, Cooke RE (1959). A test for concentration of electrolytes in sweat in cystic fibrosis of the pancreas utilizing pilocarpine by iontophoresis. Pediatrics 23: 545.
21. Röhme D, Fox H, Herrmann B, Frischauf AM, Edström JE, Mains P, Silver LM, Lehrach H (1984). Molecular clones of the mouse t complex derived from microdissected metaphase chromosomes. Cell 36: 783.
22. Monajembashi S, Cremer C, Cremer T,Wolfrum J, Greulich KO (1986). Microdissection of human chromosomes by a Laser microbeam. Exp Cell Res 167: 262.
23. Murray NE, Brammar WJ, Murray K (1977). Lambdoid phages that simplify the recovery of in vitro recombinants. Mol Gen Genet 150: 53.
24. Bartels J, Grzeschik KH, Cooper DN, Schmidtke J (1986). Regional mapping of six cloned DNA sequences on human chromosome 7. Am J Hum Genet 38: 280.
25. Carle GF, Frank M, Olson MV (1986). Electrophoretic separation of large DNA molecules by periodic inversion of the electric field. Science 232: 65.
26. Boucher RC, Stutts MJ, Knowles MR, Cantley L, Gatzy JT (1986). Na^+-Transport in cystic fibrosis respiratory epithelia. J Clin Invest 78: 1245.
27. Yankaskas JR, Cotton CU, Knowles MR, Gatzy JT, Boucher RC (1985). Culture of human nasal epithelial cells on collagen matrix supports. Am Rev Respir Dis 132: 1281.
28. Takahashi T, Neher E, Sakmann B (1987). Rat brain serotonin receptors in Xenopus oocytes are coupled by intracellular calcium to endogeneous channels. Proc Natl Acad Sci (USA) 84: 5063.

Gene Transfer and Gene Therapy, pages 293–306

IDENTIFICATION OF MUTATIONS PRODUCING HEMOPHILIA A[1]

H.H. Kazazian, Jr., C. Wong, H. Youssoufian[2],
P. Woods Samuels, D.G. Phillips,
A.F. Scott, S.E. Antonarakis

Genetics Unit, Department of Pediatrics,
The Johns Hopkins University School of Medicine,
Baltimore, Maryland 21205 USA

ABSTRACT The cloning of the factor VIII gene has provided a great impetus to the molecular analysis of hemophilia A. The large number of deletions and point mutations characterized to date have produced further basic insights into the variety and nature of mutations. In particular, the first instances of mutation via retrotransposition have been observed in the factor VIII gene. In addition, the knowledge gained from these studies has improved the accuracy of prenatal diagnosis and carrier detection in hemophilia. Future studies will include the identification of further mutations and polymorphisms, the correlation of mutations with the severity of disease and inhibitor production, a complete characterization of _de novo_ mutations and their parental origins, and the use of heterologous cell systems to study abnormalities of RNA processing.

INTRODUCTION

Hemophilias are relatively common inherited disorders of blood coagulation due to deficiency of two different clotting factors VIII and IX. Hemophilia A, or classic hemophilia, is associated with abnormality of factor VIII and affects about 1 in every 10,000 males; hemophilia B or Christmas disease is associated with an abnormality of factor IX and affects about

---

[1]This work was supported by NIH, MOD, JHU Institutional grants.

[2]Present address: Hematology-Oncology Unit, Cox 6, Massachusetts General Hospital Boston, MA 02114

1 in every 50,000 males (1). Both factors are involved in intermediate steps of the intrinsic clotting cascade, which consists of several inactive proteases and cofactors that are serially activated in response to an initial stimulus. The end product of this cascade is the production of an insoluble protein, fibrin, from its soluble precursor, fibrinogen. Fibrin then forms a filamentous network and stabilizes the platelet plug. Factor VIII in its "activated" form, VIII:Ca, then serves as a cofactor in the activation of factor X (2).

Both hemophilias (A and B) exhibit similar phenotypes (clinical pictures) characterized mainly by prolonged bleeding after minor trauma. Males with the disease have frequent hematomas of soft tissue and joints. The differential diagnosis depends on laboratory clotting tests which distinguish between these two defects (3). Both disorders are inherited as X chromosome-linked conditions with males affected and females carriers of the abnormal genes. The disease is treated with factor VIII concentrates which are given at the first sign of bleeding. Because these Factor VIII concentrates were contaminated with the HTLV-III virus between 1980 and 1985, a high fraction (approaching 90%) of affected males over age 6 are now seropositive for the AIDS virus.

The cloning and characterization of factor VIII gene, the description of DNA polymorphism markers associated with this gene, and recent advances in the molecular pathology of hemophilia A have provided a better understanding of the pathogenesis of this disorder and facilitated the detection of carriers and the prenatal diagnosis.

In 1984, researchers in two biotechnology companies (Genentech, Inc. and Genetics Institute, Inc.) reported the cloning of human factor VIII gene (4-6). Both groups used synthetic oligonucleotides as probes to clone and characterize the factor VIII gene from genomic and cDNA libraries. The oligonucleotide sequences used were deduced from small sequenced peptides of the human and porcine factor VIII (7,8). The entire gene spans 186 kb (kilobases) of DNA or to a first approximation about 0.1% of the human X chromosome. It is divided into 26 exons and 25 introns (Figure 1). The coding DNA (exon length) is 9kb, which codes for 2351 amino acids. The complete nucleotide sequence of the coding regions, the promoter elements, and the intron-exon boundaries has been determined, and the amino acid sequence of the protein has been deduced (4-6).

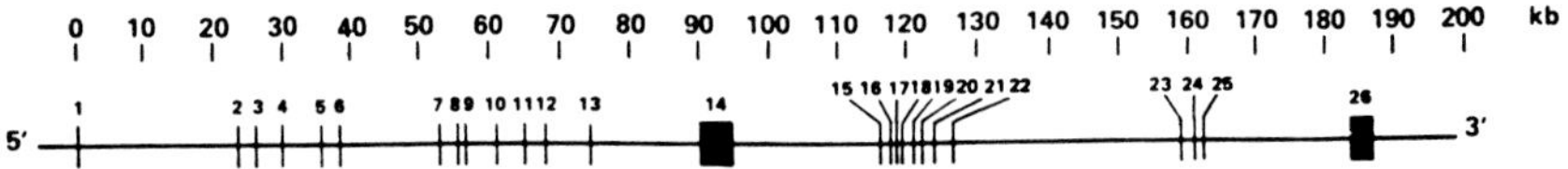

FIGURE 1. Schematic representation of the factor VIII gene. Kilobase number from the first exon of the gene is shown on top. Each exon is represented by a vertical line or filled box. Exons are numbered from 1 to 26.

The first 19 amino acids of the protein sequence comprise the secretory leader peptide of the precursor FVIII and, therefore, the mature excreted polypeptide consists of 2332 amino acids. A striking feature of the FVIII protein is that it contains domains of internal homology (7). Computer analysis of the FVIII protein sequence revealed that there are three homologous sequences (A domain) found at amino acid positions 1-329, 380-711, and 1649-2019 of the mature polypeptide. The A domains have ~30% amino acid homology. The second and third A domains are separated by the B domain of 983 amino acids which is extremely rich in potential asparagine-linked glycosylation sites. After the third A domain there are two C domains of 150 amino acids with ~40% homology to each other. Most of the 23 cysteine residues of the mature FVIII are located in the A and C domains. The mature polypeptide has the structure A_1-A_2-B-A_3-C_1-C_2 from the amino terminus to the carboxy terminus. It is of interest that the A domains show striking homology (about 30%) with the three domains of the copper binding protein ceruloplasmin, and that the B domain is encoded by the unusually long (3106 nt) exon 14.

Factor VIII circulates in plasma in conjunction with von Willebrand factor, a large polymer of a polypeptide encoded by an autosomal gene in human chromosome 12 (9,10). Factor VIII isolated from plasma is usually degraded because it undergoes proteolytic cleavages during its activation (11). It is thought that the "activated" form of factor VIII is a 90,000 Da polypeptide composed of the first two A domains (N-terminus) and an 80,000 Da polypeptide which contains the third A and both C domains (C-terminus). These polypeptides are the result of thrombin cleavage of the whole factor VIII molecule. Furthermore, the large B domain is cleaved off during the activation of factor VIII, and it appears to have no role in procoagulation (7,12).

DNA POLYMORPHISMS IN THE FACTOR VIII GENE

Since each family with hemophilia A usually has a different mutation in the factor VIII gene, it is almost impossible to detect directly the molecular defect using restriction endonuclease analysis and, therefore, provide accurate carrier detection and prenatal diagnosis. On the other hand, indirect detection of the abnormal factor VIII gene (regardless of the nature of the abnormality) can be achieved using as markers DNA polymorphisms within or adjacent to the factor VIII gene. In the past three years there has been considerable effort to identify DNA polymorphisms within the factor VIII gene, but the yield has been relatively poor. The following high frequency polymorphic sites within the factor VIII gene have been identified to date and shown in Figure 2.

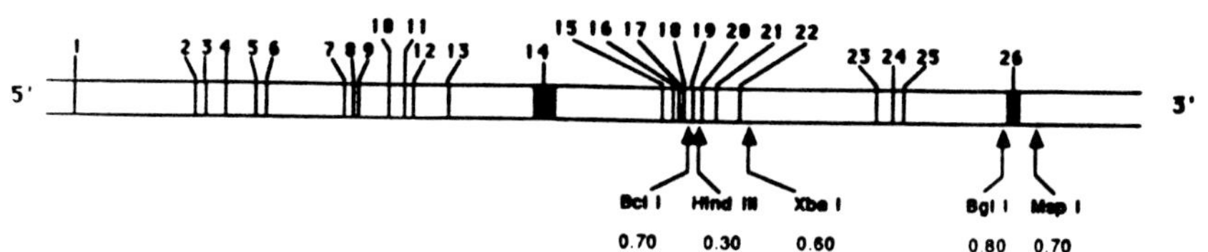

FIGURE 2. DNA polymorphisms within the factor VIII gene. A Bcl I site in IVS-18, Hind III site in IVS-19, Xba I site in IVS-22, Bgl I site at the 3′ end of IVS-25, and an Msp I site 3′ to exon 26 are depicted. The Hind III and Msp I sites are in nearly complete linkage disequilibrium with the Bcl I site and do not add new information over that obtained from the Bcl I site for prenatal diagnosis and carrier detection.

1. Bcl I site within intron 18 of the factor VIII gene comprising a two-allele system (13). This site is an excellent marker for carrier detection and prenatal diagnosis (14-17) because about 50% of Caucasian women are heterozygotes for this polymorphic site.

2. Bgl I site in intron 25 just 2 kb 5′ of exon 26 of the factor VIII gene, comprising a two allele system. The frequency of the presence of the Bgl I polymorphic site is about 90% in Caucasians and 75% in American Blacks (14). It is, therefore, a relatively good marker for carrier detection and prenatal diagnosis in the American Black population.

There is a low level of linkage disequilibrium between the Bcl I and Bgl I sites and, therefore, combined use of both sites does not dramatically increase the yield of informative pedigrees (heterozygous females for at least one marker) (18).

3. Xba I site in intron 22 of the factor VIII gene. The frequency of the presence of this polymorphic site in a mixed population from San Francisco was about 60% (19). In addition, there is some but not a high degree of linkage disequilibrium with the Bcl I polymorphic site and, therefore, about 25% of females who are homozygous for the Bcl I site are heterozygous for the Xba I polymorphic site.

If one uses these polymorphic sites as markers for the normal and abnormal factor VIII alleles in families, the error in carrier detection and/or prenatal diagnosis is negligible because the recombination rate between a given marker and the actual site of mutation will be extremely low.

Other extragenic DNA polymorphic markers have been described which are tightly linked with the factor VIII gene and the hemophilia A phenotype. These markers are:

1. DNA polymorphisms Taq I and Msp I associated with probe ST14 or DXS52. This is an extremely useful polymorphic system in which several alleles from the same locus can be recognized (20,21). There is about 3-5% recombinational distance between DXS52 and factor VIII (22,23) and, therefore, an unavoidable 3-5% error in the DNA diagnosis of the presence or absence of the mutant allele in a given family.

2. Bgl II site adjacent to DNA probe DX13 or DXS15. This latter probe also maps to the same chromosomal band as DXS52 and factor VIII and is tightly linked with factor VIII (24). However, several recombinants have been recognized between these two markers, and the recombination distance is on the order of 3-5%.

MUTATIONS IN THE FACTOR VIII GENE IN HEMOPHILIA A

Hemophilia A patients can be divided according to clinical severity into mild, moderate, and severe, and this clinical classification relates closely to factor VIII clotting activity. In the great majority of patients, the biological activity of factor VIII closely parallels the amount of protein in plasma as measured by an immunological method (25). In addition, about 6-12% of patients with hemophilia A develop antibodies against factor VIII (inhibitor patients) after therapy with exogenous factor VIII

(26,27). Inhibitors develop almost exclusively in patients that have no detectable factor VIII in their plasma.

In the past three years we have examined the DNA of 250 different patients with hemophilia A and found molecular defects in 29 patients. Deletions, single nucleotide changes, and, more recently, insertions in the factor VIII gene were found as causes of hemophilia A. These mutations have provided new insights into the pathogenesis of hemophilia A, including the existence of "hotspots" for mutation and further evidence that a considerable number of mutations occur _de_ novo. In addition, they have provided evidence for a new mechanism of mutation in man, i.e., retrotransposition of L1 or LINE sequences.

Deletions of the factor VIII gene

We have observed 14 different deletions within the factor VIII gene (Figure 3) (14,28,29). The ends of two of these deletions were cloned and sequenced. In neither case did the deletion involve an Alu element or unequal crossing over within homologous sequences.

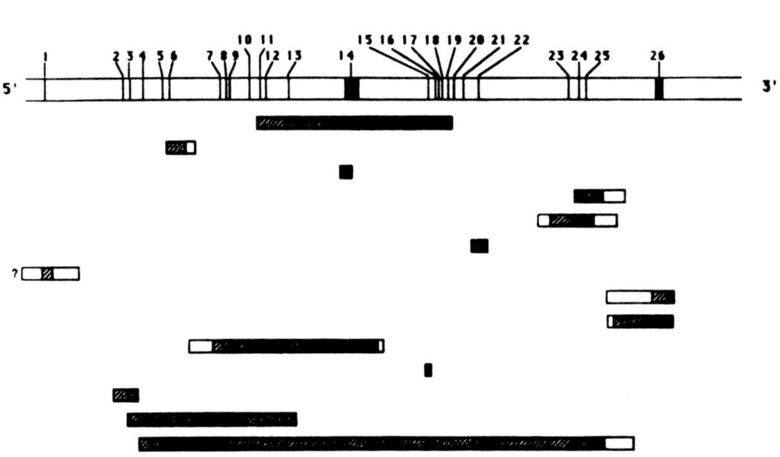

FIGURE 3. Deletions of the factor VIII gene in Hemophilia A. The factor VIII gene is shown on top. Each horizontal bar represents a different deletion within the factor VIII gene. Open bars at the ends of some deletions denote the uncertainty of the extent of the deletion. A question mark denotes that the extent of the deletion is unknown.

All but one of the partial factor VIII gene deletions described to date are associated with severe hemophilia A. Two of 14 deletions were associated with the presence of antibodies against factor VIII. No definitive conclusions can be drawn concerning the association of inhibitor formation and the size or position of the deletion, but it appears that there is little, if any, correlation between partial gene deletions and the presence of inhibitors (29). Finally, it is of interest that 14/250 patients (6%) with hemophilia A examined have sizable deletions within the factor VIII gene which can be recognized by simple restriction analysis.

Single nucleotide mutations within the factor VIII gene

Although the gene for factor VIII is very large and it seemed unlikely that single nucleotide changes would be identified using restriction analysis, a large number of point mutations have actually been identified by us and others (Figure 4) (14,30-35).

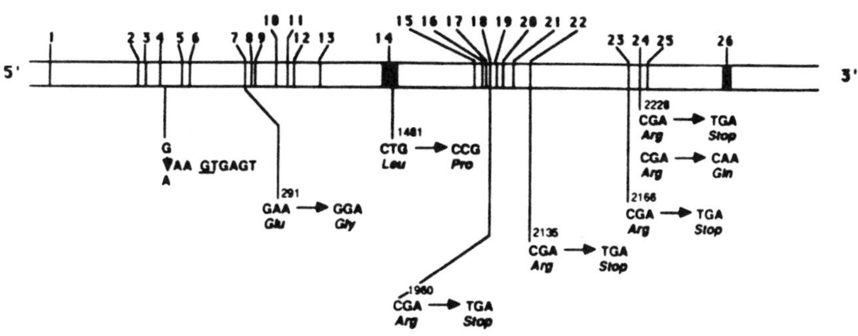

FIGURE 4. Single nucleotide mutations within the factor VIII gene identified at Johns Hopkins. The nucleotide changes and the amino acid changes are shown. Arg: Arginine, Glu: glutamine, stop: nonsense codon.

Screening the factor VIII gene using restriction analysis with Taq I has turned out to be fruitful in discovering point mutations, many of which are CpG-TpG substitutions. Ten different substitutions of this type were discovered among the 250 patients at 5 Taq I sites in exons. Nine of these mutations were CpG-TpG substitutions and the tenth was not in the CpG dinucleotide. CpG dinucleotides are thought to be "hotspots" for mutations because C can be methylated at the 5′ position of the pyrimidine ring and subsequently deaminated spontaneously to thymine. This accounts for CG-TG and CG-CA mutations. That CpG dinucleotides are "hotspots" for mutations is also supported by the fact that a considerable number of DNA polymorphic sites are in restriction endonuclease recognition sequences which contain CpG (Taq I and Msp I for example) (36). The factor VIII gene contains 7 Taq I sites in its coding region, 5 of which have CGA as a codon for arginine. Mutations have been observed in all of these five Taq I sites (exons 18,22,23,24,26).

It is of interest that examples of recurrent mutation have been observed for five different mutations among the first 500 defective factor VIII genes examined, and one can calculate that up to a few thousand recurrences of these same exact mutations may have occurred in man in the last 2000 years. The estimated mutation rate of C-T in CG dinucleotides in the factor VIII gene is calculated to be 10-20 times more than the average mutation rate for a single nucleotide. It appears that, on average, a mutation in each of the five Taq I sites in exons is observed in every 170 hemophilia A patients examined.

Insertion of L1 sequences

L1 sequences are a human-specific family of long interspersed repetitive elements present in about 10^5 copies dispersed throughout the genome (37). The full length L1 sequence is 6.1 kb, but the majority of L1 elements are truncated at the 5′ end resulting in a five-fold higher copy number of 3′ sequences (37). The nucleotide sequence of L1 elements includes an A-rich 3′ end and two long open reading frames (ORF-1 and ORF-2), the second of which encodes a potential polypeptide with homology to reverse transcriptases (37-40). This structure suggests that L1 elements represent a class of non-viral retrotransposons (37,38). A number of L1 cDNAs, including a nearly full-length element, have been isolated from an undifferentiated teratocarcinoma cell line (41). We have found insertions of L1 elements into exon 14 of the factor VIII gene in two of 240 unrelated patients with hemophilia A (42). Both of these insertions (3.8 kb and 2.3 kb) contained 3′ portions of the L1 sequence, including the poly A tract, and created target site duplications of at least 12 and 13 nucleotides of the factor VIII gene (figure 5). In addition, their 3′-trailer sequences following ORF-2 are nearly identical to the consensus sequence of L1 cDNAs (6) (figure 6). The data indicate that in man certain L1 sequences can be dispersed presumably via an RNA intermediate and cause disease by insertional mutation.

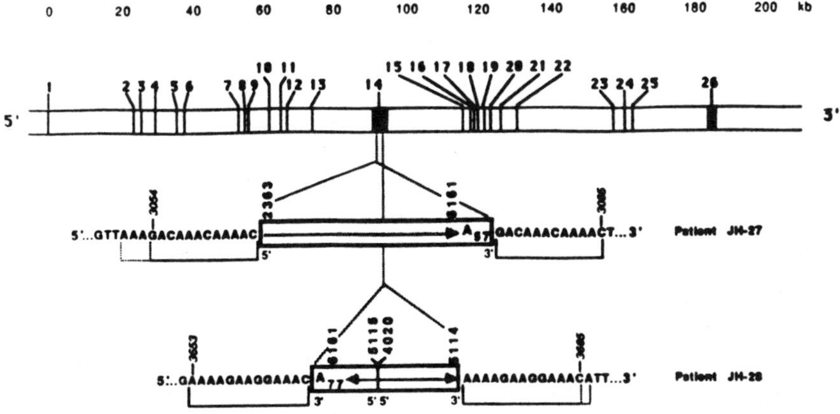

FIGURE 5. Diagram of L1 insertions in exon 14 of the factor VIII gene. The 3.8 kb L1 insertion from patient JH-27 is flanked by a 12 bp target site duplication of factor VIII cDNA sequence (nucleotides 3054-3065 where nucleotide 1 is the A of the initiator codon). Residues 3051-3053 of the factor VIII cDNA are adenylic acids and could also be duplicated (shown by the hatched bracket). The rearranged 2.3 kb insertion from patient JH-28 is shown and is flanked by a 13 bp target site duplication. Residue 3666 of the factor VIII cDNA is an A and could also be duplicated. Filled boxes represent the L1 elements, and the arrows within the boxes point toward the 3′ end of the L1 sequence. Both L1 insertions showed 98% similarity to the consensus genomic L1 sequence outside of the 3′-trailer region.

```
L1 GENOMIC          ...AGGAAGGGGAACATCACACACTGGGGCCTGTTGTGGGGTGGGGGGNGGGGGGAGGGATAGCA 6032
L1 cDNA                            T       T      A   G       C      A
JH-27 Insert                       T       T      A   G       C      A
JH-28 Insert                       T       T      A   G       C      A

                       TTAGGAGATATACCTAATGCTAAATGACGAGTTAATGGGTGCAGCACACCAACATGGCACAT 6092
                         G                    G    ACA  G    O   G        G
                         G                    G    ACA  G    G   G        G
                         G                    G    ACG  G    G   G        G

                       GTATACATATGTAACAAACCTGCACGTTGTGCACATGTACCCTAGAACTTAAAGTATAATAA..6152
                                      T         AA                  A     A
                                      T         AA                  A     G
                                      T         AA                  A     G
```

FIGURE 6. Comparison of 3′ trailer region sequence (186 nt) of the L1 element from genomic consensus (30) (top line), cDNA consensus (43) (middle line) and FVIII L1 insertion sequences of JH-27 and JH-28. The cDNA consensus shown is that of subset Ta (43) and the additional four nucleotides which differ from the L1 cDNA and genomic consensus sequences are underlined.

Insertion of L1 elements, involving retrotransposition of DNA sequences through an RNA intermediate into a new and distant location in the genome, represents a fundamentally different mechanism of mutation producing human disease from those previously described. Because we do not know when these L1 insertion events occur, whether in the sperm or ovum, after fertilization, or during early stages of embryogenesis, the proportion of such insertions that are heritable is unknown. Yet finding two L1 insertions among 240 patients with hemophilia A suggests that this mechanism of mutation is not uncommon.

HEMOPHILIA A: POTENTIAL FOR GENE THERAPY

Hemophilia A represents a good potential candidate for gene therapy. Present therapy of the disease is imperfect; in general, the physician waits for the hemorrhagic event prior to administration of factor VIII. All the long-term effects of the disorder, including crippling, could be prevented by continuous expression of a transferred factor VIII gene. Among the questions to be answered are: Can the gene be expressed appropriately in lymphocytes after transfer into hemopoietic stem cells? Must it be transferred into hepatocytes, its normal site of expression? What regulatory sequences are necessary for appropriate expression? Does the large cDNA size (9 kb) pose a problem for effective gene therapy? Gene therapy studies are facilitated by availability of an animal model, a dog colony with hemophilia A.

ACKNOWLEDGEMENTS

The authors thank Drs. R Lawn, J. Gitschier, I. Peake, L.W. Hoyer, for sharing unpublished data, helpful discussions and critical comments. We also thank Drs. J. Toole, J. Wozney, S. Aronis, G. Tsiftis, G. Stamatoyannopoulos, D. Fass, G. Bowie, and many physicians and genetic counselors for their help in the collection of the data. We also thank Ms. Emily Pasterfield for expert secretarial assistance.

REFERENCES

1. McKee PA (1983). Haemostasis and disorders of blood coagulation. In Stanbury JB, Wyngaarden JB, Fredrickson DS, Goldstein JL, Brown MS (eds): "The Metabolic Basis of Inherited Disease," 5th edition, New York: McGraw-Hill, p. 1531.
2. Jackson CM, Nemerson Y (1980). Blood coagulation. Ann Rev Biochem 49:765.
3. Ratnoff OD (1960). "Bleeding Syndromes: A Clinical Manual." Springfield: CC Thomas Publications.
4. Gitschier J, Wood WI, Goralka TM, Wion KL, Chen EY, Eaton DH, Vehar GA, Capon DJ, Lawn RM (1984). Characterization of the human factor VIII gene. Nature 312:326.
5. Toole JJ, Knopf JL, Wozney JM, Sultzman LA, Buecher JL, Pittman DD, Kaufman RJ, Brown E, Shoemaker C, Orr EC, Amphlett GW, Foster WB, Coe ML, Knudson GJ, Fass DN, Hewick RM (1984). Molecular cloning of a cDNA encoding human anti-haemophilic factor. Nature 312:342.
6. Wood WI, Capon DJ, Simonsen CC, Eaton DL, Gitschier J, Keyt B, Seeburg PH, Smith DH, Hollingshead P, Wion KL, Delwart E, Tuddenham EGD, Vehar GA, Lawn RM (1984). Expression of active human factor VIII from recombinant DNA clones. Nature 312:330.
7. Vehar GA, Keyt B, Eaton D, Rodriguez H, O'Brian DP, Rotblat F, Oppermann H, Keck R, Wood WI, Harkins RN, Tuddenham EGD, Lawn RM, Capon DJ (1984). Structure of human factor VIII. Nature 312:337.
8. Fass DN, Knutson GJ, Katzmann JA (1982). Monoclonal antibodies to porcine factor VIII:C and their use in the isolation of active coagulant protein. Blood 59:594.

9. Hoyer LW (1981). The factor VIII complex: Structure and function. Blood 58:1.
10. Ginsburg D, Handin RI, Bonthron DT, Donlon TA, Bruns GA, Latt SA, Orkin SH (1985). Human Von Willebrand factor (VWF): Isolation of cDNA clones and chromosomal localization. Science 228:1401.
11. Rotblat F, O'Brien DP, O'Brien FJ, Goodall AH, Tuddenham EG (1985). Purification of human factor VIII:C and its characterization by western blotting using monoclonal antibodies, Biochem 24:4294.
12. Toole JJ, Pittman DD, Orr EC, Mortha P, Wasley LC, Kaufman RJ (1986). A large region (95 KDa) of human factor VIII is dispensable for in vitro procoagulant activity. Proc Natl Acad Sci USA 83:5939.
13. Gitschier J, Drayna D, Tuddenham EGD, White RI, Lawn RM (1985). Genetic mapping and diagnosis of haemophilia A achieved through a Bcl I polymorphism in the factor VIII gene. Nature 314:738.
14. Antonarakis SE, Waber PG, Kittur SD, Patel AS, Kazazian HH, Jr., Mellis MA, Counts RB, Stamatoyannopoulos G, Bowie EJW, Fass DN, Pittman DD, Wozney JM, Toole JJ (1985). Hemophilia A: Molecular defects and carrier detection by DNA analysis. N Eng J Med 313:842.
15. Antonarakis SE, Copeland KL, Carpenter RJ, Carta CA, Hoyer LW, Caskey CT, Toole JJ, Kazazian HH, Jr. (1985). Prenatal diagnosis of hemophilia A by factor VIII gene analysis. Lancet 1:1407.
16. Gitschier J, Wood WI, Tuddenham EGD, Shuman MA, Goralka TM, Chen EY, Lawn RM (1985). Detection and sequence of mutations in the factor VIII gene of haemophiliacs. Nature 315:427.
17. Din N, Schwartz M, Kruse TA, Vestergaard SR, Ahrens P, Caput D, Herzog K, Quiroga M (1985). Factor VIII gene specific probe for prenatal diagnosis of hemophilia A. Lancet 1:1446.
18. Phillips DG, Kazazian HH, Jr., Scott AF, Toole JJ, Antonarakis SE (1985). Hemophilia A: Experience with prenatal diagnosis using DNA analysis. Am J Hum Genet 37:A224 (Abstract).
19. Wion KL, Tuddenham EGD, Lawn RM (1986). A new polymorphism in the factor VIII gene for prenatal diagnosis of hemophilia A. Nucl Acids Res 14:4535.
20. Oberle I, Camerino G, Heilig R, Grunebaum L, Cazenave J-P, Crapanzano C, Mannucci P, Mandel JL (1985). Genetic screening for hemophilia A (classic hemophilia) with a polymorphic DNA probe. N Eng J Med 312:682.

21. Oberle I, Drayna D, Camerino G, White R, Mandel JL (1985). The telomeric region of the human X chromosome long arm: presence of a highly polymorphic DNA marker and analysis of recombination frequency. Proc Natl Acad Sci USA 82:2824.
22. Peake IR, Bloom AL (1986). Recombination between genes and closely linked polymorphisms. Lancet 1:1335.
23. Driscoll MC, Miller CH, Goldberg JD, Aledort LM, Hoyer LW, Golbus MS (1986). Recombination between factor VIII:C gene and ST14 locus. Lancet 2:279.
24. Harper K, Winter RM, Pembrey ME, Hartley D, Davies KE, Tuddenham EGD (1984). A clinically useful DNA probe closely linked to hemophilia A. Lancet 2:6.
25. Lazarchick J, Hoyer LW (1978). Immunoradiometric measurement of the factor VIII procoagulant antigen. J Clin Invest 62:1048.
26. Brinkhous KM, Roberts HR, Weiss AE (1972). Prevalence of inhibitors in hemophilia A and B. Thromb Diath Haemorrh 51:315.
27. Gill FM (1984). The natural history of factor VIII inhibitors in patients with hemophilia A. In Hoyer LW (ed): "Factor VIII inhibitors," New York: Alan R. Liss, p 19.
28. Youssoufian H, Antonarakis SE, Phillips DG, Aronis S, Tsiftis G, Kazazian HH, Jr. (1987). Characterization of five partial deletions of the factor VIII gene. Proc. Natl. Acad. Sci. USA 84:3772.
29. Youssoufian H, Kasper CK, Phillips DG, Kazazian HH, Jr., Antonarakis SE (1988) Restriction endonuclease mapping of six novel deletions of the factor VIII gene in hemophilia A. Hum. Genet., in press.
30. Youssoufian H, Kazazian HH, Jr., Phillips DG, Aronis S, Tsiftis G, Brown VA, Antonarakis SE (1986) Recurrent mutations in hemophilia A: Evidence for CpG dinucleotides as mutation hotspots. Nature 324:380.
31. Youssoufian H, Antonarakis SE, Bell W, Griffin AM, Kazazian HH, Jr. (1988) Nonsense and missense mutations in hemophilia A: Estimate of the relative mutation rate at CG dinucleotides. Amer. J. Hum. Genet., in press.
32. Youssoufian H, Wong C, Aronis S, Platokoukis H, Kazazian HH, Jr., Antonarakis SE (1988) Moderately severe hemophilia A resulting from Glu-Gly substitution in exon 7 of the factor VIII gene. Amer. J. Hum. Genet., in press.

33. Youssoufian H, Kazazian HH, Jr., Patel A, Aronis S, Tsiftis G, Hoyer LW, Antonarakis SE (1988) Mild hemophilia A associated with a cryptic donor splice site mutation in intron 4 of the factor VIII gene. Genomics, in press.
34. Gitschier J, Wood WI, Tuddenham EGD, Shuman MA, Goralka TM, Chen EY, and Lawn RM (1985) Detection and sequence of mutations in the factor VIII gene of haemophiliacs. Nature 315:427.
35. Levinson B, Janco R, Phillips J III, Gitschier J (1987) A novel missense mutation in the factor VIII gene identified by analysis of amplified hemophilia DNA sequences. Nucl. Acids Res. 15:9797.
36. Barker D, Schafer M, White R (1984). Restriction sites containing CpG show a higher frequency of polymorphism in human DNA. Cell 36:131.
37. Fanning T, Singer MF (1987). A mammalian transposable element I. Biochim. Biophys. Acta, 910:203.
38. Scott AF, Schmeckpeper BJ, Abdelrazik M, Theisen Comey C, O'Hara B, Pratt Rossiter J, Cooley T, Heath P, Smith KD, Margolet L (1987). Origin of the human L1 elements: Proposed progenitor genes deduced from a consensus DNA sequence. Genomics 1:113-125.
39. Hattori M, Kuhara S, Takenaka O, Sakaki Y (1986). L1 family of repetitive DNA sequences in primates may be derived from a sequence encoding a reverse transcriptase-related protein. Nature 321:625.
40. Sakaki Y, Hattori M, Fujita A, Yoshioka K, Kuhara S and Takenaka O (1986). The LINE-1 family of primates may encode a reverse transcriptase-like protein. Cold Spring Harbor Symp Quant Biol 51:465.
41. Skowronski J, Singer MF (1985). Expression of a cytoplasmic LINE-1 transcript is regulated in a human teratocarcinoma cell line. Proc Natl Acad Sci USA 82:6050.
42. Kazazian HH, Jr., Wong C, Youssoufian H, Scott AF, Phillips D, Antonarakis SE (1988) A novel mechanism of mutation in man: Hemophilia A due to _de novo_ insertion of L1 sequences. Nature, in press.
43. Skowronski J, Singer MF (1986). The abundant LINE-1 family of repeated DNA sequences in mammals: Genes and Pseudogenes. Cold Spring Harbor Symp Quant Biol 51:457.

Gene Transfer and Gene Therapy, pages 307–314

GENOMIC AMPLIFICATION WITH TRANSCRIPT SEQUENCING (GAWTS) AND ITS APPLICATION TO THE RAPID DETECTION OF MUTATIONS AND POLYMORPHISMS IN THE FACTOR IX GENE[1]

D.D. Koeberl, J-M. Buerstedde, E.S. Stoflet, C.D.K. Bottema, G. Sarkar, and S.S. Sommer

Department of Biochemistry and Molecular Biology
Mayo Clinic, Rochester, Minnesota 55905

ABSTRACT A sequencing method called genomic amplification with transcript sequencing (GAWTS) is described that is based on amplification with the polymerase chain reaction (PCR). GAWTS bypasses cloning and increases the rate of sequence acquisition by at least fivefold. The method involves the attachment of a phage promoter onto at least one of the PCR primers. The segments amplified by PCR are transcribed to further increase the signal and to provide an abundance of single-stranded template for reverse transcriptase-mediated dideoxy sequencing. An end-labeled reverse transcriptase primer complementary to the desired sequence generates the additional specificity required to generate unambiguous sequence data. GAWTS can be performed on as little as a nanogram of genomic DNA. The utility of GAWTS was demonstrated by sequencing more than 60,000 base pairs (bp) from eight regions of the factor IX gene in a hemophiliac and in multiple normal individuals of defined ethnicity. The regions contain those sequences most likely to be mutated in individuals with hemophilia B. A transition at a CpG sequence in the calcium binding region (glutamine substitution at arginine 29) was found in the hemophiliac. In the normal individuals, one novel amino acid polymorphism was found in the signal sequence.

[1]This work was aided by March of Dimes Birth Defect Grant 5-647.

INTRODUCTION

In contrast to autosomal recessive mutations, deleterious X-linked mutations are eliminated within a few generations because the affected males reproduce sparingly if at all. Thus, each family with an X-linked disease such as hemophilia B represents an independent mutation. From the perspective of efforts to understand the expression, processing, and function of factor IX, this is useful since a large number of mutations are potentially available for analysis. Recently, a rapid method of sequencing an allele in a region of known sequence was developed (1). The method involves amplification with polymerase chain reaction (PCR) and cloning into M13 phage.

Using PCR, we have developed a method known as genomic amplification with transcript sequencing (GAWTS) which should facilitate structure-function correlations and make it practical to perform direct carrier testing and prenatal diagnosis on at-risk individuals (2).

MATERIALS AND METHODS

The PCR, transcription, and sequencing reactions were performed as follows: PCR was performed manually or with the Perkin Elmer Cetus automated thermal cycler. In brief, 30 µl of 10 ng/µl genomic DNA, 1 µM of each primer, 50 mM KCl, 10 mM Tris HCl (pH 8.3), 2 mM magnesium chloride, 0.01% (w/v) gelatin, and 200 µM of each dNTP was incubated at 94°C for 10 min (Perkin Elmer Cetus protocol). Subsequently 1 U of _Taq_ polymerase was added and 30 cycles of PCR were performed (annealing: 2 min at 50°C; elongation: 3 min at 72°C; denaturation: 1 min at 94°C).

After a final 10 min elongation, 3 µl of the amplified material was added to 17 µl of the RNA transcription mixture: 40 mM Tris-HCl pH 7.5, 6 mM magnesium chloride, 2 mM spermidine, 10 mM sodium chloride, 0.5 mM of the four ribonucleoside triphosphates, RNasin (1.6 U/µl), 10 mM DTT, 10 U of T7 RNA polymerase, and diethylpyrocarbonate treated water. Samples were incubated for 1 hr at 37°C and the reaction was stopped by adding 0.5 µl of 0.5 M EDTA pH 7.4.

For sequencing, 2 μl of the transcription reaction and 1 μl of ^{32}P end-labeled (see below) reverse transcriptase primer were added to 10 μl of annealing buffer (250 mM KCl, 10 mM Tris-HCl pH 8.3). The samples were heated at 80°C for 3 min and then annealed for 45 min at 45°C (approximately 5°C below the denaturation temperature of the oligonucleotide). Microfuge tubes were labeled with A, C, G, and T. The following was added: 3.3 μl reverse transcriptase buffer (24 mM Tris-HCl pH 8.3, 16 mM magnesium chloride, 8 mM DTT, 0.8 mM dATP, 0.4 mM dCTP, 0.8 mM dGTP, and 1.2 mM dTTP) containing 5 U of AMV reverse transcriptase, 1 μl of a dideoxyribonucleoside triphosphate (1 mM ddATP or 0.25 mM ddCTP or 1 mM ddGTP or 1 mM ddTTP), and 2 μl of the primer RNA template solution. The sample was incubated at 55°C for 45 min and the reaction was stopped by adding 2.5 μl of 100% formamide with 0.3% bromophenol blue and xylene cyanol FF. Samples were boiled for 3 min and 1.5 μl were loaded onto a 100-cm sequencing gel separated by electrophoresis for about 15,000 V-h. Subsequently, autoradiography was performed.

End-labeling of the reverse transcriptase primer was performed by incubating a 0.1 μg sample of oligonucleotide in a 13-μl volume containing 50 mM Tris-HCl (pH 7.4), 10 mM $MgCl_2$, 5 mM DTT, 0.1 mM spermidine, 100 μCi [α-^{32}P]ATP (5,000 Ci/mmole) and 7 U of polynucleotide kinase for 30 min at 37°C. The reaction was heated to 65°C for 5 min and 7 μl of water was added for a final concentration of 5 ng/μl of oligonucleotide per μl. One μl of labeled oligonucleotide was added per sequencing reaction without removal of the unincorporated mononucleotide.

RESULTS AND DISCUSSION

Figure 1 outlines GAWTS (2). The method is generally applicable and has been used by the laboratory to rapidly obtain sequence in multiple genes.

To test the sensitivity of GAWTS, the amount of genomic DNA was incrementally decreased. With the aid of an intensifying screen, a sequence could be discerned with 1 ng of input DNA (the amount of DNA contained in 150 diploid cells). At this level, PCR is possible in a crude cell lysate (3).

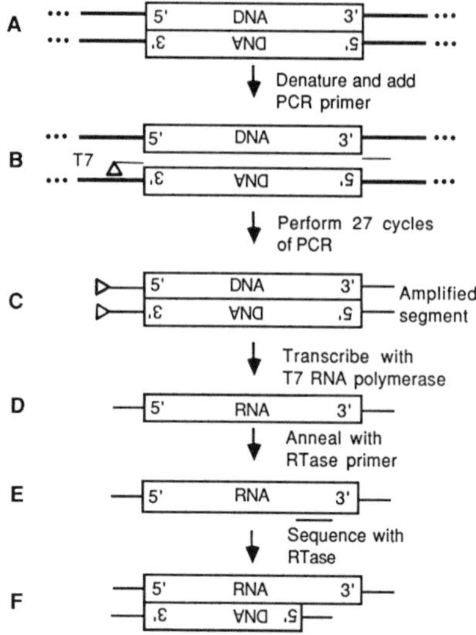

FIGURE 1. Schematic of GAWTS. (A) The region of genomic DNA to be amplified is indicated by the open rectangle. Two strands with their 5' to 3' orientation are shown. The darkened regions represent flanking sequences. (B) The oligonucleotides anneal to sites just outside the sequence to be amplified. One of the oligonucleotides has a T7 promoter sequence. (C) PCR consists of repetitive cycles of denaturation, annealing with primers, and DNA polymerization. Since the number of fragments with defined ends increases much faster than the number with undefined ends, virtually all the fragments are of defined size after 30 cycles. However, since the oligonucleotides anneal to other sites in the genome, spurious fragments can also be amplified. The segment pictured is the specifically amplified sequence. (D) RNA is transcribed from the T7 promoter. This provides a convenient source of single-stranded nucleic acid for dideoxy sequencing. (E) Because of the complexity of the mammalian genome, the amplified and transcribed sequences contain other genomic segments whose flanking sequences cross-hybridize with the PCR primers at the stringency generated by the DNA polymerization reaction. As a result, another level of specificity is crucial to obtaining interpretable sequences. That specificity is provided by utilizing a nested oligonucleotide sequencing primer which lies in the region of interest. (F) Reverse transcriptase is used to generate sequence data by the dideoxy method.

GAWTS was applied to the factor IX gene which is encoded in 34 kb that contains eight exons (1.4 kb of coding sequence, 2.8 kb total) and seven introns that account for over 90% of the sequence (4,5). For this study, eight regions encompassing 2.6 kb of sequence were chosen. Region A contains the putative promotor, exon a, and the adjacent splice junction. Region B/C contains exon b, intron b, exon c, and the flanking splice junctions. Regions D through G contain the appropriate exon and flanking splice junctions. Region H-5' contains a splice junction, the amino acid coding sequence of exon h, and the proximal 3' untranslated segment of the mRNA. Region H-3' contains the distal 3' untranslated region in exon h (including the poly A addition sequence) as well as the sequence immediately following the gene. It is anticipated that the overwhelming majority of causative mutations will be in these regions.

These regions were sequenced in HB2, an individual with hemophilia B of European descent who has mild disease (factor IX coagulant = 30%). Two sequence changes were found, a G->A transition in exon b which substitutes a glutamine for arginine 29 (Fig. 2B) and a silent T->A transversion at the third position of valine 227 which is located more than .23 kb downstream in exon g. Had the G->A transition originated in the complementary strand in exon b, it would have been a C->T transition at CpG. This and other data (Buerstedde et al., unpublished) indicates that CpG is a major hotspot for mutation in the factor IX gene.

As the above were the only sequence changes found, it is inferred that the change in exon b constitutes the causative mutation. However, the inference is less compeling than in some situations such as a nonsense mutation or splice junction defect. Therefore it was important to better characterize the polymorphisms in these eight regions. Genomic DNA from 18 individuals of European descent, one Asian Indian, and one Lebanese Arab were sequenced with GAWTS. Examination of 52 kb of sequence has revealed only one new polymorphism, a phenylalanine substitution for isoleucine at amino acid -40 in the signal sequence in individual E91 (Fig. 2A). Sequence from region A of an additional 10 individuals failed to reveal another example of the substitution indicating that this signal sequence alteration is not a high frequency polymorphism as defined by a minor allele frequency of greater than 20% ($p < .02$ by the Poisson distribution).

A.

```
                                        REGION A

           5'(-139)-16D        5'(-120)-15D
         AA TAATGACCAC TGCCCATTCT CTTCACTTGT CCCATTCTCT TCACTTGTCC CAAGAGGCCA TTGGAAATAG TCCAAAGACC CATTGAGGGA
       -139                 -120             -105

                                                                                                           -46->
                                                                   1                                         Met
GATGGACATT ATTTCCCAGA AGTAAATACA GCTCAGCTTG TACTTTGGTA CAACTAATCG ACCTTACCAC TTTCACAATC TGCTAGCAAA GGTT ATG

                        Phe
Gln Arg Val Asn Met Ile Met Ala Glu Ser Pro Gly Leu Ile Thr Ile Cys Leu Leu Gly Tyr Leu Leu Ser Ala Glu Cys
CAG CGC GTG AAC ATG ATC ATG GCA GAA TCA CCA GGC CTC ATC ACC ATC TGC CTT TTA GGA TAT CTA CTC AGT GCT GAA TGT
                        T(E91)
-18
Thr                             239            257
ACA G GTTT GTTTCCTTTT TTAAAATACA  TTGAGTATGC  TTGCCT...
                                GT  AACTCATACG  AACGGA
                                                      AGAGGGATATCACTCAGCATAATCCATGG
                                  (T7-29)I1(257)-47U              T7 promoter
```

FIG. 2A.

B.

```
                                        REGION B/C

                                        (T7-29)I1(6248)-47D
       GGTACCTAA TACGACTCAC TATAGGGAGA
                                ...AAAGACT TTCTTAAGAG ATGTAAAATT TTCATGATGT TTTCTTTTTT GCTAAAACTA AAGAATTATT
                                  6248                6265
           -17                                                                 -1  +1
           Val Phe Leu Asp His Glu Asn Ala Asn Lys Ile Leu Asn Arg Pro Lys Arg Tyr Asn Ser Gly Lys Leu
CTTTTACATT TCAG TT TTT CTT GAT CAT GAA AAC GCC AAC AAA ATT CTG AAT CGG CCA AAG AGG TAT AAT TCA GGT AAA TTG

                                                                                   Gln
Glu Glu Phe Val Gln Gly Asn Leu Glu Arg Glu Cys Met Glu Glu Lys Cys Ser Phe Glu Glu Ala Arg Glu Val Phe Glu
GAA GAG TTT GTT CAA GGG AAC CTT GAG AGA GAA TGT ATG GAA GAA AAG TGT AGT TTT GAA GAA GCA CGA GAA GTT TTT GAA
                                                                                    A(HB2)
                38
Asn Thr Glu Arg Thr                                           6535
AAC ACT GAA AGA ACA G TGAGTATTTC CACATAATAC CCTTCAGATG CAGAGCATAG AATAGAAAAT CTTTAAAAAG ACACTTCTCT TTAAAATTTT
                                                      AC GTCTCGTATC TTATC
                                                          I2(6535)-17U

AAAGCATCCA TATATATTTA TGTATGTTAA ATGTTATAAA AGATAGGAAA TCAATACCAA AACACTTTAG ATATTACCGT TAATTTGTCT TCTTTTATTC

        39                       46
        Thr Glu Phe Trp Lys Gln Tyr Val              6720                  6739                  6759
TTTATAG ACT GAA TTT TGG AAG CAG TAT CTT G G TAAGCAATTC ATTTTATCCT CTAGCTAATA TATGAAACAT ATGAGAATTA TGTGGG...
                                                           GGA GATCGATTAT ATACTT  TA TACTCTTAAT ACACCC
                                                             I3(6739)-19U          I3(6759)-18U
```

FIG. 2B.

As the silent base change in exon g might well have represented an additional polymorphism, that region was sequenced in another 20 individuals. No sequence changes were seen indicating that the change in HB2 represents either a rare polymorphism/variant or a second mutation as commonly occurs in cells exposed to mutagens that act at the replication fork or in cells with nucleotide pool abnormalities (6).

With GAWTS and other PCR-based methods of direct sequencing (7,8), the routine sequencing of relevant regions of a gene such as Factor IX has become technically feasible. In the future, automation of these methods should provide another dramatic increase in the rate of genomic information retrieval.

LEGEND TO FIGURE 2. Sequence of Regions A and B/C showing the sequence changes found and the location of the PCR and reverse transcriptase primers (boxed sequences). The numbering system (5) and the notation for primers (2) can be found elsewhere.

A. T->A transversion substitutes Phe for Ile at amino acid -40. The genomic sequence generated by the reverse transcriptase primer begins at base -105 and ends at base 239. Due to technical difficulties the first 10 bases of this and other regions were not obtained in all individuals. PCR primers: 5'(-139)-16D and (T7-29)I1(257)-47U. Reverse transcriptase primer: 5'(-120)-15D.
B. Amino acid changes found in Region B/C in the hemophiliac, HB2. Due to the orientation of the primers, the sequence generated is of the complementary RNA, beginning at 6720 and ending at 6265. Two reverse transcriptase primers were used: I2(6535)-17U and I3(6739)-19U.

ACKNOWLEDGMENTS

We thank E.J.W. Bowie, M.D. for his generous support.

REFERENCES

1. Scharf SJ, Horn GT, Erlich HA (1986) Direct cloning and sequence analysis of enzymatically amplified genomic sequences. Science 233:1076.
2. Stoflet ES, Koeberl DD, Sarkar G, Sommer SS (1988) Genomic amplification with transcript sequencing. Science 239:491.
3. Saiki RK, Bugawan TL, Horn GT, Mullis KB, Erlich HA (1986) Analysis of enzymatically amplified β-globin and HLA DQα DNA with allele-specific oligonucleotide probes. Nature (London) 324:163.
4. Anson DS, Choo KH, Rees DJG, Giannelli F, Gould JA, Huddleston JA, Brownlee GG (1984) The gene structure of human anti-haemophilia factor IX. EMBO J 3:1053.
5. Yoshitake S, Schach BG, Foster DC, Davie EW, Kurachi K (1985) Nucleotide sequence of the gene for human factor IX (antihaemophilic factor B). Biochemistry 24:3736.
6. Phear G, Nalbantoglu J, Meuth M (1987) Next-nucleotide effects in mutations driven by DNA precursor pool imbalances at the aprt locus of Chinese hamster ovary cells. Proc Natl Acad Sci USA 84:4450.
7. Wong C, Dowling CE, Saiki RK, Higuchi RG, Erlich HA, Kazazian HH (1987) Characterization of B-thalassaemia mutations using direct genomic sequencing of amplified single copy DNA. Nature 330:384.
8. Engelke DR, Hoener PA, Collins FS (1988) Direct Sequencing of Enzymatically Amplified Human Genomic DNA. Proc Natl Acad Sci USA 85:544.

Gene Transfer and Gene Therapy, pages 315–323

LEUKOCYTE ADHESION DEFICIENCY: AN INHERITED DEFECT IN THE MAC-1, LFA-1, AND P150,95 GLYCOPROTEINS

Donald C. Anderson

Departments of Pediatrics & Cell Biology,
Baylor College of Medicine,
Houston, Texas 77054.

INTRODUCTION

Leukocyte adhesion deficiency (LAD) is a recently recognized autosomal recessive trait characterized by recurrent bacterial infections, impaired pus formation and wound healing, and abnormalities in a wide spectrum of adherence-dependent functions of granulocytes, monocytes, and lymphoid cells. Features of this disease are attributable to deficiency (or absence) of cell surface expression of a family of functionally and structurally related glycoproteins. These include Mac-1 (complement receptor type 3), lymphocyte function-associated antigen-1 (LFA-1), and p150,95 (termed the CD11/CD18 complex by the World Health Organization).

RESULTS AND DISCUSSION

Structure and Function of the CD11/CD18 Leukocyte Adherence Glycoproteins.

The structure, cell distribution and function of Mac-1 and the two related glycoproteins that have identical 95,000 Mr β (CD18) subunits are summarized in Table I. Mac-1, LFA-1, and p150,95 molecules are $\alpha_1 \beta_1$ complexes defined with monoclonal antibodies (MAbs) specific for their αM (CD11b), αL (CD11a), and αX (CD11c) subunits, respectively (1). The amino acid sequences of the αM, αL, and αX subunits show 33 to 50% identity. Because of this homology, the α subunits are considered a protein family. It is likely that a primordial gene duplication event led to the evolution of this family. These leukocyte glycoproteins are also evolutionarily related to extracellular matrix receptors or integrins, and thus are members of a supergene family of adhesion molecules (2).

Mac-1, LFA-1, and p150,95 have distinct as well as common functions in leukocyte adhesion. MAC-1 is a complement receptor (CR-3) that binds iC3b (3). Mab to Mac-1α inhibit binding and phagocytosis of iC3b-opsonized particles by granulocytes and macrophages (4). A distinct nonspecific adhesive function for Mac-1 has also been defined in Mab inhibition experiments as well as by observations in functional assessments of LAD cells (4,5). Among these include the adhesion of myeloid or lymphoid cells to vascular endothelium and homotypic aggregation (6). These cellular reactions are Mg^{++} dependent. LFA-1 participates in Mg++ dependent lymphocyte and monocyte adhesion to a variety of cells (7). It can mediate antigen-independent adhesion, and is required for T-lymphocyte antigen-dependent adhesion to and killing of some target cells. It is also important in natural killing, antibody-dependent killing by K cells and granulocytes, and in T-lymphocyte helper cell interactions (8). Purified p150,95 has been shown to bind iC3b, and anti-p150,95α Mab appear to impede neutrophil or monocyte adherence to surfaces, but the overall functional role of p150,95 is less well defined than that of Mac-1-LFA-1 (9).

TABLE 1
THE MAC-1, LFA-1 FAMILY

	Mac-1	LFA-1	p150,95
Subunits ($M_r \times 10^{-3}$)	αM (170) β (95)	αL (180) β (95)	αX (150) β (95)
Cell distribution	Monocytes Macrophages Granulocytes Large granular lymphocytes	Lymphocytes Monocytes Granulocytes Large granular lymphocytes	Monocytes Macrophages Granulocytes
Chemotactic or secretory stimulation increases surface expression	Yes	No	Yes
Functions inhibited by monoclonal antibodies	Complement Receptor type 3 function (iC3b binding, phagocytosis, and intracellular killing of C3-opsonized microorganisms) Granulocyte, adherence, spreading, aggregation, chemotaxis, and antibody-dependent cellular cytotoxicity	Cytolytic T-lymphocyte-mediated killing and T helper cell responses Natural killing Antibody-dependent cellular cytotoxicity Phorbol ester-stimulated lymphocyte aggregation	Granulocyte adherence and aggregation

Common features: The β subunits appear identical. The α subunits αM and αL are 35% homologous in sequence. The α and β subunits are noncovalently associated in $\alpha_1\beta_1$ complexes. Both α and β subunits are glycosylated and expressed on the cell surface. All functions shown require divalent cations.

Biosynthesis and Intracellular Stores.

Biosynthesis of CD11/CD18 glycoproteins has been studied both in the mouse and in humans (1,10). Each of their α subunits and the common β subunit is encoded by a separate mRNA. These subunits are synthesized as precursors that are cotranslationally glycosylated with N-linked high mannose carbohydrate groups. After α and β subunit association, most high mannose groups are converted to complex-type carbohydrates in the Golgi apparatus, and the subunits increase slightly in M_r. Mature glycoproteins are then transported to the cell surface or to storage sites in intracellular secretory vesicles.

In unstimulated neutrophils and monocytes, Mac-1 and p150,95 are present in an intracellular, vesicular compartments as well as on the cell surface (11,12). Inflammatory mediators including C5α and f-Met-Leu-Phe stimulate a 5 - 10-fold increase in Mac-1 and p150,95 (but not LFA-1) on the cell surface. This "upregulation" is rapid (5-10 min), does not require protein synthesis, and appears to be of great importance in regulating granulocyte or monocyte adhesiveness.

Clinical and Histopathologic Features of LAD.

Clinical features among a large cluster of LAD patients studied in Houston, Texas is shown in Table 2. Recurrent, necrotic, and indolent infections of soft tissues primarily involving skin, mucous membranes, and intestinal tract as well as impaired healing of surgical or traumatic wounds represent the clinical hallmarks of LAD (9). These features appear to reflect the profound impairment of leukocyte mobilization into extravascular inflammatory sites. Biopsies of infected cutaneous, periodontal, umbilical cord, or other soft tissues demonstrate inflammatory infiltrates totally devoid of neutrophilic granulocytes (5). These histopathologic features are particularly striking considering that marked peripheral blood granulocytosis (5 - 20-fold higher than normal) is a constant characteristic feature of this disorder. The severity of infectious complications among LAD patients is directly related to the degree of glycoprotein deficiency (5). Patients with "severe" phenotypes (total deficiency of CD11/CD18 surface expression) are susceptible to life-threatening systemic infections (peritonitis, septicemia, pneumonitis), while individuals with moderate LAD phenotypes (5 - 10% of normal surface CD11/CD18 expression) commonly survive to the third or fourth decade of life, and they infrequently experience systemic progression of superficial infections.

Table II Clinical features of Mac-1, LFA-1, p150,95 deficiency syndrome in Texas patients

Clinical features	Severe deficiency				Moderate deficiency					
	#1 18M/0 F[a]	#2 16M/0 F[a]	#3 6 Y/0 F	#9 6 Y/0 M	#4 11Y/0 M	#5 16Y/0 M	#6 38Y/0 M	#7 9Y/0 M	#8 12Y/0 F	#10 13Y/0 M
Delayed umbilical cord severance	+	+	+	−	−	−	−	−	−	−
Persistant granulocytosis (15–161,000/mm^3)	+	+	+	+	+	+	+	+	+	±
Recurrent soft tissue infections										
necrotic cutaneous abscess or cellulitis	+	+	+	+	+	+	+	+	+	+
perirectal cellulitis/sepsis	+	+	+	+	−	−	−	−	−	−
stomatitis/facial cellulitis	+	+	+	+	+	+	+	−	−	−
gingivitis/periodontitis	+	+	+	+	+	+	+	+	+	+
pneumonitis	+	+	+	+	−	−	+	+	−	−
necrotizing enterocolitis peritonitis/septicemia	+	−	+	−	−	−	−	−	−	−
Impaired wound healing	+	+	+	−	+	+	++	−	−	−
Parental consenguinity	−	+	−	+	−	−	±	±	±	−
Ethnic background	Iranian	Hispanic	Anglo-saxon	Hispanic	Hispanic	Hispanic	Hispanic	Hispanic	Hispanic	Canadian

[a] Deceased.

Functional Abnormalities.

More profound abnormalities of tissue leukocyte infiltration and *in vitro* chemotaxis, hyperadherence, phagocytosis of iC3b-opsonized particles and complement or antibody-dependent cytotoxicity are observed in "severe" as compared to "moderate" deficiency individuals (5). Some heterogeneous results between reporting laboratories may reflect methodologic differences (13). However, as first observed by Crowley (13), abnormalities of granulocyte (and monocyte) adherence to substrates and adhesion-dependent functions including chemotaxis, and aggregation have been observed among almost all patients studies. Leukocyte chemotaxis is affected because it requires adhesion (14). CR3-dependent binding and phagocytosis of iC3b-opsonized particles is consistently deficient, in agreement with the identity of the CR3 and Mac-1 (3). Opsonized particles are ingested poorly and hence, fail to trigger the respiratory burst associated with intracellular killing. Abnormalities of granulocyte or mononuclear leukocyte ADCC have been frequently observed in "severe" patients. In contrast, adherence-independent cellular functions including f-Met-Leu-Phe receptor-ligand binding, cell bipolarization, and oxidative metabolism or degranulation when mediated by soluble stimuli are generally normal (5).

Impaired hyperadherence of stimulated granulocytes or monocytes of LAD patients is causally related to impaired mobilization of intracellular stores of Mac-1 and p150,95 to the cell surface (and/or functional activation of these surface molecules). Whereas a 5 - 10-fold increase in Mac-1 and p150,95 surface expression is consistently demonstrated when normal granulocytes or monocytes are exposed to chemotactic factors, little or no increase is observed on LAD patients cells. These findings and those derived from MAb blocking studies indicate that hyperadherence responses are mediated by "up regulation" of these surface glycoproteins. Thus, we propose that *in vivo*, chemoattractants or other inflammatory byproducts diffusing from sites of inflammation into the circulation elicit enhanced surface Mac-1 and p150,95 which facilitates homotypic aggregation and "directed" adherence of granulocytes and monocytes to endothelial cells at the inflammatory site. The profound inability of patient granulocytes to migrate into extravascular inflammatory sites appears to be related to a lack of Mac-1 glycoprotein "up regulation" and/or activation (5,15).

Molecular Basis for LAD.

The molecular basis for LAD has thus far been studied via two methods: biosynthesis and human x mouse lymphocyte hybrids. Biosynthesis experiments have utilized EBV-transformed B-lymphocyte and mitogen-stimulated T-lymphocyte cell lines, which from healthy individuals synthesize the LFA-1 α subunit and the common β subunit and express the LFA-1 α β complex on the cell surface. Early studies showed that patient cell lines synthesized apparently normal LFA-1 α-subunit precursor, but the αL precursor did not undergo carbohydrate processing, did not associate in an αβ complex, and neither subunit was expressed on the cell surface (15). In human x mouse lymphocyte hybrids, human LFA-1 α and β subunits from normal cells could associate with mouse LFA-1 subunits to form interspecies hybrid αβ complexes. Surface expression of the αL, but not the β subunit of patient cells was rescued by the formation of interspecies complexes. These findings showed that the LFA-1 α subunit in genetically deficient cells is competent for surface expression in the presence of an appropriate mouse β subunit; this suggests that the genetic lesion affects the β subunit. Moreover, these findings confirm that α subunits must associate with the β subunit in an αβ complex in order to mature and to be expressed in intracellular pools or on the cell surface (Fig. 1). The LFA-1 α subunit is degraded in the absence of the β subunit (16).

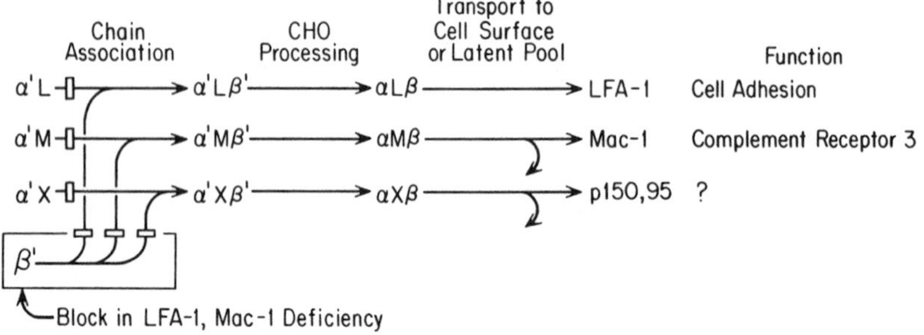

Figure 1 Biosynthesis of the Mac-1, LFA-1 glycoprotein family. The biosynthetic pathway in normal cells is as described in Reference 1. The evidence for a primary block in β-subunit synthesis, a secondary block in α'L biosynthesis due to a lack of β-subunit association, and hypothetically similar blocks in α'M and α'X biosynthesis, is discussed in the text. Reprinted by permission from *The Journal of Experimental Medicine*.

The development of a rabbit anti-human β-subunit serum has allowed β-subunit precursors in both healthy and deficient cell lines to be immunoprecipitated and examined as to size in SDS-PAGE (17) These studies have identified heterogeneous mutations of the β subunit among LAD kindreds. Among seven LAD kindreds studied, one patient has an aberrantly large β precursor, two have β precursors of normal size, and one has no β precursor at all. Four individuals in the same family ("moderate" phenotype) have an identical abberantly small β precursor. Of ten relatives of this family, nine have been typed as heterozygote carriers and one as a noncarrier by quantity of surface expression (5).

All heterozygoses show both a normal and an abnormally small β precursor; the noncarrier shows only the normal β precursor. These data provide conclusive evidence that the defect is in the β-subunit gene.

The recent availability of a complementary cDNA for the human β subunit has facilitated further definition of LAD β-gene mutations employing Northern blot analysis. Total RNA for patient cell lines has been probed with a nick translated β subunit cDNA. Of seven LAD patients studied, four expressed relatively normal quantities of β subunit mRNA, one had diminished mRNA expression, and two others produced no detectable message (17). Studies in four patients of one kindred demonstrated an identical aberrantly small β precursor, but normal levels mRNA expression. Further studies in this kindred have identified a 90 nucleotide deletion in β mRNA, which in turn is due to a single base pair alteration in genomic DNA (17). Thus, at least four classes of mutations of the β gene in LAD have been characterized as shown below.

TABLE III
CLASSIFICATION OF LAD MUTATIONS

Class of Mutation	Phenotype	β Precusor	β mRNA Levels
I	Severe	None detectable	None detectable
	Severe	None detectable	None detectable
II	Moderate	Trace amounts	Low
III	Moderate	Aberrantly small	Normal
IV	Severe	Aberrantly large	Normal
V	Severe	Normal size	Normal
	Normal	Normal size	Normal

Prospects for Gene Therapy

The identification of the genetic lesion in LAD in the common β subunit of the Mac-1, LFA-1 glycoprotein family predicts that introduction of a normal β-subunit gene into hematopoietic cells should cure the disease as has bone marrow transplantation (18). The mouse β-subunit has been shown to complex and rescue surface expression of the patient LFA-1 α-subunit in patient x mouse lymphocyte hybrids (16).

Efforts are now directed into introducing (via retroviral or other vectors) the β-subunit gene into bone marrow cells with the goal of curing this disease through gene therapy.

REFERENCES

1. Sanchez-Madrid F, P Simon, S Thompson, and TA Springer (1983). Mapping of antigenic and functional epitopes on the alpha and beta subunits of two related glycoproteins involved in cell interactions, LFA-1 and Mac-1. J Exp Med 158:586.
2. Kishimoto TK, K O'Connor, A Lee, TM Roberts, and TA Springer (1987). Cloning of the beta subunit of the leukocyte adhesion proteins: Homology to an extracellular matrix receptor defines a novel supergene family. Cell 48:681.
3. Beller DI, TA Springer, and RD Schreiber (1982). Anti-Mac-1 selectively inhibits the mouse and human type three complement receptor. J Exp Med 156:1000.
4. Anderson DC, LJ Miller, FC Schmalstieg, R Rothlein, and TA Springer (1986). Contributions of the Mac-1 glycoprotein family to adherence-dependent granulocyte functions: Structure-function assessments employing subunit-specific monoclonal antibodies. J Immunol 137:15.
5. Anderson DC, FC Schmalstieg, AS Goldman, WT Shearer, and TA Springer (1985). The severe and moderate phenotypes of heritable Mac-1, LFA-1, p150,95 deficiency: Their quantitative definition and relation to leukocyte dysfunction and clinical features. J Infec Dis 152:668.
6. Smith CW, R Rothlein, BJ Hughes, MM Mariscalco, FC Schmalstieg, and DC Anderson (1987). Recognition of an endothelial determinant for CD18 - dependent human neutrophil adherence and transendothelial migration. J Clin Inv
7. Springer TA, ML Dustin, TK Kishimoto, and SD Marlin (1986). The lymphocyte function associated (LFA-1,CA2 & LFA-3) molecules: Cell adhesion receptors of the immune system. Annu Rev Immunol 5:223.
8. Kohl S, TA Springer, and FC Schmalstieg (1984). Defective natural killer cytotoxicity and polymorphonuclear leukocyte antibody dependent cellular cytotoxicity in patients with LFA-1/OKM-1 deficiency. J Immunol 133:2942.
9. Anderson DC, and TA Springer (1987). Leukocyte adhesion deficiency: An inherited defect in the Mac-1, LFA-1 and p150,95 glycoproteins. Ann Rev Med 38:175.
10. Ho MK, and TA Springer (1983). Biosynthesis and assembly of the alpha and beta subunits of Mac-1, a macrophage glycoprotein associated with complement receptor fraction. J Biol Chem 258:2766.

11.Jones DH, DC Anderson, BL Burr, H.E. Rudloff, CW Smith, and FC Schmalstieg (1987). Subcellular location of Mac-1 (CR-3) in human neutrophils: effects of chemotactic factors and PMA. Pedi Res 21:312a.

12.Miller LJ, DF Bainton, N Borregaard, and TA Springer (1987). Stimulated mobilization of monocyte Mac-1 and p150,95 adhesion proteins from an intracellular vesicular compartment to the cell surface. J Clin Inv 80:185.

13.Crowley CA, JT Curnutte, RE Rosin, J Andre-Schwartz, JI Gallin, M Klempner, R Snyderman, FS Southwick, TP Stossel, and BM Babior (1980). An inherited abnormality of neutrophil adhesions: Its genetic transmission and its association with a missing protein. N Eng J Med 302:1163.

14.Schmalstieg FC, HE Rudloff, GR Hillman, and DC Anderson (1986). Two dimensional and three dimensional movement of human polymorphonuclear leukocytes: Two fundamentally different mechanisms of location. J Leuk Biol 40:677.

15.Springer TA, WS Thompson, LJ Miller, and DC Anderson (1984). Inherited deficiency of the Mac-1, LFA-1, p150,95 glycoprotein family and its molecular basis. J Exp Med 160:1901.

16.Marlin SD, CC Morton, DC Anderson, and TA Springer (1986). LFA-1 immunodeficiency disease: Definition of the genetic defect and chromosomal mapping of alpha and beta subunits by complementation in hybrid cells. J Exp Med 164:855.

17.Kishimoto TK, N Hollander, TM Roberts, DC Anderson, and TA Springer (1987). Heterogenous mutations of the beta subunit common to the LFA-1, Mac-1, and p150,95 glycoproteins cause leukocyte adhesion deficiency. Cell 50:193.

18.Fischer A, S Blanche, F Veber, F LeDeist, I Gerota, M Lopez, A Durandy, and C Griscelli (1986). In Recent Advances in Bone Marrow Transplantation, Gale, RP (eds): "Correction of immune disorders by HLA matched and mismatched bone marrow transplantation," New York: Alan R. Liss

Gene Transfer and Gene Therapy, pages 325–334

MOLECULAR APPROACHES TO INHERITED DEFICIENCY OF PROXIMAL UREA CYCLE DEFECTS

Arthur L. Horwich[1]

Yale University School of Medicine
Department of Human Genetics
New Haven, CT 06510

I wish to discuss two mitochondrial enzymes of the liver whose inherited deficiency leads to severe disease. I'll cover four different areas: clinical phenotype of the diseases, biogenesis of the enzymes, DNA diagnosis, and therapeutic considerations. A portion of this work has been carried out in collaboration with Leon Rosenberg's laboratory, with whom we've had a close relationship over the past four years and I'll indicate the particular investigators as the work is presented.

First, I wish to present a clinical vignette: Alana P. was the 7-pound product of a normal pregnancy and delivery, who was normal, nursing and alert, until the third day of life when she was noted to be breathing rapidly. A blood count and chest film taken in the Well Baby Nursery were normal. Late in the third day she became less responsive and fed poorly. She was transferred to the Newborn Intensive Care Unit where blood culture and spinal tap were performed and antibiotics commenced. A blood gas revealed a respiratory alkalosis (pCO_2 20, pH 7.6). In the ensuing 12 hours, she became unable to respond to a bell or to noxious stimuli, developed disconjugate gaze, respiratory failure and shock. Following a call to the Genetic Consultation service, a blood ammonia was obtained and measured 2,176 μg/dl, elevated 10-fold above the upper limits of normal. Shortly thereafter, the baby suffered a cardiac arrest and could not be resuscitated.

---

[1]This work was supported by GM 34433

Following death, an extract of liver was prepared and revealed complete deficiency of activity of the enzyme carbamyl phosphate synthetase I (CPS). As shown below in Figure 1, this enzyme catalyzes the first step of the urea cycle, a metabolic pathway that converts ammonia, a toxic end-product of amino acid metabolism, to the excretable molecule urea. CPS and its neighbor in the pathway, ornithine transcarbamylase (OTC), are both enzymes localized to the mitochondrial matrix of hepatocytes, while the remaining three enzymes of the pathway are cytosolic and "constitutively" expressed in all tissues. Deficiency of OTC, like that of CPS, usually results in severe disease.

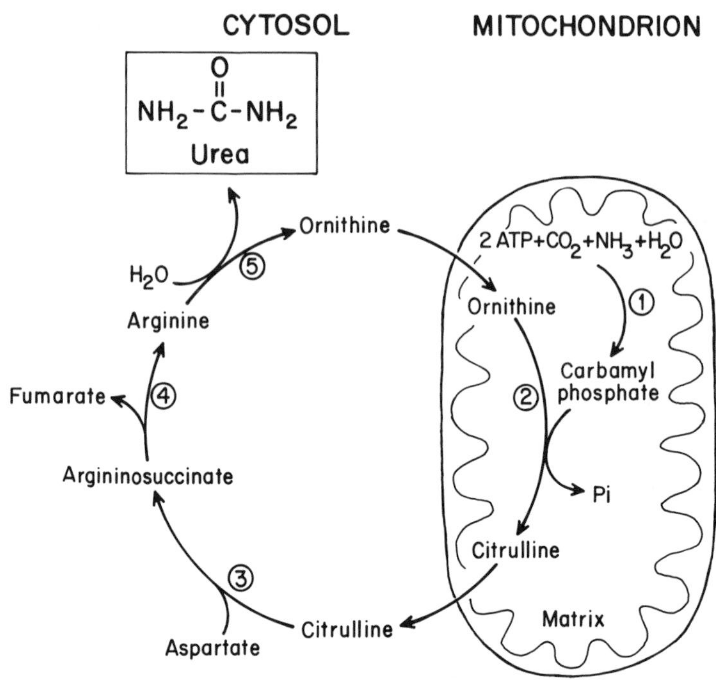

ENZYMES

1. Carbamyl phosphate synthetase I
2. ORNITHINE TRANSCARBAMYLASE (OTC)
3. Argininosuccinate synthetase
4. Argininosuccinate lyase
5. Arginase

Figure 1. Urea cycle pathway.

CPS deficiency is inherited as an autosomal recessive trait. The parents of the affected baby, Alana P., went on to have another affected baby, Ashley. Deficiency was detected in the immediate postnatal period by inability to detect citrulline in plasma, and by development of mild hyperammonemia even in the presence of protein-restricted dietary intake. DNA analysis with a Bgl I restriction fragment length polymorphism (RFLP), detected at the CPS locus by John Phillips of Vanderbilt, using cloned human cDNA, revealed the same haplotype in Ashley as in her affected sib. While Ashley has not suffered any severe hyperammonemic episodes while on a regimen of protein-restricted diet and conjugating agents, the latter to be described below, she has suffered mild episodes associated with intercurrent illnesses, that is, a cold or fever lead to a stress response of catabolic breakdown of endogenous protein and increased production of ammonia. One such episode required hospitalization. She is thus extremely fragile, a veritable "clinical timebomb."

Figure 2 shows a pedigree of a family followed in New Haven with OTC deficiency:

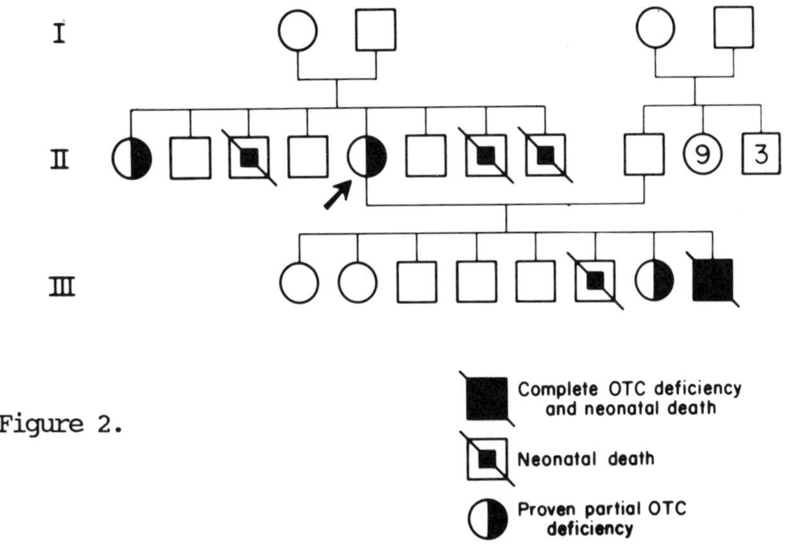

Figure 2.

Here, multiple male family members, in two generations, have died of neonatal hyperammonemia, indicative of X-linked inheritance. In addition, female III-7 is a symptomatic female heterozygote who develops hyperammonemia following a challenge of dietary protein. This little girl's liver is presumed to have undergone lyonization in such a way that most of her active X chromosomes bear the mutant OTC allele. Her total hepatic OTC activity is likely to measure less than approximately 15%, the level below which clinical symptomatology develops. Thus, such female heterozygotes require dietary protein restriction and the use of the two conjugating agents sodium benzoate and sodium phenylacetate. These two agents covalently conjugate the amino acids glycine and glutamine, respectively, to the excretable molecules hippurate and phenylacetylglutamine. The benzoate reaction removes one amino nitrogen per molecule while phenylacetate removes two nitrogen atoms per molecule.

The use of dietary restriction and conjugating agents has had an appreciable positive effect on mortality and morbidity of CPS and OTC deficiencies but the overall outcome of these two proximal urea cycle enzymopathies is poor. As shown in Figure 3, 40% of CPS deficient patients have died by the age of 5 years, while 70% of OTC deficient patients are dead by the age of 3.

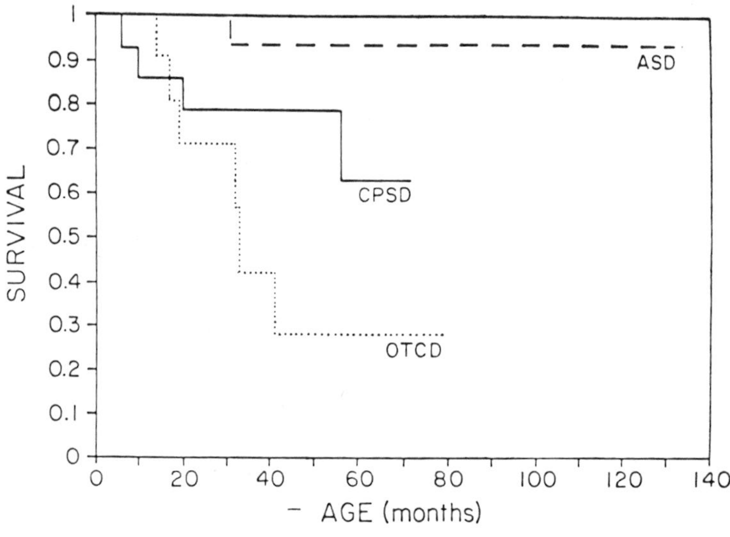

Figure 3.

The mortality figures are perhaps even worse than illustrated, as the survival curve incorporates data on a number of patients with "atypical", milder, deficiency of the two enzymes, who are long-term survivors. One can only guess that perhaps 80% of the more typical severely affected hemizygous OTC-deficient males are dead by one year of age. Concerning morbidity, survivors with OTC deficiency have suffered much more substantially than those with CPS deficiency, exhibiting substantial developmental delay. Perhaps CPS deficient patients are relatively spared because in autosomal recessive disease there is the possibility, in a setting where a compound of two different mutations is usually present, of making a small amount of functional enzyme, whereas in hemizygous X-linked disease no such opportunity exists.

Using the cloned human OTC cDNA, Uta Francke's lab has mapped the gene, using somatic cell hybrids and in situ hybridization, to Xp2.1. Rima Rozen, Joyce Fox, and Wayne Fenton in Lee Rosenberg's lab have used the cDNA to examine the OTC gene in normal and affected individuals. The normal gene is approximately 70 kb in size and contains 9 introns. Analysis to date of approximately 50 probands reveals deletion in five of them, with apparently different breakpoints in each case, some completely deleting the gene and others partially or internally deleting it. These deletions have arisen in two cases de novo, from mothers who by biochemical testing, described below, are determined not to be heterozygotes. The remaining 90% of probands are presumed to exhibit point mutations which, to date, have only been characterized in three pedigrees, by Bob Nussbaum and colleagues, where alteration of a Taq I restriction site at codon 109 of the mature portion results in nonsense or missense.

While other point mutations are not yet characterized, it seems clear that mutation at the OTC locus, like that at other loci examined in this meeting, is heterogeneous. Some patients are CRM positive while others are CRM negative. At the RNA level, studies of mutations have been daunted by the fact that the gene is normally expressed only in the liver, and this tissue is difficult to obtain in a fresh state from deceased probands given the usually catastrophic circumstances of their demise.

A particular class of OTC mutations is of special interest to us, those affecting the signal that targets the OTC subunit precursor to mitochondria. The OTC subunit is normally translated on free cytosolic ribosomes as a 40 K precursor containing an NH_2-terminal cleavable portion called a leader peptide. Following entry into mitochondria the leader is cleaved and three mature-sized 36 K subunits then assemble into the active enzyme. In particular, the OTC precursor exhibits no enzyme activity: elegant studies in Lee's lab revealed that cleavage and assembly are required to produce activity.

The cloned cDNA allowed prediction of the primary structure of the leader peptide:

↓
MLFNLRILLNNAAFRNGHNFMVRNFRCGQPLQ NKVQ

In contrast with the hydrophobic signals of secreted proteins, the 32-residue leader is basic in overall composition, containing four arginine residues and devoid of acidic residues. We note also that it is cleaved after a glutamine residue, differing from the short side chain residues after which secretory signals are cleaved. To examine the critical functional elements of the leader, both gene fusions and mutations were programmed. Import of altered precursors could be studied using an in vitro reconstitution system in which SP6-programmed transcripts are translated in reticulocyte lysate, and the lysate is then directly incubated with isolated mitochondria. We first demonstrated that the leader contains sufficient information to direct mitochondrial localization. It was fused with a small protein, DHFR, which normally localizes to the cytosol. The fusion protein was observed to localize to mitochondria of both intact cells and in the in vitro reconstituion systme. Franta Kalousek, Bob Pollock, Krystyna Furtak and I next examined a battery of single residue leader substitutions and defined two functional elements in the OTC leader peptide, net positive charge, conferred by the basic residues, and midportion structure, particularly critical at position 23. Positive charge may play a role in membrane translocation, via an electrostatic interaction of the positively charged leader peptide with the electrochemical gradient across the inner mitochondrial membrane, which is relatively charge-negative at the inner aspect; secondary structure may play a role in recognition by a putative outer membrane receptor molecule.

During the past decade, a host of experiments in a number of labs have provided a phenomenological description of the import pathway.

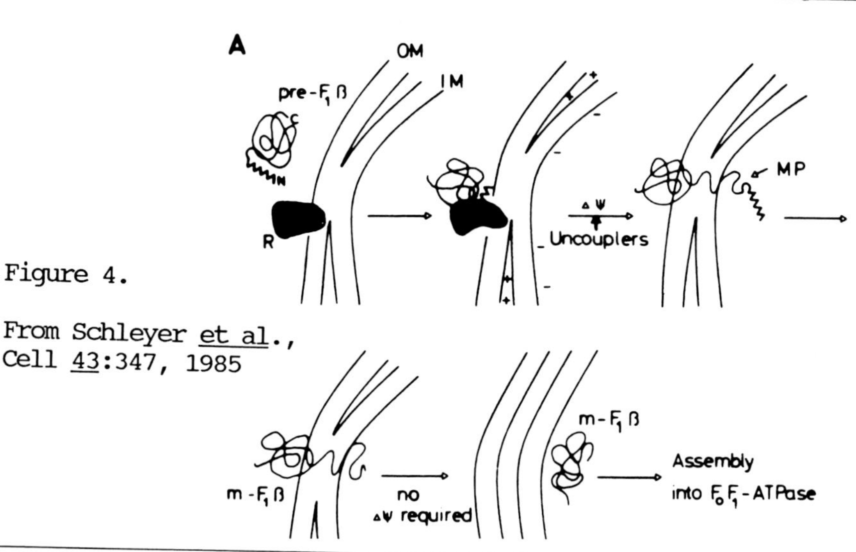

Figure 4.

From Schleyer et al., Cell 43:347, 1985

Following translation on free ribosomes, precursors are recognized by the mitochondria, most likely by an outer membrane receptor molecule like that recently identified in the envelope of pea chloroplasts by the Blobel lab using antiidiotypic antibodies developed against a synthetic peptide derived from the leader peptide of a nuclear-coded pea chloroplast precursor. After recognition, the leader peptide is translocated across a point of contact between outer and inner membrane, in a step that requires an intact electrochemical potential gradient. The leader is then proteolytically removed by an enzyme in the mitochondrial matrix space. The remaining portion of the precursor is then translocated, in a step that does not require an electrochemical gradient. Assembly of mature subunits then occurs in the matrix space.

To address in molecular terms the components of the import pathway, we wished to employ a genetically manipulable system and thus turned to Saccharomyces cerevisiae.

Ming Cheng, in my lab, programmed expression of the human OTC precursor in yeast from a galactose operon promoter, and found that, following induction with galactose, the newly-synthesized human OTC precursor was recognized by the mitochondria of yeast, translocated to the mitochondrial matrix and cleaved to a mature form that assembled into active enzyme. Thus, the mitochondrial import pathway of Saccharomyces is capable of recognizing a human precursor protein, and the pathway of import appears to be conserved in evolution. To produce nuclear mutations affecting this pathway we analyzed a collection of temperature-sensitive lethal mutants for their ability to produce OTC enzyme activity following simultaneous shift to nonpermissive temperature and induction with galactose. Here we make the assumption that obstruction of import is a lethal condition due to failure to constitute enzymes of mitochondrial pathways involved with amino acid and fatty acid metabolism. Mutants that failed to produce activity in the shift assay were examined by immunoblot analysis, and a number were found to accumulate both OTC precursor and the precursor of a yeast mitochondrial matrix protein. Four complementation groups were identified, two of which have been previously identified by Yaffe and Schatz. The various genes and products are now under study.

To return to analysis of human OTC deficiency, Brusilow and colleagues have recently developed a metabolic test that reliably identifies OTC heterozygotes. It relies on shuttling behind the partial OTC block of carbamyl phosphate, a substrate of OTC, out of mitochondria and into the pyrimidine pathway. In the presence of the drug allopurinol, which blocks the distal pyrimidine enzyme OMP decarboxylase, the metabolites orotidine and orotic acid accumulate, and are measurable in the urine. Interestingly, when mothers of OTC probands are examined with this test, virtually all exhibit heterozygote status, suggesting that "new mutation", i.e., first cases in a pedigree, do not arise from the egg. Haldane's hypothesis would have predicted a 33% frequency of such occurrence. Rather, it appears that new mutations at this locus arise commonly in sperm, as can be demonstrated by DNA analysis, carried out by the Rosenberg lab, of affected pedigrees using 4 RFLPs (2 Msp I, 1 BamH I, 1 Taq I) at the locus. Such a pedigree is shown in Figure 5 below, where new mutation arises on the D haplotype of the maternal grandfather. Further pedigree studies are needed to confirm this hypothesis.

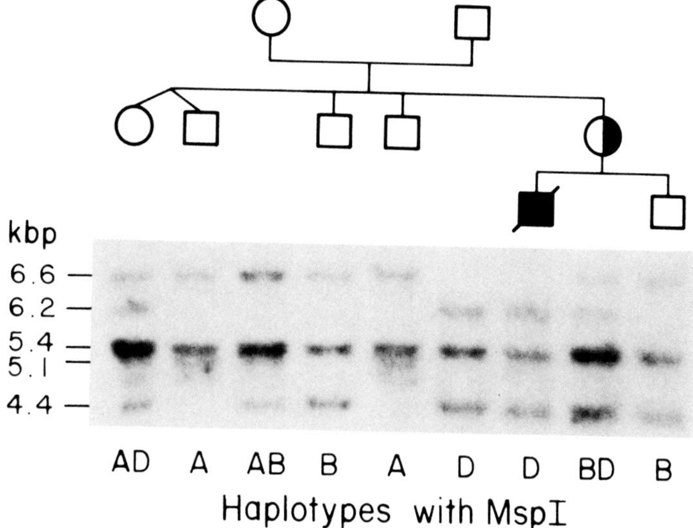

Figure 5.

Together, the carrier test and RFLPs make prenatal DNA diagnosis, using either chorionic villi or amniotic fluid, possible for approximately 80% of female carriers. The question arises whether additional therapeutic strategies are available for affected OTC males. A first line strategy that can currently be considered for severely affected males is the use of hepatic transplantation. This might be implemented even during infancy as was recently demonstrated with a patient of ours with another liver enzymopathy called hereditary tyrosinemia. This disease results in liver necrosis and failure and our patient, in extremis, at age 3 months was transplanted. She has done well postoperatively, cured of the metabolic disorder, but obligated to immunosuppressive treatment. Are other therapeutic avenues available?

Three such avenues arise as possibilities. First, one might reconstitute a pathway outside the liver. This presents difficulty in the case of OTC as one would require transfer also of CPS, the other liver-specific enzyme. Our efforts to carry this out in fibroblasts have been unsuccessful to date. A second approach is the transfer of liver cells containing the desired pathway, to be discussed by Jim Wilson. A third approach is direct vector-mediated delivery. Here we have attempted to employ retroviruses. Because these viruses can only infect a dividing cell population we resorted to infecting fetal liver, an organ that is actively mitotic. For this study, Jennifer Rasmussen and I employed chick embryos and replication-competent Rous sarcoma virus. The top of the chick egg was excised at day 6 or 7 of development, the liver was directly visualized, and several microliters of high-titer virus stock injected. In all cases, when the chicks were allowed to proceed to day 19 and DNA from the livers examined by blot analysis, the viral src gene was found to be introduced. No such signal was observed in mock-infected embryos. Evidence for infection of hepatocytes as opposed to Kupffer or other liver cell types was garnered from positive blot analysis of cell fractions substantially enriched for hepatocytes. We have attempted similar experiments using both replication-competent and defective Moloney viruses injected into liver of E9 to E11 mouse embryos but have failed to detect viral sequences in the liver at the time of birth.

In closing, it should be pointed out that to reconstitute a deficient enzymatic activity in liver, one needs to consider both the number of cells to be rescued/inserted and the level of enzyme activity obtained. The liver probably contains approximately 10 billion hepatocytes, so to reach a level of total OTC activity 10% that of normal, a level required to avert clinical symptomatology, one needs either to hit 1 billion cells with 100% activity or 10 billion cells with 10% normal activity. In the case of OTC, this level of expression is a tall order because the normal level of OTC mRNA is .1% of total mRNA. Normal levels of OTC activity were obtained in fibroblasts only when an SV40-driven cDNA was coamplified with a mutant DHFR to a copy level of several hundred. Clearly, both delivery to the liver and expression in the organ need to be addressed in a systematic fashion.

Gene Transfer and Gene Therapy, pages 335–344

MOLECULAR GENETIC STUDIES IN METHYLMALONIC ACIDEMIA

Fred D. Ledley[1]

Howard Hughes Medical Institute
Department of Cell Biology
Baylor College of Medicine
Houston, TX 77030.

ABSTRACT A cDNA for human methylmalonyl CoA mutase (MCM) has been cloned from a human liver cDNA library by antibody screening. The identity of the clone was established by gene transfer of the cDNA into COS cells and the observation of increased MCM enzymatic activity in transformed cells. Fibroblasts from patients with MCM apoenzyme deficiency exhibit variable levels of hybridizable mRNA with several CRM$^-$ cell lines having severely decreased or non-detectable MCM mRNA. The present results demonstrate the feasibility of constituting MCM holoenzyme activity by gene transfer and raise the possibility of gene therapy for this disorder which causes considerable morbidity and mortality.

INTRODUCTION

Methylmalonic acidemia (MMA) is an inborn error of organic acid metabolism caused by deficiency of the enzyme methylmalonyl CoA mutase (MCM) (1). MCM catalyzes the isomerization of methylmalonyl CoA to Succinyl CoA which is an essential step in the degradation of various branch chain amino acids and odd chain fatty acids. Newborn

---

[1]Fred D. Ledley is an Assistant Investigator of the Howard Hughes Medical Institute.

screening has demonstrated that 1:29,000 newborns have increased methylmalonate excretion (2) due either to deficiency of the MCM apoenzyme (disorders designated _mut_) or enzymes required for generation of adenosylcobalamin which is an obligate cofactor for MCM activity (disorders designated _cbl_) (1). Deficiency of this enzyme causes methylmalonic acidemia (MMA), an often fatal disorder of organic acid metabolism in which the precursors and abnormal metabolites of methylmalonic acid acummulate leading to widespread disruptions of metabolic homeostasis (1,3). Individuals with milder enzymatic defects may have methylmalonic aciduria without clinical symptoms (2).

MCM is a homodimer with identical subunits of 72-77,000 daltons (4,5). It is a mitochondrial protein, encoded by a nuclear gene, which is synthesized in the cytoplasm and must be transported into the mitochondria in order to express biological activity (6,7).

We have recently reported identification of a cDNA for human MCM (8). This clone has been used for gene transfer of human MCM (8), analysis of MCM mRNA expression in _mut_ fibroblasts (8), and chromosomal mapping of the human MCM gene and MUT locus (9).

RESULTS

Identification of human MCM clones.

A human MCM cDNA (MCM105) was isolated from a human liver cDNA library (10) in the vector λgt11 by antibody screening using a chicken anti-human placental MCM antiserum (11) and methods (10) described previously. This antibody detected several distinct cDNA species which did not cross hybridize. The authentic MCM clone was identified by epitope selection (12) in which the λgt11 fusion protein is used to affinity purify hapten specific antibodies which are then tested for cross reactivity with purified MCM protein. One of six clones isolated from the human liver library tested positive by epitope selection. This clone hybridized to a mRNA in human liver of approximately 2500 bases which was sufficiently large to encode the 72-77kD MCM protein. A full length clone (MCM26) was isolated from the human liver cDNA library by hybridization with the (partial) MCM105 cDNA clone. The full length clone contains three EcoRI fragments (designated 5'-_a_-_b_-_c_-3') and approximately 2500 bases.

Expression of recombinant human MCM in COS cells.

The full length MCM cDNA was subcloned into the expression vector 91023(B) provided by Dr. Randy Kaufman of the Genetics Institute, Cambridge MA. This vector contains the SV40 origin of replication, adenovirus major late promotor, tripartite leader sequence, and Va genes. This clone was introduced into COS cells (13) by calcium phosphate co-precipitation (14), and 48 hours after transfection, cells were harvested for MCM enzymatic assay performed as described (15). Low levels of MCM are present in COS cells (a derivative of the monkey kidney cell line CV-1). The level of activity was increased 2-5 fold by transfection with subclone MCM60 containing two EcoRI fragments (a and b) from clone MCM26 comprising approximately 2000 bp (table 1). No increase in activity occurred following transfection with subclones containing fragments b and c in different orientations, or human PAH cDNA as a control (table 1).

This result indicated that the a and b fragments represented the 5' end of the cDNA and contained a full length open reading frame.

TABLE 1.
MCM ACTIVITY IN COS CELLS FOLLOWING DNA MEDIATED GENE TRANSFER[1]

	MCM activity nmoles succinate formed/mg protein (15)	
Sample	Experiment 1	Experiment 2
HUMAN LIVER	37	80
COS	--	5.6
COS + PAH	7.1	--
COS + MCM60	16.9	37
COS + MCM62	3.0	--
COS + MCM75	7.3	10.3

1 COS cells were transfected with partial EcoRI fragments from clone MCM26 or a human PAH cDNA in the vector 91023B. Clone MCM60 contains fragments a and b. Clones MCM75 and MCM62 contain fragments b and c in opposite orientations. The results from two independent transfections are shown. Data from ref 8.

MCM expression in mut fibroblasts.

The human MCM cDNA was used as a probe to examine the MCM gene structure and mRNA expression in fibroblasts from individuals with mut MMA. Fibroblasts with cbl MMA, which have normal MCM apoenzyme were used as controls.

Expression of MCM mRNA was analyzed by northern blot analysis of total cellular RNA probed with the MCM26b fragment as a probe. Hybridizable mRNA was present in all cell lines from individuals deficient in cobalamin metabolism (figure 1A. lanes 1,6) which contain normal amounts of MCM apoenzyme activity. Hybridizable MCM mRNA

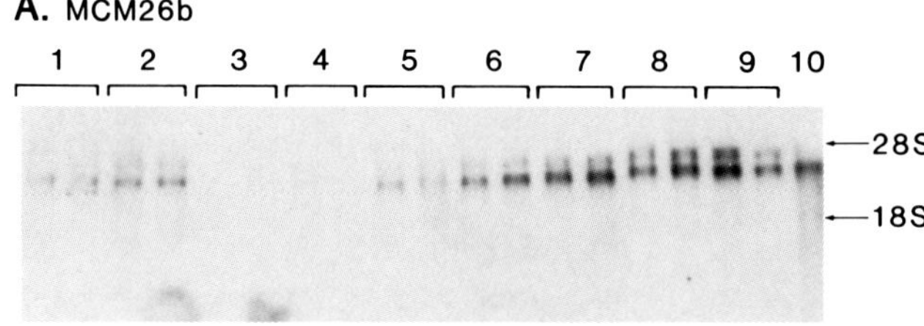

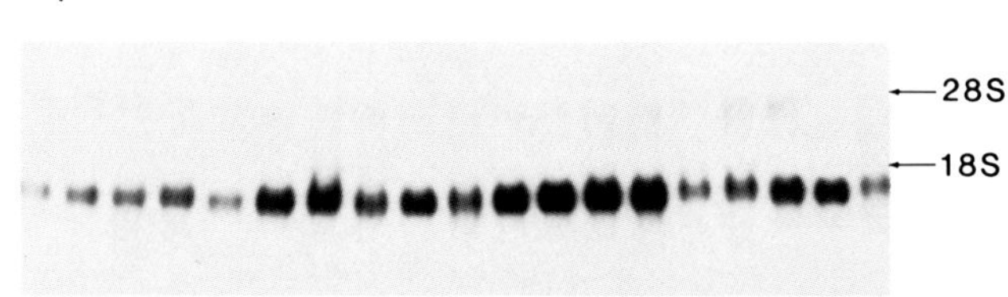

FIGURE 1. Northern blot showing mMCM mRNA in fibroblasts from nine patients with MMA (lanes 1-9) and human liver (lane 10). Panel A: Total RNA probed with MCM26b. Panel B. The same filter reprobed with β-actin cDNA. Legend: 1. GM2452; 2. GM930; 3. GM1673; 4. GM50; 5. 79-865; 6. 87-1645; 7. 82-1211; 8. 81-1085; 9. 87-1640; 10. human liver. Duplicate samples representing independent isolations of RNA are shown.

was present in several mut cell lines (figure 1A. lanes 2,5,6,7,8) though in two other cell lines (figure 1A. lanes 3,4) hybridizable mRNA was absent or present at extremely low levels. In order to demonstrate that the difference in hybridizable MCM mRNA in these cell lines was specific, the same filter was reprobed with a cD A to ß-actin (figure 1B). ß-actin mRNA was present in all lanes at levels equivalent to, or higher, than levels in the control cbl fibroblast cell lines. This suggests that there is a specific decrease in the amount of hybridizable MCM mRNA in the cell lines GM1673 and GM50 (lanes 3, 4). Thus the mutations in these cells presumably represent defects in transcription or processing of the MCM mRNA.

MCM gene structure and restriction fragment length polymorphism.

Southern blotting of human genomic DNA with the MCM probe and a variety of restriction enzymes reveals multiple hybridizing bands (figure 2) suggesting that the MCM locus is large, with multiple exons spanning at least 40-50 kb. A polymorphic HindIII site was identified with

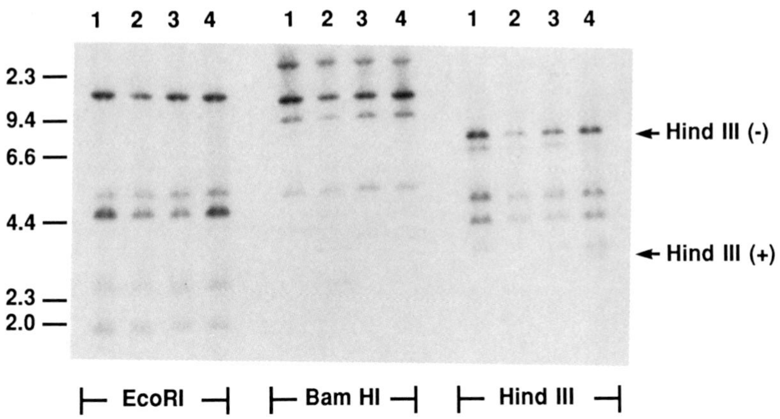

FIGURE 2. Southern blot of DNA from normal and mut fibroblast cell lines digested with EcoRI, BamHI, or HindIII. The polymorphic bands observed with HindIII are indicated (+/-).

the MCM26b probe which results in restriction fragments of 7kb or 4kb (figure 2). Of 68 normal alleles examined, 31 contained HindIII(+) polymorphic site (46%). Identical restriction fragments were observed in DNA from 15 mut cell lines digested with EcoRI or BamHI, suggesting that none of the mutations in these cell lines involve large deletions or insertions of DNA.

Chromosomal localization of MCM and MUT locus.

The MCM gene locus was identified by hybridization of DNAs from panels of hamster-human somatic cell hybrid cell lines and in situ hybridization (9). Somatic cell panels identified the MCM locus on chromosome 6 in the region 6p23-q12. *In situ* hybridization was performed to further refine the MCM locus. Of 200 grains observed, 28 (14%) were on chromosome 6. The distribution of 28 grains localized on chromosome 6 is shown in figure 3. Sixteen (58%) were in the region 6p12-21.2.

Mut MMA constitutes a single complementation group in somatic cell hybrid studies and is thus thought to represent mutations within a single locus, designated MUT. The fact that some individuals with *mut* deficiency have absent or decreased mRNA hybridizable to the MCM26b clone (8) indicates that the MCM26b clone recognizes the mRNA product of the MUT locus. Thus the MCM gene locus at 6p12-21.2 is properly designated MUT (9).

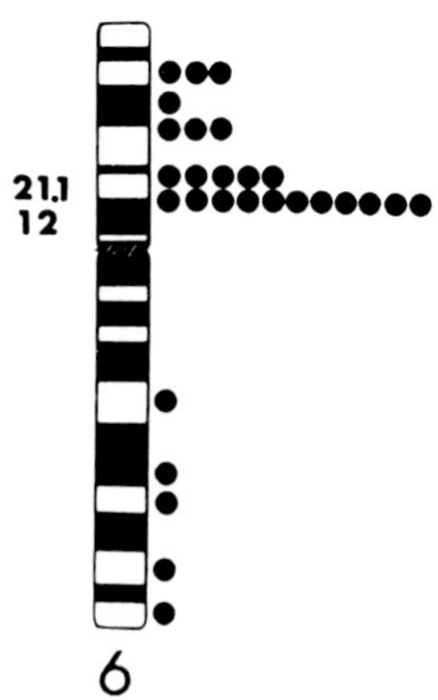

FIGURE 3. Chromosomal location of MUT locus determined by in situ hybridization with MCM probe.

DISCUSSION

This report summarizes the cloning of a full length cDNA for methylmalonyl CoA mutase and preliminary experiments with DNA mediated gene transfer of the MCM cDNA and analysis of MCM deficienct cell lines. The initial MCM clone was identified using a relatively impure antibody against placental MCM. The authentic MCM clones were identified by epitope selection of antibodies against purified mutase (8). The identity of the MCM clone was verified by DNA mediated gene transfer of MCM into COS cells which resulted in an increase in MCM activity in these cells. The observation that several mut (MCM deficient) cell lines have a specific decrease in the amount of hybridizable MCM mRNA supports the authenticity of the MCM clones (8). The MCM locus has been mapped to chromosome 6p12-21.2 (9) and restriction fragment length polymorphisms have been identified which will be useful in demonstrating linkage between the mut genotype and the MCM locus as well as between the MUT locus and other loci on chromosome 6.

Considerations for gene therapy of MMA

The preliminary gene transfer experiments described here demonstrate the feasibility of constituting L-methylmalonyl CoA mutase activity by gene transfer for biochemical analysis or gene therapy.

MMA represents an interesting model for somatic gene replacement therapy. Most cases of mut MMA present come to clinical attention in the neonatal period with acute methylmalonic acidosis associated with secondary metabolic abnormalities such as specific amino acidemia, ketoacidosis, hyperammonemia, and hypoglycemia (3). MMA can also present with neurological or mental retardation or failure to thrive. Despite therapy, most children suffer recurrent episodes of life threatening acidosis. Even the most carefully treated children generally suffer mental retardation and often die during episodes of fulminant acidosis (3).

Because of the protean clinical manifestations of severe MMA, this disorder may be an important candidate for experiments in somatic gene therapy (16). In considering gene therapy for MMA, however, it is necessary to consider the multisystem pathology associated with this disorder. MCM is ubiquitously expressed in human and rodent cells. MCM deficiency, however, is often considered a disorder of hepatic metabolism, with secondary disruptions of systemic metabolism and organ functions resulting from the action of circulating metabolites. If this understanding of the pathology of MMA is true, gene therapy directed at easily accessible sites such as bone marrow or fibroblasts might clear circulating precursors of methylmalonate but might not reverse disruptions of hepatic metabolism. Thus gene therapy might need to be targeted to the liver in order to prevent disruptions of hepatic metabolism by local (intracellular) excesses of abnormal organic acids.

It remains unclear, however, whether restitution of normal MMA activity in liver or remote sites will reverse the CNS or bone marrow pathology associated with MCM deficiency, or whether MCM activity within these cells may be important. These questions must be addressed before rational experiments in somatic gene replacement therapy can be designed. Unfortunately such data is difficult to elucidate from clinical studies given the complex phenotype of MMA deficiency. Gene transfer experiments in which MCM activity is eliminated by recombinant techniques in specific cell types in culture, or ideally in animals, may address these questions. The present experiments represent only a first step in these directions.

ACKNOWLEDGEMENTS

The cloning of MCM was performed in collaboration with Drs. Fred Kolhouse and Robert Allen of the University of Colorado Health Sciences Center, Denver CO. This work was performed with the expert technical assistance of Michele Lumetta, Ruud Jansen, and Nikki Nguyen. I thank Drs. Mark Batshaw, Lawrence Sweetman, Flemming Güttler, and David Valle, for providing cell lines and Dr. Robert Schwartz, Randy Kaufman, and Savio Woo for providing clones and libraries used in the present work. This manuscript was prepared with the assistance of Cyndi Mitchell.

REFERENCES

1. Rosenberg, L.E. 1983. Disorders of propionate and methylmalonate metabolism. Pp. 474 in J.B. Stanbury, J.B. Wyngaarden, D.S. Fredrickson, J.L. Goldstein and M.S. Brown, eds. Metabolic Basis of Inherited Disease 5th edition. McGraw Hill, New York.
2. Ledley, F.D., H.L. Levy, V.E. Shih, R. Benjamin, and M.J. Mahoney. 1984. Benign methylmalonic aciduria. New Eng. J. Med. 311:1015-1018.
3. Matsui, S.M., M.J. Mahoney, and L.E. Rosenberg. 1983. The natural history of the inherited methylmalonic acidemias. N. Engl. J. Med. 308:857-861.
4. Fenton, W.A., A.M. Hack, F.H. Willard, A. Gertler, and L.E. Rosenberg. 1982. Purification and properties of methylmalonyl coenzyme A mutase from human liver. Arch. Biochem. Biophys. 214:815-823.
5. Kolhouse, J.F., C. Utley, and R.H. Allen. 1980. Isolation and characterization of methylmalonyl-CoA mutase from human placenta. J. Biol. Chem. 255:2708-2712.
6. Fenton, W.A., M.A. Hack, D. Helfgott, and L.E. Rosenberg. 1984. Biogenesis of the mitochondrial enzyme methylmalonyl-CoA mutase. Synthesis and processing of a precursor in a cell-free system and in cultured cells. J. Biol. Chem. 259:6616-6621.
7. Fenton, W.A., A.M. Hack, J.P. Kraus, and L.E. Rosenberg. 1987. Immunochemical studies of fibroblasts from patients with methylmalonyl-CoA mutase apoenzyme deficiency: detection of a mutation interfering with mitochondrial import. Proc. Natl. Acad. Sci. USA 84:1421-1424.
8. Ledley, F.D., M.R. Lumetta, P.N. Nguyen, J.F. Kolhouse, and R.H. Allen. Molecular cloning of methylmalonyl CoA mutase: gene transfer and analysis of _mut_ cell lines. Proc Natl Acad Sci USA (in press).
9. Ledley, F.D., M.R. Lumetta, H.Y. Zoghbi, P. VanTuinen, S.A. Ledbetter, and D.H. Ledbetter. Mapping of human methylmalonyl CoA mutase (MUT) locus on chromosome 6. Am. J. Hum. Genet. (in press).
10. Kwok, S.C.M., F.D. Ledley, A.G. DiLella, K.J.H. Robson, and S.L.C. Woo. 1985. Nucleotide sequence of a full-length cDNA clone and amino acid sequence of human phenylalanine hydroxylase. Biochem. 24:556-561.

11. Kolhouse, J.F., C. Utley, W.A. Fenton, and L.E. Rosenberg. 1981. Immunochemical studies on cultured fibroblasts from patients with inherited methylmalonic acidemia. Proc. Natl. Acad. Sci. USA 78:7737-7741.

12. Weinberger, C., S.M. Hollenberg, E.S. Ong, J.M. Harmon, S.T. Brower, J. Cidlowski, E.B. Thompson, M.G. Rosenfeld, and R.M. Evans. 1985. Identification of human glucocorticord receptor complementary DNA clones by epitope selection. Science 228:740-742.

13. Mellon, P., V. Parker, Y. Gluzman, T. and Maniatis, T. 1981. Identification of DNA sequences required for transcription of the human alpha$_1$-globin gene in a new SV40 host-vector system. CELL 27:279-288.

14. Graham, F.L., and A.J. VanderEb. 1973. A new technique for the assay of infectivity of human adenovirus 5 DNA. Virology 52:456-467.

15. Kolhouse, J.F., S.P. Stabler, and R.H. Allen. 1987. Methylmalonyl CoA mutase. Method in Enzymol. (in press).

16. Ledley, F.D. 1987. Somatic gene therapy for human disease: background and prospects. J. Pediatr. 110:1-8;167-174.

Gene Transfer and Gene Therapy, pages 345–354

GENETIC HETEROGENEITY IN GAUCHER DISEASE: IDENTIFICATION OF MULTIPLE ALLELIC MUTATIONS AND IMPLICATIONS FOR THERAPEUTIC APPROACHES UTILIZING GENE TRANSFER[1]

Shoji Tsuji[2], Brian M. Martin, and Edward I. Ginns

Molecular Neurogenetics Section
Clinical Neuroscience Branch, DIRP, NIMH
Bethesda, Maryland 20892

ABSTRACT A single base mutation (A to G) was found in exon 9 of a glucocerebrosidase gene from an Ashkenazic type 1 Gaucher patient. This results in the amino acid substitution of Ser for Asn at position 370 in the enzyme. Allele-specific hybridization with oligonucleotide probes demonstrated that this mutation was exclusively found in the type 1 patients in the population studied. Some Ashkenazic Jewish and non-Jewish type 1 patients had only one allele with this mutation. Furthermore, a type 1 Gaucher patient had one allele with the substitution of Ser for Asn at position 370, and the previously described substitution of Leu for Pro at amino acid 444 encoded by the other allele. These findings demonstrate that there are multiple allelic mutations responsible for type 1 Gaucher disease in both Ashkenazic and non-Ashkenazic individuals. The results also suggest that the presence of mutant glucocerebrosidase encoded by an allele having the Asn to Ser mutation at amino acid 370 might prevent the development of central nervous system abnormalities.

---

[1]This work was supported in part by grants from the National Gaucher Foundation.

[2]Present address: Department of Neurology, Brain Research Institute, Niigata University, 1 Asahimachi-dori, Niigata 951, JAPAN

INTRODUCTION

The sphingolipidosis called Gaucher disease results from mutations in the gene for the lysosomal enzyme glucocerebrosidase (EC 3.2.1.45, β-D-glucosyl-N-acylsphingosine glucohydrolase) (1,2). Although the biochemical defect is present in all tissues, the accumulation of the neutral glycosphingolipid glucosylceramide occurs predominantly within the lysosomes of reticuloendothelial cells. As a consequence of this accumulation, a prominent feature of this disorder is the presence of Gaucher cells within tissue. These cells are derived from Kupffer cells in the liver, alveolar macrophages in the lung, osteoclasts and macrophages within the bone marrow, and histiocytes in the spleen. On the basis of clinical signs and symptoms, Gaucher disease has been divided into three major phenotypes (3): type 1, non-neuronopathic; type 2, acute neuronopathic; and type 3, chronic neuronopathic. Type 1, the most common form, is characterized by hepatosplenomegaly, thrombocytopenia, anemia, and bone complications. Type 1 occurs in increased frequency in the Ashkenazic Jewish population (4). In contrast, there is no ethnic predilection of types 2 or 3 Gaucher disease. Type 2 patients usually have symptom onset by six months of age and manifest cranial nerve and brainstem abnormalities, as well as the other symptoms observed in type 1 patients. Death occurs in type 2 patients between two and three years of age and usually results from respiratory complications. Type 3 Gaucher patients have the systemic symptoms found in type 1 disease, in addition to neurologic abnormalities that appear during childhood or adolescence. Type 1 and type 3 patients have a much more variable symptom onset and severity than the more stereotypic course seen in type 2 patients. The absence of functional complementation between phenotypes in somatic-cell hybridization studies (5), and the ethnic predilection seen only in type 1 Gaucher disease (4), suggested that the phenotypes of this disorder were each a result of different allelic mutations.

The isolation of cDNA for human glucocerebrosidase (6-8) permitted the identification of the chromosomal locus at 1q21 (9) and the isolation of genomic clones for human glucocerebrosidase (10,11). The normal glucocerebrosidase gene has 11 exons and 10 introns contained within approximately 7 kb, with consensus sequences for the putative promoter and splice junctions. In addition, at locus 1q21 there is another region that is highly homologous to the glucocerebrosidase gene but lacks parts of several exons and splice junctions. These sequence deletions, as well as Southern blot analyses of total genomic DNA from control subjects and types 1, 2 and 3 Gaucher patients suggest that this latter region is a pseudogene (12).

We previously reported the isolation and characterization from a type 2 Gaucher patient of a glucocerebrosidase genomic clone having a mutation (Pro for Leu at amino acid 444) that is frequently found in types 2 and 3 Gaucher disease (12), often only in one allele. We have now identified a single base mutation (A to G) in exon 9 of a glucocerebrosidase genomic clone from an Ashkenazic Jewish patient with type 1 Gaucher disease (13). Other type 1 patients had one or both alleles with this mutation, indicating, as is the case for type 2 disease, that multiple allelic mutations are also responsible for type 1 Gaucher disease in both the Jewish and non-Jewish populations. The occurrence of multiple genotypes within phenotypes of Gaucher disease is discussed.

MATERIALS AND METHODS

Tissues and Cells

All patients, except two with type 2 Gaucher disease, were seen at the National Institutes of Health under informed consent. The GM1260 and GM877 type 2 Gaucher fibroblast cell lines were obtained from the Human Genetic Mutant Cell Repository (Camden, NJ). The ethnic origins and ages of patients and controls have been previously described (12,13).

Preparation of DNA

High-molecular weight DNA was prepared from autopsy or biopsy tissues, mononuclear blood cells, or cultured skin fibroblasts as previously described (14).

Genomic Library Construction, Clone Isolation and Sequence Analysis

Genomic libraries were constructed from EcoR1 digested DNA extracted from cultured fibroblasts both from an Ashkenazic type 1 and a type 2 Gaucher patient (12,13). Recombinant bacteriophage carrying the glucocerebrosidase gene were identified by plaque hybridization using glucocerebrosidase cDNA as the probe. The BamHl endonuclease fragments of the human genomic DNA inserts from lambda clones were subcloned and subsequently sequenced by the dideoxynucleotide chain-terminator technique (15). Nucleotide sequence of all exons and splice junctions, as well as the 5'- and 3'-flanking regions of both the type 1 and type 2 glucocerebrosidase genes was determined.

Southern Blot Analysis

For Southern blot analysis, 5 micrograms of DNA was digested with a restriction enzyme, fragments were fractionated electrophoretically in an agarose gel, and transferred to a nitrocellulose filter. In order to demonstrate the NciI mutation, the filters were hybridized to a 1.1 kb KpnI cloned genomic DNA subfragment of a normal genomic clone as previously described (12).

Allele Specific Hybridization Analysis

The presence of the single base mutation in exon 9 in type 1 patients was demonstrated using oligonucleotide probes because the mutation did not result in the alteration of an endonuclease restriction site (13). Using a 19-mer oligonucleotide with the normal sequence (5'TACCCTA-GAACCTCCTGTA3', probe A), a 19-mer oligonucleotide with the type 1 mutant sequence (5'TACCCTAGAGCCTCCTGTA3', probe B), and a 9-mer primer oligonucleotide (5'TACAGGAGG3') high specific radioactivity probes were synthesized by primer extension reactions (12,13).

In-vitro Mutagenesis and Transient Expression

The type 1 mutation identified in exon 9 of the glucocerebrosidase gene from the Ashkenazic Jewish patient and the type 2 mutation found in exon 10 were each separately introduced into a normal human glucocerebrosidase cDNA clone isolated from an Okayama-Berg pcDX library constructed from simian virus 40 transformed normal human fibroblasts (16). Glucocerebrosidase activity was determined and Western blots (17) done on extracts from COS cells transfected with both normal and mutagenized clones.

RESULTS

Identification of a Mutation in the Glucocerebrosidase Gene in an Ashkenazic Individual with Type 1 Gaucher Disease

The genomic clone from the type 1 Ashkenazic patient had a 14 kb insert and a restriction pattern that was identical to that of the normal genomic clone. Nucleotide sequence analysis of the 5'- and 3'-flanking regions, splice junctions, and all the exons revealed only a single base change (A to G) in exon 9 of this type 1 glucocerebrosidase gene (13). This mutation results in the replacement of Asn by Ser at amino acid 370 as shown in Table 1.

TABLE 1
TYPE 1 MUTATION IN EXON 9

```
           Asn Leu Leu Tyr His Val Val Gly Trp
..cttaccctag AAC CTC CTG TAC CAT GTG GTC GGC TGG..
            ↓
           AGC
           Ser
```

Allele specific hybridization analysis of the family of the type 1 Ashkenazic patient whose genomic clone was sequenced demonstrates that the mutation has been transmitted from the paternal grandfather through three generations. The mother, however, does not have an allele with this A to G mutation in exon 9 and therefore must have an allele with an as yet unidentified mutation in the glucocerebrosidase gene.

Using oligonucleotide probes that identify either the normal or mutant sequence in exon 9, 24 type 1 patients, 6 type 2 patients, and 11 type 3 patients, as well as 12 normal controls were studied. As seen in Table 2, none of the normal controls, type 2 or type 3 patients studied had this mutant allele. In contrast, 18 type 1 patients had at least one allele with this mutation.

A single base substitution (T to C) in exon 10 has previously been demonstrated in the cloned glucocerebrosidase gene from a patient with type 2 Gaucher disease (12). This mutation produces a new cleavage site for the NciI restriction endonuclease. Although this mutation

TABLE 2
OCCURRENCE OF THE A TO G MUTATION IN EXON 9

Subjects	Genotype +/+	+/-	-/-
Normal controls	12	0	0
Patients with Gaucher disease			
Type 1	6	15	3
Type 2	6	0	0
Type 3	11	0	0

occurs in high frequency in patients with types 2 and 3 Gaucher disease, twenty percent of the type 1 patients studied had one allele with this T to C change in exon 10. Interestingly, one type 1 patient had one allele with the Asn to Ser mutation in exon 9 and another allele with the Leu to Pro mutation in exon 10.

Full length mutant cDNA clones containing either the A to G (exon 9) or T to C (exon 10) mutation were constructed as previously described (12,13). COS cells transfected with the mutant cDNAs had no significant increase in glucocerebrosidase activity, but the appropriate human protein polymorphism was demonstrated on immunoblot analysis (12,13,18-20).

DISCUSSION

The maturation and correct targeting of human glucocerebrosidase to lysosomes depends on multiple cotranslational and posttranslational modifications (18). Glucocerebrosidase is initially synthesized on polyribosomes with a nineteen amino acid signal peptide that is cotranslationally removed (3,19). Posttranslational processing of the four Asn-linked oligosaccharide chains, ultimately resulting in the formation of complex type oligosaccharides, occurs within the endoplasmic reticulum and Golgi (3,19). The enzyme is then translocated to the lysosome by a mechanism other than the mannose-6-phosphate pathway. After the signal peptide is removed, all of the different species observed by either immunoprecipitation of in vitro cell-free translation products or pulse-chase analysis are due to oligosaccharide remodelling (19).

The allelic mutations causing Gaucher disease result in the synthesis of catalytically deficient glucocerebrosidases that differ from the normal enzyme in posttranslational processing (17,20), compartmentalization (21), and/or stability (22). It has previously been demonstrated that discrimination between the phenotypes of Gaucher disease is possible from either Western blot (17,20) or pulse-chase analyses (19,22) using polyclonal or monoclonal antiserum against homogeneous human placental glucocerebrosidase. Western blot analysis of Percoll-gradient fractionated normal human fibroblast extracts has demonstrated that the lysosomal form of glucocerebrosidase is that having the lowest apparent molecular mass (23).

Since type 1 Gaucher disease is the most prevalent form, investigators have sought to identify the mutations in the type 1 gene that might provide a basis for diagnostic tests useful for genetic counseling. It has previously been suggested that the Ashkenazic Jews with type 1 Gaucher disease might have a single gene mutation (24), but linkage analysis and biochemical studies have recently provided evidence for genetic heterogeneity in this population (25). Our present study clearly demonstrates that there are at least four genotypes of the glucocerebrosidase gene in type 1 Gaucher disease: (a) Ser-370/Ser-370, (b) Ser-370/Pro-444, (c) Ser-370/X, and (d) X/X, where X is an as yet unidentified mutation. We found three of these genotypes (a,b, and c) in Ashkenazic Jewish individuals, clearly indicating that there is more than one allelic mutation in the glucocerebrosidase gene in Ashkenazic Jews. Furthermore, this A to G mutation in exon 9 is found in both Jewish and non-Jewish type 1 Gaucher patients.

We previously reported that the Leu to Pro mutation in exon 10 occurred frequently in neuronopathic (types 2 and 3) Gaucher disease, and that all patients homozygous for this mutation had neurologic abnormalities (12). Although twenty percent of the type 1 Gaucher patients studied had this mutant NciI site in exon 10, all of these type 1 individuals were heterozygous for this mutation. In contrast, the Asn to Ser mutation in exon 9 has been found only in type 1 Gaucher patients (13). Furthermore, the fact that an adult with type 1 Gaucher disease had both the Asn to Ser (in exon 9 of one allele) and Leu to Pro (in exon 10 of one allele) mutations suggests that the presence of mutant glucocerebrosidase coded by the allele having the exon 9 Asn to Ser mutation might prevent patients from developing central nervous system abnormalities. These findings also suggest that the glucocerebrosidase mRNA and protein within a cell may be heterogeneous. Thus, caution must be exercised in the interpretation of both biochemical and kinetic data on mutant glucocerebrosidase in a patient's cells.

In Gaucher disease where clinical symptomatology is the result of lipid storage in cells derived from hematopoietic progenitors (3), both enzyme replacement and gene transfer are potential therapeutic approaches (11). The prevention of neurological abnormalities by the type 1 allele (Asn to Ser 370) in individuals also having the allele containing the Leu to Pro mutation in exon 10 suggests that retroviral mediated transfer of the normal glucocerebrosidase gene into a patient's macrophage cell line would be beneficial (11,13).

Two mutations in the glucocerebrosidase gene have thus far been identified in Gaucher patients. One mutation is exclusively found in type 1 patients, while the other is frequently found in types 2 and 3 Gaucher disease. Since approximately eighty percent of Gaucher patients appear informative with respect to these mutations, testing for these mutations should be useful for diagnosis and genetic counseling. Identification of additional mutations in the glucocerebrosidase gene should provide a more complete understanding of this disorder.

ACKNOWLEDGMENTS

The authors thank Ms. Kara Maysak, Ms. Barbara Stubblefield, Ms. Mary LaMarca, Ms. Suzanne Winfield, and Mr. William Eliason for their technical assistance and suggestions, and Ms Andrea Hobbs for help in manuscript preparation.

REFERENCES

1. Brady RO, Kanfer JN, Shapiro D (1965). Metabolism of gluccoerebrosides. II. Evidence of an enzymatic deficiency in Gaucher's disease. Biochem Biophys Res Commun 18:221.
2. Patrick AD (1965). A deficiency of glucocerebrosidase in Gaucher's disease. Biochem J 97(Suppl):17c.
3. Barranger JA, Ginns EI (1988). Glucosylceramide lipidoses: Gaucher disease. In Scriver CR, Beaudet AL, Sly WS, Valle D (eds): "The Metabolic Basis of Inherited Disease," New York: McGraw-Hill (in press).

4. Kolodny EH, Ullman MD, Mankin HJ, Raghavan SS, Topol J, Sullivan JL (1982). Phenotypic manifestations of Gaucher disease: Clinical features in 48 biochemically verified type 1 patients and comments on type 2 patients. Prog Clin Biol Res 95:33.
5. Gravel RA, Leung A (1983). Complementation analysis in Gaucher disease using single cell microassay techniques. Evidence for a single "Gaucher gene." Hum Genet 65:112.
6. Ginns EI, Choudary PV, Martin BM, Winfield S, Stubblefield B, Mayor J, Merkle-Lehman D, Murray GJ, Bowers LA, Barranger JA (1984). Isolation of cDNA clones for human β-glucocerebrosidase using the λgt11 expression system. Biochem Biophys Res Commun 123:57.
7. Sorge J, Gelbart T, West C, Westwood B, Beutler E (1985). Molecular cloning and nucleotide sequence of human glucocerebrosidase cDNA. Proc Natl Acad Sci USA 82:7289.
8. Tsuji S, Choudary PV, Martin BM, Winfield S, Barranger JA, Ginns EI (1986). Nucleotide sequence of cDNA containing the complete coding sequence for human lysosomal glucocerebrosidase. J Biol Chem 261:50.
9. Ginns EI, Choudary PV, Tsuji S, Martin B, Stubblefield B, Sawyer J, Hozier J, Barranger JA (1985). Gene mapping and leader polypeptide sequence of human glucocerebrosidase: Implications for Gaucher disease. Proc Natl Acad Sci USA 82:7101.
10. Choudary PV, Ginns EI, Barranger JA (1986). Molecular cloning and analysis of the human β-glucocerebrosidase gene. DNA 5:78.
11. Choudary PV, Tsuji S, Martin BM, Guild BC, Mulligan RC, Murray GJ, Barranger JA, Ginns EI (1986). The molecular biology of Gaucher disease and the potential for therapy. In "Cold Spring Harbor Symposia on quantitative Biology LI," New York: Cold Spring Harbor Laboratory, p 1047.
12. Tsuji S, Choudary PV, Martin BM, Stubblefield BK, Mayor JA, Barranger JA, Ginns EI (1987). A mutation in the human glucocerebrosidase gene in neuronopathic Gaucher's disease. N Engl J Med 316:570.
13. Tsuji S, Martin BM, Barranger JA, Stubblefield B, LaMarca ME, Ginns EI (1988). Genetic Heterogeneity in type 1 Gaucher disease: Multiple genotypes in Ashkenazic and non-Ashkenazic individuals. Proc Natl Acad Sci USA (in press).

14. Maniatis T, Fritsch EF, Sambrook J (1982). "Molecular cloning: A laboratory manual." New York: Cold Spring Harbor Laboratory.
15. Biggin MD, Gibson TJ, Hong GF (1983). Buffer gradient gels and 35S label as an aid to rapid DNA sequence determination. Proc Natl Acad Sci USA 80:3963.
16. Okayama H, Berg P (1983). A cDNA cloning vector that permits expression of cDNA inserts in mammalian cells. Mol Cell Biol 3:280.
17. Ginns EI, Brady RO, Pirruccello S, Moore C, Sorrell S, Furbish FS, Murray GJ, Tager JM, Barranger JA (1982). Mutations of glucocerebrosidase: Discrimination of neurologic and non-neurologic phenotypes of Gaucher disease. Proc Natl Acad Sci USA 79:5607.
18. Ginns EI (1985). The molecular biology of the sphingolipid hydrolases. Bio Essays 2:118.
19. Erickson AH, Ginns EI, Barranger JA (1985). Biosynthesis of the lysosomal enzyme glucocerebrosidase. J Biol Chem 260:14319.
20. Ginns EI, Tegelaers FPW, Barneveld R, Galjaard H, Reuser AJJ, Tager JM, Brady RO, Barranger JA (1983). Determination of Gaucher's disease phenotypes with monoclonal antibody. Clin Chim Acta 131:283.
21. Willemsen R, Van Dongen JM, Ginns EI, Sips HJ, Schram AW, Tager JM, Barranger JA, Reuser AJJ (1987). Ultrastructural localization of glucocerebrosidase in cultured Gaucher's disease fibroblasts by immunocytochemistry. J Neurol 234:44.
22. Jonsson LMV, Murray GJ, Sorrell S, Strijland A, Aerts JFGM, Ginns EI, Barranger JA, Tager JM, Schram AW (1987). Biosynthesis and maturation of glucocerebrosidase in Gaucher fibroblasts. Eur J Biochem 164:171.
23. Ginns EI, Lazlow DJ, Moore C, Brady RO, Barranger JA (1985). Characterization of lysosomal glucocerebrosidase in normal and Gaucher's disease tissues. Fed Proc 44:709.
24. Motulsky AG (1979). In Goodman RM, Motulsky AG (eds): "Genetic Diseases Among Ashkenazi Jews," New York: Raven Press, p 167.
25. Beutler E, Kuhl W, Sorge J (1984). Cross-reacting material in Gaucher disease fibroblasts. Proc Natl Acad Sci USA 81:6506.

Gene Transfer and Gene Therapy, pages 355–363

RETROVIRAL-MEDIATED GENE TRANSFER OF HUMAN PAH INTO MOUSE PRIMARY HEPATOCYTES

D. Armentano[1], H. Peng[1], L. MacKenzie-Graham[1], M. Seh[1], R.F. Shen[1], F.D. Ledley[1], G.J. Darlington[1], and S.L.C. Woo[1]

Howard Hughes Medical Institute
Department of Cell Biology
Institute for Molecular Genetics
Baylor College of Medicine
Houston, TX 77030

ABSTRACT Phenylketonuria is caused by a deficiency of the hepatic enzyme, phenylalanine hydroxylase. A vector, pNASPAH was constructed by cloning a human PAH cDNA with a hybrid liver-specific promoter into the retroviral vector, N2. Recombinant virus was produced by introducing pNASPAH into the packaging cell lines, Ψ2 and PA317. Primary mouse hepatocytes were infected with recombinant virus and abundant expression of human PAH RNA was observed. The results demonstrated the feasibility of retroviral-mediated gene transfer into hepatocytes and the potential of this approach for somatic gene therapy of PKU.

INTRODUCTION

Defects at the gene and protein level have been identified for many heritable disorders. Mutations that define the molecular basis for the deficiency of phenylalanine hydroxylase (PAH) in the metabolic

---

[1]This work was partially supported by NIH grants HD-21452 and HD-17711. S.L.C. Woo is an Investigator and F.D. Ledley is and Assistant Investigator of Howard Hughes Medical Institute

disorder, phenylketonuria (PKU), have been characterized (1-3). PAH is a hepatic enzyme that converts phenylalanine to tyrosine, thus, in PKU, phenylalanine accumulates and results in abnormalities of aromatic amino acid metabolism. If left untreated, such abnormalities cause postnatal brain damage and severe mental retardation (4,5). PKU is transmitted as an autosomal recessive trait and occurs at a frequency of about 1 in 10,000 Caucasian births.

Restriction of dietary phenylalanine intake is currently the only treatment for PKU. The diet must be implemented vigorously during the first few years of life and followed through for as long as the patients can bear to be effective (6). Recent advances in the ability to transfer genes into mammalian cells allows consideration for an alternative therapeutic strategy for PKU (7,8). The unique properties of retroviruses have been utilized for their development as vectors for efficient gene transfer and ultimately as a potential approach for somatic gene therapy of many genetic disorders. Retroviral vectors can be engineered to contain exogenous genes and the recombinant viruses can efficiently infect a wide variety of host cells. This results in stable integration of the exogenous gene into the host genome, and in many instances, its expression. We have explored the possibility of using recombinant retroviruses for the transfer and expression of PAH and the potential of somatic gene therapy of PKU (9). In preliminary experiments we tested the ability of PAH to be transferred to two mammalian cell lines that normally lack PAH activity. We previously demonstrated that infection of NIH 3T3 cells and hepa 1-a cells with recombinant virus containing human PAH cDNA leads to expression of PAH mRNA, immunoreactive protein and enzymatic activity in vitro.

Until recently, most gene replacement therapy has focused on the correction of various disorders by in vitro infection of bone marrow with recombinant retroviruses followed by transplantation of the infected bone marrow into animals. When considering a target tissue for gene therapy of PKU it is important to take into account that the complete metabolic reaction for the hydroxylation of phenylalanine involves two steps. In addition to PAH, the first step requires a reduced pterin cofactor, tetrahydrobiopterin, which becomes oxidized concommittently with the hydroxylation of phenylalanine. The oxidized cofactor is then reduced by dihydropterin reductase which allows reutilization of the cofactor (10). The liver contains the enzymes required for synthesis and reduction of the cofactor,therefore, hepatocytes may be the desired target for gene therapy of PKU.

Early studies failed to demonstrate that hepatocytes could be infected with retroviruses in vivo after birth or following partial hepatectomy (11-15). We explored the potential of retroviral-mediated gene transfer into primary mouse hepatocytes in vitro with a variety of recombinant viruses containing the bacterial neomycin resistance gene, Tn5 (16). We were able to demonstrate that in the presence of hormonally defined medium, mouse primary hepatocytes divide in short term culture, can be infected with recombinant retrovirus and express G418 resistance. In addition, these cells continued to express liver-specific functions. In this report we describe the construction of a recombinant retrovirus containing the bacterial neomycin resistance gene and human PAH cDNA which is expressed from a liver-specific promoter. The recombinant virus was used to infect primary hepatocytes in vitro and abundant expression of human PAH RNA in the infected cells was observed. The results demonstrated the feasibly of transducing functional genes into mouse primary hepatocytes by retroviral mediated gene transfer and provide the basis for further investigation of this approach to somatic gene therapy of PKU.

RESULTS

Construction of a recombinant retroviral vector containing human PAH cDNA under the control of a liver-specific promoter.

The retroviral vector N2 was chosen as the parental vector for gene transfer of human PAH cDNA. N2 is a Moloney murine leukemia virus-based vector in which the coding regions for viral structural proteins have been deleted and replaced with the bacterial neomycin resistance gene that serves a dominant selectable marker (17). In order to direct PAH cDNA expression from the vector in hepatocytes, the transcriptional regulatory elements of the human α1-antitrypsin gene (AAT) were used in conjunction with the TATA box and mRNA cap site of the SV40 early promoter. The AAT elements were used because this gene is abundantly expressed in the liver and because they were shown to drive the expression of a reporter gene in livers of transgenic mice (18, unpublished observations). A chimeric "minigene", created by linking the hybrid promoter and the PAH cDNA, was then cloned into the XhoI site of N2 (pNASPAH).

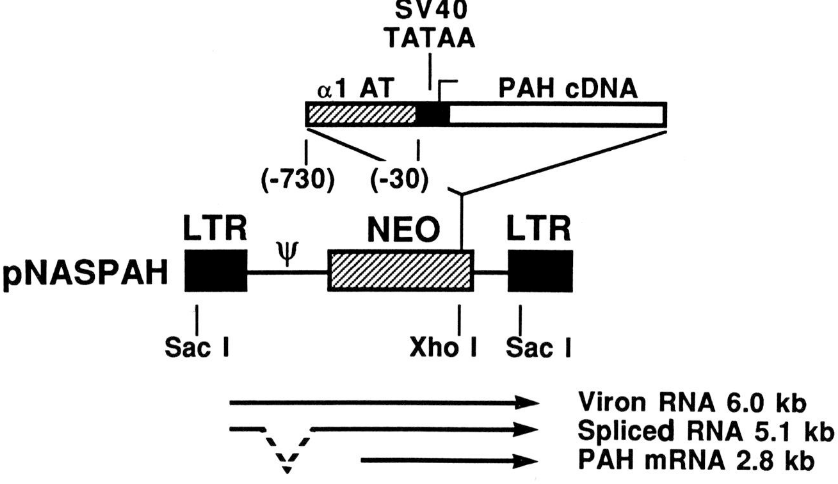

FIGURE 1. The structure of a recombinant retroviral vector containing a human PAH cDNA and a liver-specific promoter.

Production of Amphotropic recombinant virus

High titer virus was produced by infecting the amphotropic packaging cell line, PA317 (19), with transient supernatants derived from Ψ2 (20) ecotropic packaging cell lines that had been transfected with pNASPAH or N2. Single colonies were isolated and screened for the production of virus by assaying supernatants for titer of G418 resistant cfu/ml on rat 208F cells. The highest virus producer cell lines for

N2 and pNASPAH were expanded and had titers of $1x10^6$ and $2x10^5$ G418 resistant cfu/ml, respectively.

Infection of mouse primary hepatocytes and selection for G418 resistance.

Primary hepatocytes were isolated from neonatal mice and plated in defined hormonal medium without serum as previously described (16). Hepatocyte cultures were infected with high titer virus preparations of N2 or pNASPAH and were then selected with 250ug/ml G418. In the absence of G418, cultures of uninfected cells grew to confluence in 10 days (Panel A). Control plates of uninfected cells that were subjected to G418 died within 5 days (Panel B). Viable colonies with characteristic morphology of hepatocytes remained in cultures infected with N2 or pNASPAH and selected with G418 (Panels C and D, respectively). The efficiency of gene transfer was greater for N2 than was for pNASPAH which most probably reflects the difference in titer of these two viruses.

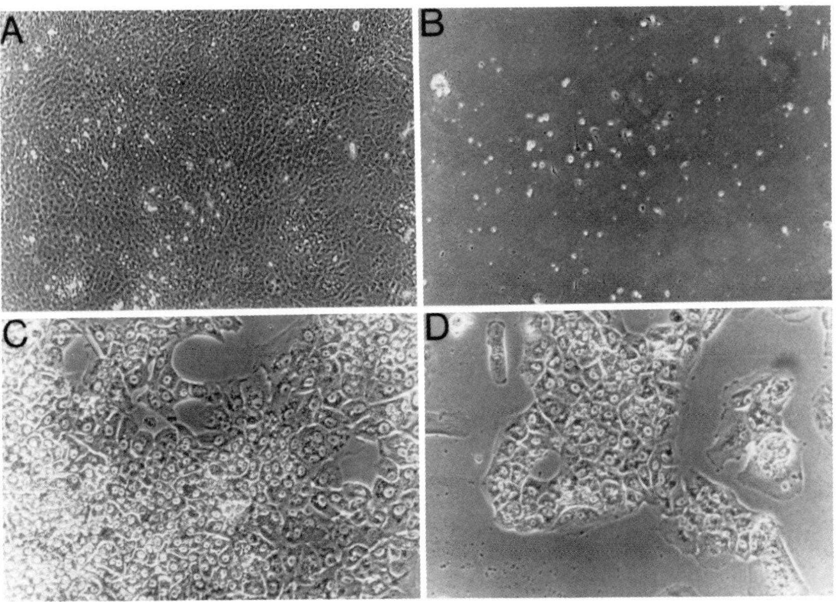

FIGURE 2. Infection of primary hepatocytes with recombinant retroviruses and selection with G418.

Expression of human PAH in virus-infected primary mouse hepatocytes.

Total RNA isolated from mouse hepatocytes infected with recombinant virus was examined by Northern blot analysis for the presence of transcripts containing human PAH sequence. Hepatocytes infected with pNASPAH and selected with G418 contain three RNA species (Fig. 3, lane 4). The longer transcripts correspond to full length genomic proviral RNA and a spliced RNA product which is generated from cryptic splice sites in N2. These two RNAs are also detected in the virus producing cell lines (Fig. 3, lane 5). The smallest RNA species (Fig. 3, lane 4) is present in hepatocytes infected with pNASPAH, but absent in the virus producing cell line. This transcript corresponds to the expected length of RNA which is initiated from the internal AAT-SV40 hybrid promoter and terminated in the 3' LTR. The uninfected hepatocytes do not contain any RNA detectable with human PAH cDNA probe under the stringent wash conditions employed (Fig. 3, lane 2). The steady-state levels of human PAH RNA in the virally infected hepatocytes are comparable to that present in normal human liver, which is shown as a 2.5kb band (Fig. 3, lane 1).

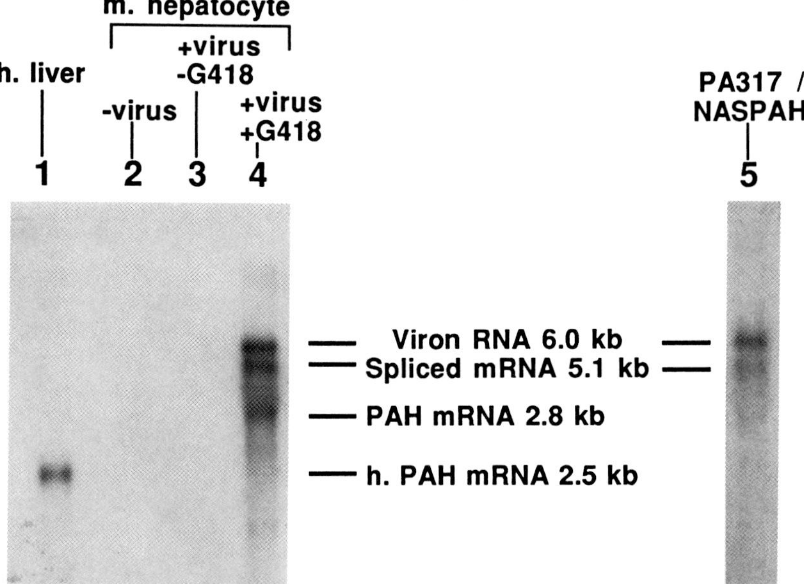

FIGURE 3. RNA analysis of hepatocytes infected with recombinant virus.

DISCUSSION

These experiments describe the construction of a recombinant retroviral vector, pNASPAH, for retroviral-mediated gene transfer of human PAH into mouse primary hepatocytes. As in the parental vector, N2 (17), the neomycin resistance gene in pNASPAH is expressed from the viral 5′ LTR, whereas the PAH cDNA is targeted for expression in hepatocytes by the transcriptional regulatory elements of the human α1-antitrypsin gene (18).

In primary hepatocyte cultures, the presence of hormonally defined medium sustains proliferation of hepatocytes but prevents the overgrowth of fibroblasts and endothelial-like cells (21). Cells that are transformed to G418 resistance by infection with the recombinant viruses, N2 and pNASPAH, exhibit hepatocyte morphology. For cells infected with pNASPAH, RNA analysis indicates that the α1-antitrypsin transcriptional regulatory elements are activated and drive the expression of human PAH mRNA. This further demonstrates that these cells are hepatocytes since this promoter is not active in the virus producing cell line which is of fibroblast origin.

For gene therapy to be clinically useful, the transferred gene must produce an enzymatically active protein that is expressed in an appropriate level for correction of the disorder. Since mouse hepatocytes produce endogenous PAH it is not possible to quantitatively measure PAH activity produced from the exogenous human PAH cDNA. However, we have previously demonstrated that this cDNA encodes for anenzymatically active protein (9). Both mouse and human PAH have a molecular weight of 54kd and are inseparable by Western blot or isoelectric focusing. We are currently analyzing protein extracts from infected hepatocytes for the presence of human PAH by analysis of CNBr cleavage products of immunoprecipitated protein. The sizes of the cleavage products for mouse and human PAH are different and should allow detection of human PAH.

Somatic cell gene therapy will also require the successful transplantation of the modified target cells into the host. It should be noted that individuals with only 1-5% of the normal hepatic PAH activity have only moderate elevations in the level of blood phenylalanine and do not suffer from mental retardation (22).

Therefore, the transplantation of a relatively fewhepatocytes that express functional PAH at a high level may be sufficient to reduce the blood level of phenylalanine for treatment of PKU.

The experiments described here, taken together with previously published studies, establish the feasibility of using recombinant retroviral vectors to transfer and express human PAH in primary hepatocytes (9,16). This approach to somatic gene therapy may provide a clinical means for correcting PKU and perhaps other diseases associated with deficiencies of hepatic proteins.

ACKNOWLEDGMENTS

We thank Dr. Eli Gilboa for providing us with N2, Dr. Dusty Miller for providing the PA317 cell line, and Dr. Richard mulligan for the Ψ2 cell line.

REFERENCES

1. DiLella, A.G., Marvit, J., Lidsky, A.S., Guttler, F., and Woo, S.L.C. (1986). Nature 322, 799-803.
2. DiLella, A.G., Marvit, J., Brayton, K. and Woo, S.L.C. (1987). Nature 327, 333-336.
3. Lichter-Konecki, U., Konecki, D.S., DiLella, A.G., Brayton, K., Marvit, J., Hahn, T.M., Trefz, F.K. and Woo, S.L.C. (1987). Biochemistry
4. Folling, A. (1934). Z. Physiol. Chem. 227, 169-176.
5. Scriver, C.R., and Clow, C.L. (1980) N. Engl. J. Med. 303, 1336-1342.
6. Seashore, M.R., Friedman, E., Novelly, R.A. and Bapat, V. (1985) Pediatrics 75, 226-262.7. Anderson, W.F. (1984). Science 226, 401-409.
8. Williams, D.A. and Orkin, S.H. (1986) J. Clin. Invest. 77, 1053-1056.
9. Ledley, F.D., Grenett, H.E, McGinnis-Shelnutt and Woo, S.L.C. (1986). Proc. Natl. Acad. Sci. USA 83, 409-413.
10. Kaufman, S. (1976). Adv. Neurochem. 2. 1-132.
11. Jaenish, R. (1980) Cell 19, 181-186.
12. Harbers, K., Jahner, D. and Jaenisch, R. (1981) Nature (London) 293, 540-542.
13. Simon, I., Lohler, J. and Jaenisch, R. (1982) Virology 120, 106-121.

14. Stuhlmann, H., Cone, R., Mulligan, R.C. and Jaenisch R. (1984) Proc. Natl. Acad. Sci. USA 81, 7151-7155.
15. Jaenisch, R. and Hoffman, E. (1979). Virology 98, 289-297.
16. Ledley, F.D., Darlington, G.J., Hahn, T. and Woo, S.L.C. (1987). Proc. Natl. Acad. Sci. USA 84, 5335-5339.
17. Armentano, D., Yu, S.F., Kantoff, P.W., von Ruden, T., Anderson, W.F. and Gilboa, E. (1987) J. Virol. 61, 1647-1650.
18. Shen, R.F., Li, Y., Sifers, R.N., Wang, H., Hardick, C., Tsai, S. and Woo, S.L.C. (1987). Nucl. Acids Res. 15, 8399-8415.
19. Miller, A.D. and Buttimore, C. (1986). Mol. Cell. Biol. 6, 2895-2902.
20. Mann, R., Mulligan, R.C. and Baltimore, D. (1983). Cell 33, 153-159.
21. Darlington, G.J., Kelly, J.H. and Buffone, G.J. (1987). In Vitro Cell Dev. Biol.
22. Knox, W.E. (1972). in Metabolic Basis of Inherited Disease, eds. Stanbury,J.B., Wyngaarden, J.B. and Fredrickson, D.S. (McGraw-Hill, New York), pp. 266-294.

Gene Transfer and Gene Therapy, pages 365–374

TRANSFER AND EXPRESSION OF THE HUMAN ADENOSINE DEAMINASE (ADA) GENE IN ADA-DEFICIENT HUMAN T LYMPHOCYTES WITH RETROVIRAL VECTORS

Donald B. Kohn[1]*, Philip Kantoff[2], James Zwiebel[2], Eli Gilboa[3], W. French Anderson[2] and R. Michael Blaese[1]

[1] Metabolism Branch, N.C.I., N.I.H., Bethesda, MD 20892

[2] Laboratory of Molecular Hematology N.H.L.B.I., N.I.H., Bethesda, MD 20892

[3] Dept. of Molecular Biology, Memorial Sloan-Kettering Cancer Center, New York, NY 10021

*Current address: Division of Research Immunology & Bone Marrow Transplantation, Childrens Hospital of Los Angeles, U.S.C. Medical School, 4650 Sunset Boulevard Los Angeles, California 90027

ABSTRACT Transfer of the human ADA gene into an HTLV-1 transformed, ADA-deficient human T lymphocyte line (TJF-2) by a retroviral vector (SAX) produces normal levels of ADA activity. We now show that non-transformed T lymphocytes from ADA-deficient patients can also be infected by SAX. This leads to increased levels of ADA activity in these primary T cells, similar to those produced in the transformed T line. To quantitate the rate of ADA gene transfer and expression by SAX, TJF-2 cells which were infected by SAX were cloned by limiting dilution. 6/27 (22%) of the clones had acquired and were expressing the SAX vector. 1-3 copies of SAX/cell produced ADA activity in the range found in normal thymocytes and T lymphocytes. Similar vectors with other promoters were also highly active. Thus, these vectors are capable of efficient transfer and expression of the human ADA gene in ADA-deficient, human T cells.

INTRODUCTION

Adenosine deaminase (ADA) deficiency, a cause of severe combined immunodeficiency (SCID), has been identified as a favorable disease candidate for initial trials of human gene therapy (1). These patients have profound lymphopenia and consequent immunodeficiency as a result of the absence of ADA, an enzyme needed in early T lymphocyte development (2). Allogeneic bone marrow transplantation can completely correct this disorder, which suggests that genetic correction of a patient's own bone marrow stem cells by insertion of a functioning ADA gene may also be curative (3). The transferred ADA gene must be expressed in T lymphocyte progenitors, where a broad range of conferred ADA activity should allow normal T cell development.

To study ADA gene transfer and expression in hematopoietic cells of ADA-deficient patients, a retrovirus (SAX) was constructed (4). The SAX vector is a derivative of the high-titer N2 vector and contains a normal human ADA cDNA regulated by the SV40 early region promoter. We have previously shown that SAX is capable of transferring the ADA gene into an HTLV-I transformed, ADA-deficient, human T lymphocyte line, TJF-2 (4). ADA gene transfer by SAX leads to normal levels of ADA activity in these genetically deficient T cells. This corrects, *in vitro*, the physiologic defect of ADA deficiency: hypersensitivity to inhibition of cell growth by 2'-deoxyadenosine.

Here we examine the ability of SAX to transduce the ADA gene in non-transformed ADA-deficient T lymphocytes, perform a clonal analysis of the rate of ADA gene transfer and expression and compare expression of ADA by vectors with various promoters.

MATERIALS AND METHODS

Cells and SAX Vector

Establishment and characterization of the HTLV-1 transformed ADA-deficient human T cell line TJF-2 has been reported previously (5). TJF-2 cells were grown in RFI medium: RPMI 1640 with 10% fetal calf serum (FCS), and 50 U/ml recombinant IL-2 (Cetus Corp.). Non-transformed, IL-2 dependent ADA-deficient T cells were obtained by stimulating peripheral blood mononuclear (PBM) cells from such patients with 50 U/ml IL-2, 1 ug/ml PHA and irradiated, allogeneic PBM. Cultures were re-stimulated weekly with fresh IL-2 and irradiated allogeneic PBM and could be expanded for 2-3 months before undergoing senescence. Immunofluorescent staining showed these cells to be mature, activated T cells.

The SAX vector has been described previously (4). Vectors containing other promoters were constructed by similar methods. The vector viruses were produced from cells of the amphotropic packaging line PA12 (6). For co-cultivation infections, vector-producing fibroblasts were plated at 2×10^6 cells/60 mm tissue culture dish in 4 ml DMEM, 10% FCS, 2mM glutamine (D-10). The next day, the fibroblast monolayers were irradiated (1500R) and 2×10^6 ADA (-) T cells were added to each plate, followed by polybrene (4 ug/ml) and rIL-2 (50 U/ml). The T cells and fibroblasts were co-cultivated for 18-24 hours. Then the T cells were harvested, pelleted and resuspended in fresh RFI. Following co-cultivations, the T cells were successively grown in new flasks for three days to permit any contaminating fibroblasts to be removed by adherence. No detectable ADA gene transfer by virus-containing supernatants alone occured, despite titers of $1\text{-}5 \times 10^6$ virus particles/ml as determined by titering on NIH 3T3 fibroblasts (multiplicity of infection of 5-25 G418-resistance units/T cell).

ADA activity was measured, in duplicate samples, as conversion of ^{14}C-adenosine to ^{14}C-inosine at least three days after vector infection or after selection by growth in G418 as described previously (5). The concentrations of G418 used represent biologically active G418 as indicated by the manufacturer. Clones of TJF-2 cells were derived by limiting dilution and dot and

Southern blots for the NeoR sequences were performed as described previously (5,7).

RESULTS

ADA Expression in ADA-deficient, Human T Cells

Using the co-cultivation method, non-transformed, IL-2 dependent T cells from four ADA-deficient SCID patients have been infected with SAX. Prior to SAX infection, these cells had ADA levels that were only 1-8% of normal. The level of ADA activity in the T cells exposed to SAX-producing fibroblasts increased in every case (Table 1) and is in the same range that has been seen in SAX-infected TJF-2 cells (i.e., 50-150U ADA, Table 2). However, absolute conclusions cannot be made about the extent of ADA gene expression by SAX in transformed cells compared to normal, as gene transfer efficiencies may differ between the different cell types.

TABLE 1
ADA levels produced by SAX in non-transformed, ADA-deficient T cells.

	ADA Activity (U) [a]	
Patient	Uninfected	SAX
1	8.1	156.4
2	28.0	185.1
3	9.8	38.2
4	5.4	41.0

[a] Nanomoles inosine/minute/10^8 cells.

A more valid index of the level of the expression of the transferred gene is the ADA activity of cells which have been selected so that all cells contain the vector. The presence of the bacterial neomycin resistance gene in SAX allows selection to be performed for cells containing the vector by growth in G418, a neomycin analogue toxic to eukaryotic cells. SAX infection of the HTLV-1 transformed, ADA-deficient T cell line TJF-2 followed by G418 selection produces a high level of ADA activity, exceeding the level of ADA of a control T cell line from a normal donor (Table 2). After SAX infection, a population of TJF-2 cells can be routinely grown in 500-1000 ug/ml G418. In contrast to the routine success with transformed TJF-2 cells, it was much more difficult to get non-transformed T cells to grow well in G418 after SAX infection. Of six attempts, this was achieved only once, by using a lower concentration of G418 (100 ug/ml). The ADA level of the SAX-infected, G418 selected, non-transformed cells was comparable to that produced in TJF-2 cells (Table 2).

TABLE 2
Comparison of ADA Levels Produced by SAX in Non-transformed versus HTLV-I Transformed ADA-deficient Human T Cells

Cells	ADA Activity (U)
HTLV-1 transformed (TJF-2):	
uninfected	4.7
SAX infected	146.8
SAX infected, G418 selected	1022.6
Normal HTLV-1 Transformed T line	413.8
Non-transformed (patient 3)	
uninfected	3.5
SAX infected	29.8
SAX infected, G418-selected	890.6
Normal non-transformed human T line	368.7

Clonal Analysis of SAX Infection in TJF-2

The extensive proliferative capacity of the HTLV-I transformed, ADA-deficient TJF-2 cells allows them to be easily cloned and expanded in large numbers. To assess the efficiency of ADA gene transfer and expression, clonal populations of TJF-2 cells infected by SAX were derived by limiting dilution and analyzed for the presence of SAX vector DNA sequences and for the level of expression of the ADA gene.

A total of 27 clones were obtained from the population of TJF-2 cells that were exposed to the SAX virus (but not G418 selected). Genomic DNA extracted from each clone was examined by dot blot hybridization for the presence of integrated SAX genomes using a ^{32}P-labelled NeoR gene probe. Six of the twenty-seven clones contained DNA which hybridized to this probe. Thus the frequency of infection of the TJF-2 cells by SAX was approximately 22%.

The ADA enzymatic activity of each clone was also measured (Figure 1). The average ADA level of the 21 clones which did not contain SAX vector DNA was similar to the ADA activity of the parental ADA-deficient TJF-2 cells (geometric mean=4.1 U). Cells from each of the six clones which had been infected by SAX (i.e.-contained SAX vector DNA by dot blot) had levels of ADA greatly exceeding that of the SAX-negative cells (geometric mean=1270 U). This number is similar to the ADA level of the bulk population of SAX-infected, G418 selected TJF-2 cells shown in Table 2.

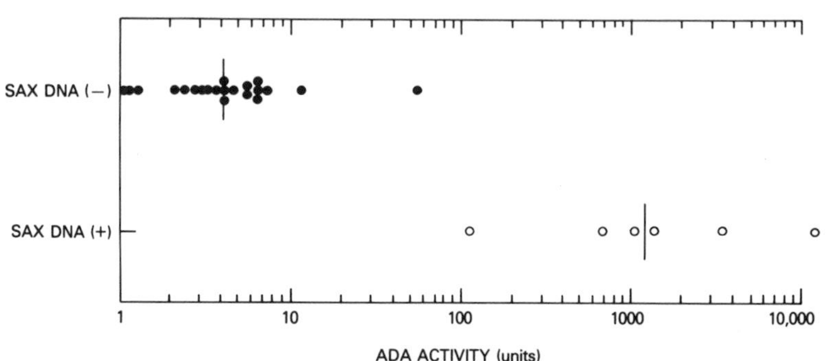

Figure 1. ADA activity of clones of TJF-2.

The ADA level of these six SAX DNA-containing clones varied over 100-fold from 110 U to 11,000 U. To determine if this wide range of gene expression was due to either rearrangements of the SAX vector or differences in the number of vector copies, Southern blot analysis was performed. The SAX DNA-containing clones with ADA levels of approximately 3000, 12,000 and 1,000 units had

1, 3 and 1 copies of SAX vector integrated within their genomic DNA, respectively (Data not shown).

ADA Activity Produced by Vectors with Other Promoters

ADA-deficient T cells provide a convenient assay system to compare the effects of heterologous promoters on the expression of the ADA gene by a series of retrovirus vectors. TJF-2 cells were infected by co-cultivation with vector-producing fibroblasts, selected in 1.0 mg/ml G418 for three weeks and then assayed for ADA activity. A control infection by cells producing the N2 vector, which does not contain an ADA gene, expressed only baseline levels of ADA (Table 3). Infection by the various ADA gene vectors produced increased levels of ADA over a wide range. The vector AX25 does not have an internal promoter for the ADA gene, so initiation of transcription of the ADA gene would have to occur from the retroviral LTR promoter. As expected, this vector produced relatively low levels of ADA.

TABLE 3
ADA activity of TJF-2 cells infected by ADA vectors

Vector	Promoter-gene	ADA (U)
N2	LTR-NeoR	3.8
AX25	LTR-hADA	17.2
TAX	Thymidine kinase-hADA	725.0
BAX	Beta-actin-hADA	942.3
CAX	Human CMV-hADA	1534
SAX	SV-40-hADA	1579
HAX	Murine H2d-hADA	1693
NAAS	Human ADA promoter-hADA	1225
	Control Cells	
Uninfected ADA (-) T Cells		2.9
Normal Human T line		495.4

In contrast, vectors with a variety of internal promoters upstream of the ADA cDNA led to high levels of ADA enzymatic activity. A promoter fragment from the 5' flanking region of the ADA gene, as well as the human CMV early promoter, SV40 virus early promoter and the murine H2d promoter all resulted in high levels of ADA in infected, G418-selected TJF-2 cells. The herpes simplex thymidine kinase and human beta-actin promoters were less active.

DISCUSSION

The high level of ADA produced by SAX in TJF-2 cells suggested that this vector would be similarly active in primary, non-transformed human T cells from ADA-deficient patients. However, it was possible that the strong expression by SAX in the TJF-2 cells resulted in part from a permissive effect on gene expression due to HTLV-1 transformation. For example, the "tat" gene product of HTLV-1 could have a stimulatory effect on the LTR or SV40 promoters of SAX. Therefore, our finding that the level of ADA activity produced by SAX in the non-transformed cells is similar to that seen in the HTLV-I transformed T line suggests that cellular transformation is not responsible for the activity of the transferred ADA gene. Rather, the SV40 early promoter appears to be an intrinsically strong promoter of the ADA gene in human T lymphocytes.

Clonal analysis by limiting dilution of SAX infected TJF-2 cells showed the gene transfer rate to be approximately 20%. Thus amphotropic-packaged retrovirus vectors are capable of high efficiency gene transfer into human T cells. The level of expression of the ADA gene by SAX in these ADA-deficient human T cells is also very high. All six clones of TJF-2 which contained SAX DNA by dot blot had normal or above levels of ADA activity, varying over a 100-fold range. In the three clones analyzed by Southern blot, this variability seemed only partly related to the number of copies of the vector integrated into the cells. Therefore, other factors, such as those imposed by the chromosomal site of integration, must also affect expression of the genes introduced by this retroviral vector. Importantly, the SAX vector appears to be quite active with a low number of copies per cell producing high levels of ADA activity.

Finally, we used this system to examine the ability of a variety of promoters to cause expression of the ADA gene. A number of viral or mammalian promoters were found to be very active in these cells. The absence of endogenous ADA activity of the TJF-2 line and the high sensitivity and quantitative nature of the radioisotopic measurement of ADA activity make this a useful model for examining promoter activity in human T cells.

The cultured ADA-deficient T cells which we have studied, both transformed and non-transformed, are phenotypically mature T lymphocytes. To achieve long-lasting correction of ADA-deficient SCID by gene therapy, a functioning ADA gene must be transferred into renewable, immature T lymphocyte progenitors which can then generate diverse, immunologically competent, mature T cells. Because <u>in vitro</u> differentiation of human T cell progenitors is not currently feasible, studies of insertion and expression of genes in T cell progenitors and their progeny will require <u>in vivo</u> systems of lymphocyte differentiation, such as animal bone marrow transplantation models. The SAX vector has successfully transferred and expressed the human ADA gene in the peripheral blood of non-human primates using an autologous bone marrow transplantation/gene transfer protocol (8). However, the specific cell type which expressed human ADA was not determined. Further experiments must be performed to learn if the SAX vector can express human ADA in T lymphocytes after infection of their progenitors.

ACKNOWLEDGEMENTS

We wish to thank Drs. R. Buckley, R.E. Stiehm, R. Parkman, and R.H. Kobayashi for providing patient samples, Jane E. Selegue for technical assistance and Debbie Carroll for preparation of the manuscript.

REFERENCES

1. Anderson WF (1984). Prospects for human gene therapy. Science 226:401-409.
2. Martin DW Jr, Gelfand EW (1981). Biochemistry of diseases of immunodevelopment. Annu Rev Biochem 50:845-877.
3. Parkman R (1986). The application of bone marrow transplantation to the treatment of genetic diseases. Science 232:1373-1378.
4. Kantoff P, Kohn DB, Mitsuya H, Armentano D, Sieberg M, Zweibel JA, Eglitis MA, McLachlin JR, Wiginton DA, Hutton JJ, Horowitz SD, Gilboa E, Blaese RM, Anderson WF (1986). Correction of adenosine deaminase deficiency in cultured human T and B cells by retrovirus-mediated gene transfer. Proc Natl Acad Sci (USA) 83:6563-6567.
5. Kohn DB, Mitsuya H, Ballow M, Selegue JE, Barankiewicz J, Cohen A, Gelfand E, Anderson WF, Blaese RM. Establishment and characterization of adenosine deaminase (ADA)-deficient human T cell lines. J Immunol-in press.
6. Miller AD, Law MF, Verma IM (1985). Generation of helper-free amphotropic retroviruses that transduce a dominant-acting methotrexate resistant DHFR gene. Mol Cell Biol 5:431-437.
7. Eglitis ME, Kantoff PW, Gilboa E and Anderson WF (1985). Gene expression in mice after high efficiency retroviral-mediated gene transfer. Science 230:1395-1398.
8. Kantoff PK, Gillio A, McLachlin JR, Bordignon C, Eglitis MA, Kernan NA, Moen RC, Kohn DB, Yu S, Karson E, Karlsson S, Zweibel JA, Gilboa E, Blaese RM, Nienhuis A, O'Reilly RJ, and Anderson WF (1987). Expression of human adenosine deaminase in non-human primates after retroviral-mediated gene transfer. J Exp Med 166:219-234.

Gene Transfer and Gene Therapy, pages 375–395

LONG TERM *IN VIVO* EXPRESSION OF HUMAN ADENOSINE DEAMINASE IN MURINE HEMATOPOIETIC CELLS[1]

G.R. MacGregor, F.A. Fletcher, K.A. Moore, S.M-W. Chang[2], J.W. Belmont, and C.T. Caskey

Institute for Molecular Genetics and Howard Hughes Medical Institute, Baylor College of Medicine, Houston, Texas 77030

ABSTRACT Somatic gene transfer offers the possibility of a new approach to the treatment of human genetic disease. Adenosine deaminase deficiency is being used as a model in which gene transfer techniques can be developed and evaluated. Multiple replication-defective retrovirus vectors were tested for their ability to transfer and express human adenosine deaminase *in vitro* and *in vivo* in a mouse bone marrow transplantation model. High titer virus production was obtained from vectors utilizing both a retrovirus long terminal repeat promoter and internal transcriptional units with human cFos and herpes virus thymidine kinase promoters. After infection of primary murine bone marrow with one of these vectors, human ADA was detected in CFU-C, CFU-S, and in the blood of reconstituted recipient animals. This system offers the opportunity to assess methods for increasing efficiency of gene transfer, for regulation of expression of foreign genes in hematopoietic progenitors, and for long term measurement of stability of expression in these cells.

---

[1]This work was supported by the Howard Hughes Medical Institute, NIH grant HD21452, and Cystic Fibrosis Foundation grant R004 7-03S1.

[2]Present address: Laboratory of Molecular Genetics, NINCDS, NIH, Bethesda, MD

INTRODUCTION

Adenosine deaminase (ADA) deficiency is a rare autosomal recessive condition which causes a form of severe combined immune deficiency (SCID, 1). It accounts for approximately 15% of all cases of SCID or one-third of autosomal recessive SCID. The pathophysiology of ADA deficiency has been intensively investigated and appears to involve selective toxicity of dATP for immature T cells (1-3). Alterations in methylation *via* inhibition of S-adenosyl homocysteine hydrolase have also been implicated (2,3). Human ADA is encoded at a single 35kb locus composed of 12 exons on chromosome 20q13 (4-10). A full length cDNA of 1437bp has been obtained which encodes a 362 amino acid protein of 40.7kd (7), *i.e.*, cloned DNA sequences are available for gene transfer. Bone marrow transplantation can be curative (11,12) and enzyme replacement has been demonstrated to improve the immune function (13). Therefore, as long as the total enzyme activity is enough to decrease the accumulation of the toxic metabolites it can be expected that the immune deficiency will be improved. Despite improvements, conventional therapies continue to be associated with significant morbidity and mortality. Thus, the risk-benefit calculation will be heavily weighted toward any potentially curative procedure. Genetic diseases, such as ADA deficiency, which directly affect blood and reticuloendothelial cells, and diseases in which accumulation of circulating toxic metabolites are involved in the pathogenesis are candidates for a novel therapy - somatic gene transfer.

Retrovirus vectors have been used extensively to transfer foreign DNA sequences into hematopoietic stem cells (14-27). Although several defective vectors have allowed expression of the introduced sequences, expression with other constructions has been problematic. Specific recognition by nuclear proteins or modification via methylation has been implicated in the failure of expression of coding sequences under transcriptional control of viral long terminal repeat (LTR) promoters both in embryonic cells and transgenic mice (28-37). We have previously reported transduction of human ADA to mouse hematopoietic progenitors *in vitro* (38). Experiments designed to test the efficacy of that particular vector in more primitive progenitors, CFU-S, were positive (39) but the low infection efficiency precluded consistent repetition. Lim *et al*. (40) have also recently shown expression of ADA in murine CFU-S but the infection efficiency was insufficient to allow long term analysis of stability of expression of the human ADA. We describe here the construction and testing of sixteen additional ADA-transducing retrovirus constructs and the results of ADA expression studies with four vectors with the capacity to produce high titer virus. These experiments represent the first demonstration of efficient gene transfer and expression of human ADA in murine hematopoietic progenitors and *in vivo* expression over an extended period of time. The results offer encouragement that the techniques of somatic gene transfer can be further improved and practically applied to gene therapy for a variety of human genetic diseases.

RESULTS

Screening of Vector Constructions

A transduction assay was used to evaluate quickly multiple vector constructions. Vectors were transfected into packaging cell lines ψ2 (ecotropic) and PA317 (amphotropic, 41-42). Since cell lines produced by infection have been shown to generate higher titer and more stable defective vector production (43), the stably transformed populations were used as a source of virus to infect the packaging cells of the opposite tropism. Clonal isolates of the infected cells were obtained and checked for expression of human ADA (39). ADA-transducing virus produced from these clonal isolates was then assayed by infection of target cells. The intensity of the human ADA enzyme activity on IEF analysis correlated roughly with the titer (data not shown). The assay was intentionally made insensitive so that vectors which either produced low titers (less than 10^4 CFU/ml) or which expressed ADA poorly could be identified and set aside. Individual cell lines giving the highest level of transduction were then used for subsequent experiments on bone marrow infection. Sixteen retrovirus vectors were tested in this way. These are illustrated in Figure 1. These vectors were designed to evaluate the effect on titer and expression of three aspects of structure: (a) sequences 3′ of the classical ψ packaging site - so called gag^+; (b) deletions in the 3′U3 region; and (c) internal transcriptional units in both orientations with respect to the vector LTR. The gag^+ region has been shown to enhance viral RNA packaging but has been associated with a high propensity for recombination to form replication competent virus (44,45). Several groups have reported the use of vectors bearing alterations in the 3′U3 region to make self-inactivating LTR's (46,47). This may avoid negative regulatory effects caused by specific recognition of LTR enhancer sequences in primitive stem cells. Use of an internal promoter with or without enhancer elements might also overcome deficiencies in expression arising from constraints on LTR function in specific tissues or in specific developmental contexts. The results of the transduction assays are indicated in Figure 1. Only vectors which contain

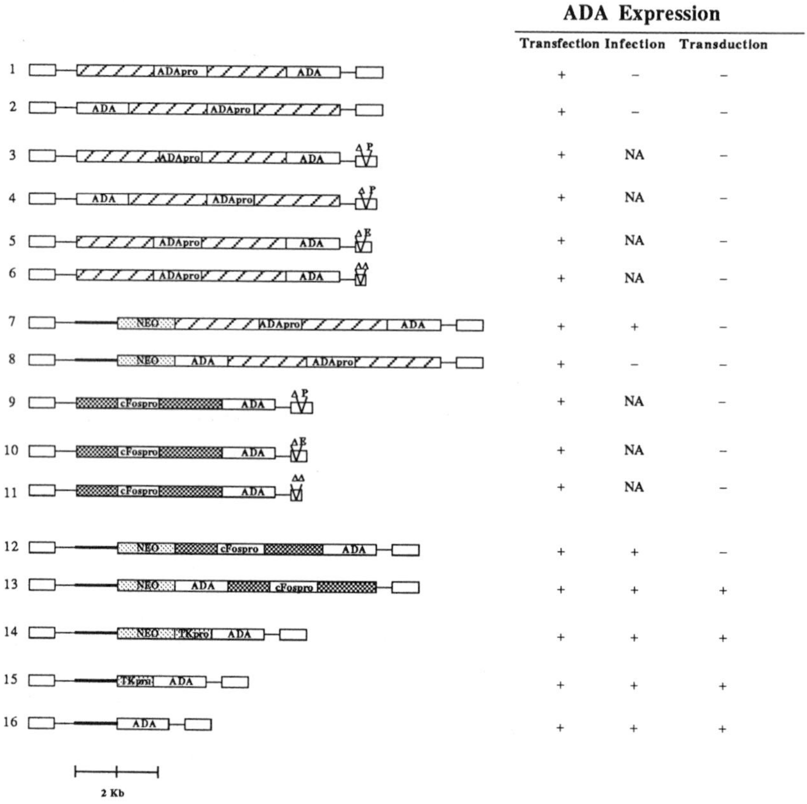

	ADA Expression		
	Transfection	Infection	Transduction
1	+	–	–
2	+	–	–
3	+	NA	–
4	+	NA	–
5	+	NA	–
6	+	NA	–
7	+	+	–
8	+	–	–
9	+	NA	–
10	+	NA	–
11	+	NA	–
12	+	+	–
13	+	+	+
14	+	+	+
15	+	+	+
16	+	+	+

2 Kb

Figure 1. Construction of defective retrovirus vectors. A human ADA promoter-ADA cDNA minigene was constructed by cloning a 1237bp *Eco*RI-*Dde*I fragment from pADA211 (7) into the *Hinc*II site of pUC18. A *Hind*III-*Nco*I genomic DNA fragment containing 3.9kb 5′ of exon 1 was ligated to the cDNA

the gag$^+$ region gave positive results in the transduction assay. Two internal promoter constructs using a human cFos and herpes simplex virus thymidine kinase (HSVTK) promoters worked well in the fibroblasts in the assay. The human ADA promoter-containing constructs proved to be unstable such that although very high titers of *neo* transducing virus (1 - 5 x 10^6 G418CFU/ml) were found, no transduction of ADA activity was observed. The orientation of the cFos minigene appeared to be important. Infection of ψ2 cells with the forward orientation construct (same direction of transcription as the LTR) did not produce virus capable of ADA transduction although the reverse orientation did. The TK promoter constructions were all successful in their ability to transduce ADA to target fibroblasts. Because all the vectors with 3' LTR deletions were made in the non-gag$^+$-containing type vector (48), it is impossible to assess negative effects of these deletions on virus production. (Titers from all these vectors were too low to proceed with bone marrow experiments).

---

subclone at the *Nco*I site (within the initiator ATG codon). This 5.1kb minigene was then blunt end-ligated in both orientations into the *Xho*I site of the N2 vector (pXM5) and the *Bgl*II site of the FVXM (48, M. Botchan) Δe, Δp, and Δe Δp vectors (A. Bernstein). The human cFos minigene was constructed by ligating a 2.25kb *Hind*III fragment from pFC1 (53, kindly provided by R.Vincent) containing the cFos cap site into the *Nae*I site of the ADA cDNA (56bp 5' of the ATG). This 3.5kb minigene was subcloned into each vector in a manner similar to the ADA minigene. pXT1ADA was constructed by ligating a *BamH*I partial cut ADA cDNA fragment from p7ADAΔ (a 1237bp *EcoR*I-*Dde*I fragment subcloned into the *Hinc*II site of pUC7) to *Bgl*II cut and Calf Intestinal Phosphatase-treated pXT1 (pXT1, kindly provided by Dr. E. Wagner, contains the 5'LTR and gag region from pXM5, a *neo* gene, and a 925bp Herpes virus thymidine kinase promoter). pΔXT1ADA was made by subcloning the *Sal*I-*Hind*III fragment from pXT1ADA (containing the TK-ADA minigene and the 3'LTR region) into *Sal*I-*Hind*III cut p5'N2 (a subclone of the 5' LTR and gag-containing *EcoR*I fragment from pXM5 in pUC18). pΔNN2ADA was made by reconstructing the N2 vector without *neo*. A *Sal*I-*Hind*III fragment from pXM5 (3' LTR region) was subcloned into *Sal*I-*Hind*III cut p5'N2, then the ADA cDNA was ligated into the unique *BamH*I site. The heavy line indicates the gag$^+$ region. Open boxes denote the Moloney Murine Leukemia Virus long terminal repeat (LTR).

Stability of ADA containing vectors.

Studies were conducted to establish the integrity of the transferred viruses. Individual cell lines from the four vectors giving positive results in the transduction assay were used to infect Rat 208F target cells. Genomic DNA purified from target cells and parental cell lines was cut with *Nhe*I (cuts in the LTR's only) and following electrophoresis, Southern blotted DNA's were probed with the human ADA cDNA to determine if internal rearrangements had taken place (Figure 2a). Virus produced from three of the four cell lines did not show detectable rearrangements in their integrated forms. Within the parental pΔXT1ADA virus-producing cell line and in one of the target cell clones a minor deletion variant was detected. In addition, total cellular RNA was isolated and subjected to Northern analysis using ^{32}P-labelled ADA cDNA as probe (Figure 2b). RNA species corresponding to the predicted transcripts from the viral LTR and the internal promoters cFos or TK were readily detected. Taken together, these results suggest that the process of selection did not engender rearranged defective viruses.

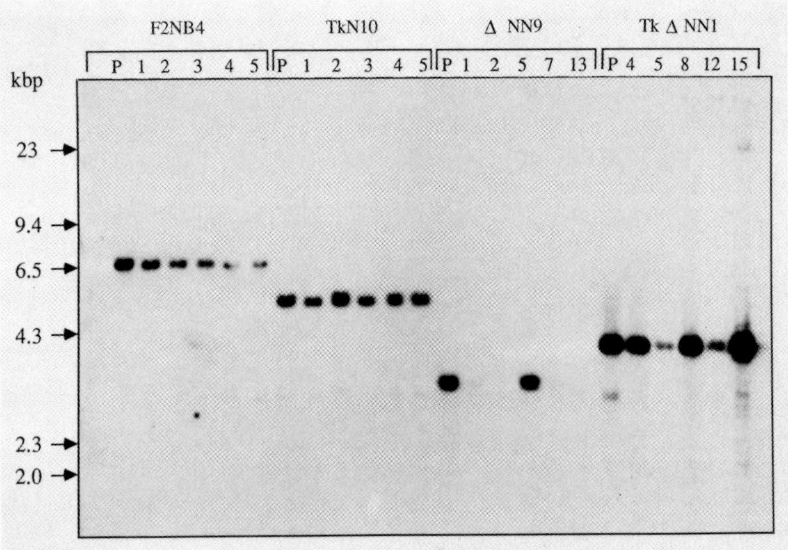

Figure 2a. Southern analysis of infected target cells.

High molecular weight DNA was isolated by standard techniques from the Parental ψ2 producer cell clones (P) and from infected Rat208F clones (numbered lanes). Five μg of genomic DNA was cut with *Nhe*I, electrophoresed through a 0.7%TAE agarose gel and blot-transferred to nylon membrane (Gene Screen Plus, Dupont, Boston, MA) by the method of Southern. ADA cDNA was labeled with ^{32}P dCTP and hybridized as described (38). Filters were washed to a final stringency of 0.1X SSC, at 65°C. The blots were exposed to Kodak XAR-5 film overnight with an intensifying screen. F2NB4 - parental cell line producing pN2Fos2ADA vector; TKN10 - pXT1ADA vector; ΔNN9 - pΔNN2ADA vector; and TKΔNN1 - pΔXT1ADA vector. ADA retroviral-specific hybridization identical to that seen in lane "P" in the ΔNN9 samples were observed in lanes 1, 2, 7, and 13 following longer exposure of the filter (data not shown).

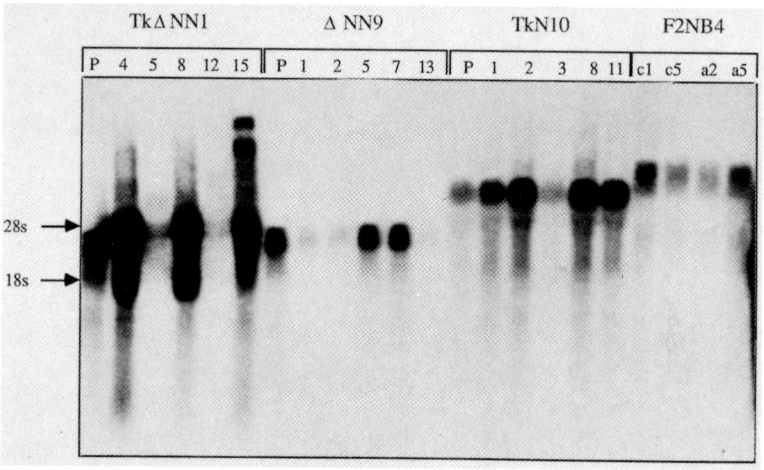

Figure 2b. Northern analysis of infected target cells. Total cellular RNA was isolated (54) from identical cultures analyzed in Figure 2a. Ten μg of RNA were electrophoresed through a 0.9% formaldehyde-MOPS agarose gel. The nucleic acids were transferred, hybridized and washed as in Figure 2a. High molecular weight species in lane 15 may indicate read-through transcription at a unique site of integration.

Because of the known propensity for gag^{+}-containing vectors to recombine in the ψ2 cells, we examined these cell lines for the presence of replication competent virus. A marker rescue assay was employed using an NIH3T3 line transfected with the defective vector SV(B) (49) as the intermediate target cell. Supernatants from the ADA vector in ψ2 cell lines all contained helper virus capable of rescuing the defective *neo*-containing virus from the SV(B) line.

Infection of murine hematopoietic cells.

The four high titer viruses (N2Fos2, XT1, ΔXT1,and ΔNN2) were subsequently used to infect primary mouse bone marrow cells. Infection was performed by 24hr co-cultivation with irradiated virus producer lines. Infected marrow was either plated in semisolid culture media for enumeration of G418-resistant CFU-C or injected into irradiated recipients. Some recipients received graded doses of the marrow cells so that discrete CFU-S could be studied. Other animals received fully reconstituting doses to allow long term studies of human ADA expression. The results of these experiments are summarized in Table 1. Two vectors, N2Fos2 and XT1, containing the cFos promoter and the TK promoters respectively fail

Table 1

Vector[a]	In Vitro		In Vivo[c]	
	% G418^{R} CFU[b]	IEF	Southern	IEF
pN2Fos2ADA	8-20%	neg	2/35 (6%)	0/2
pXT1ADA	1.6-8.5%	neg	11/18 (61%)	0/11
pΔTKADA	NA	pos	3/6 (50%)	3/3
pΔNN2ADA	NA	pos	31/45 (69%)	24/31

[a] See Figure 1 for description of vectors.
[b] Infected bone marrow cells were plated in supplemented IMDM with 0.5% methyl cellulose as previously described (38). CFU were counted at 10 days.
[c] Discrete (CFU-S) or whole spleens from fully reconstituted mice were analyzed as described in Figures 3 and 4. Survival splenectomy was performed using standard techniques (56) with Avertin anesthesia.

to express human ADA in CFU-C or in CFU-S despite these viruses ability to transduce ADA efficiently to tissue culture fibroblasts and despite relatively efficient infection of CFU-C and CFU-S. (Both vectors express *neo* from the LTR promoter and confer G418-resistance to between 1 and 20% of CFU-C. However, when those colonies were picked and assayed for human ADA expression none was found.) When individual CFU-S or spleens from fully reconstituted animals were studied we again failed to observe expression of human ADA. In contrast, bone marrow cells infected with the virus using the Moloney LTR promoter (ΔNN2) or the TK promoter and lacking the neo gene (ΔXT1) expressed human ADA. This enzyme activity could be detected in pooled CFU-C in the absence of selection indicating very high infection efficiency. Likewise, when individual CFU-S were analyzed infection efficiencies of between 50 and 85% were observed based on Southern analysis (Figure 3) and expression of human ADA (Figure 4).

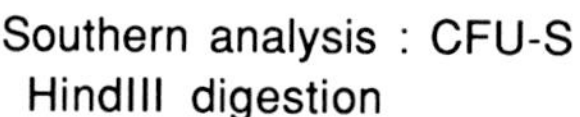

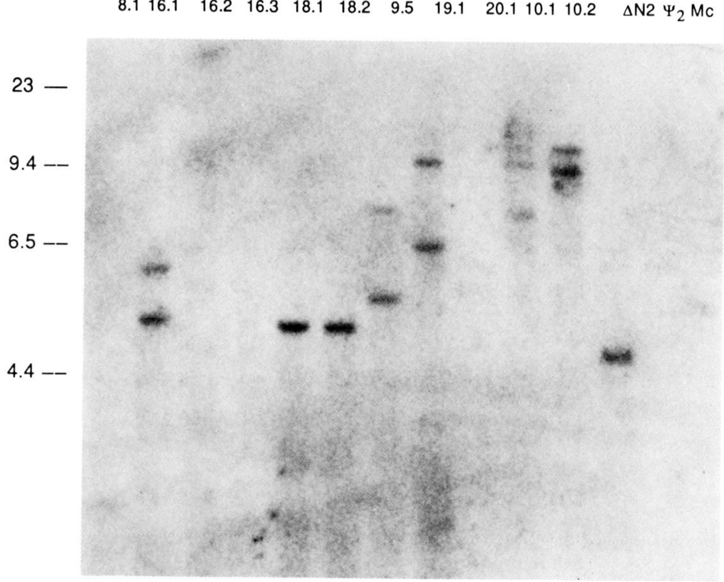

Figure 3. Southern analysis of infected CFU-S. Six x 10^6 fresh bone marrow cells harvested from the tibias and femurs of healthy 8-12wk C57Bl/6J mice (Jackson Laboratories, Bar

Harbor, ME) were co-cultivated with 6 x 10^6 irradiated (1500R, Cs137 source, Gammacell 1000, Atomic Energy Ltd. Canada) virus producer cells in Iscove's Modified Dulbecco's Medium (IMDM, Gibco) supplemented with BSA, Fe-saturated transferrin, soybean lipids (Boehringer-Mannheim Biochemicals, Indianapolis, IN), 10%FCS, 10% WEHI3B(D-) (provided by R. Phillips) conditioned medium (interleukin 3), and 10% 5637 (ATCC) conditioned medium (interleukin 1α). After 24hr the nonadherent cells were removed into petri dishes for an additional 48hr of culture. With vectors carrying the *neo* gene some of the cells were preselected in 1.50mg/ml G418. One x 10^5 viable cells were plated into 1ml semisolid methylcellulose cultures in supplemented IMDM (38). CFU-C were counted and picked after 10d. Irradiated (1000R) syngeneic recipient animals were injected IV with between 5 x 10^4 and 5 x 10^5 viable cells for study of distinct CFU-S. Recipients were sacrificed at 14d post transplantation and individual CFU-S were dissected from the surrounding stroma. The splenic tissue was disrupted with several strokes in a micro-Dounce homogenizer in lysis buffer. Genomic DNA was extracted from the nuclear pellet after centrifugation by standard methods and 10 μg cut with *Hind*III (cuts outside the vector). The cut DNA was run through a 0.8% TAE agarose gel, blotted to nylon membrane and hybridized with random hexamer ^{32}P-labelled ADA cDNA probe (38). The filter was washed in 0.1X SSC at 60°C and exposed for autoradiography (Kodak XAR).

Figure 4A

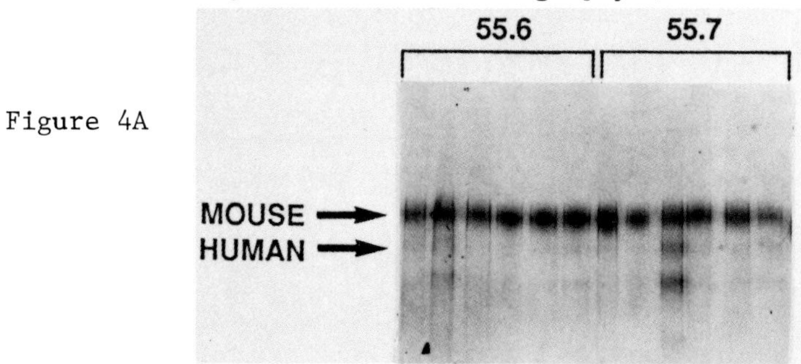

4B

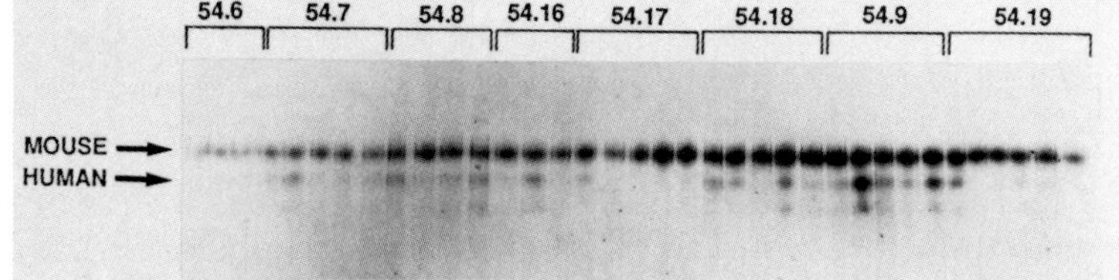

There was a 100% correlation between integration events and expression with both of these vectors in the first experiment. To determine if expression of human ADA in the spleens of animals transplanted with the ΔXT1 vector-infected marrow was obtained from transcription from the LTR or the HSVTK promoter, an RNase A mapping study was performed (Figure 5).

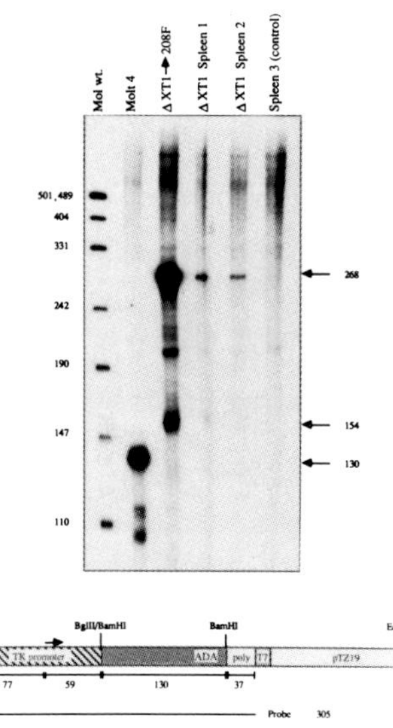

Figure 5. RNase A protection mapping of viral transcripts from the pΔXT1ADA vector. The analysis was carried out as described by Gibbs and Caskey (55). The template plasmid was constructed by subcloning an EcoRI (-136) - BamHI (+150) fragment containing both the HSVTK promoter and ADA cDNA sequences-into pTZ19R (Pharmacia, Piscataway, NJ). Survival splenectomy was performed using standard techniques (56) with Avertin anesthesia. Lane 1 - MW markers; 2 - MOLT4 (human T cell leukemia, ATCC); 3 - pΔXT1ADA infected Rat 208F; 4 - Spleen from reconstituted animal #1 14d post transplantation with pΔXT1ADA-infected marrow; 5 - pΔXT1ADA animal #2; 6 - control spleen.

This strongly suggests that the predominant RNA species originated in the LTR and that the HSVTK transcript contributes little to production of human ADA. Southern analysis of the individual CFU-S showed multiple integrations with an average of two provirus per cell. Since CFU-S have been shown to be clonally-derived (50), this result suggests that the virus titer was not limiting but rather implies that there are limitations on the receptivity of some CFU-S

Figure 4. Expression of human ADA in CFU-S. Protein extracts from individual CFU-S from animals injected with infected marrow were run on IEF gels. A.) Infection by pΔXT1ADA virus. Each lane is from one CFU-S, the boxed areas from single animals. B.) Infection by ΔNN2ADA virus. The arrows indicate the relative migration of human and murine ADA.

progenitors to infection. In a series of three experiments (Table 2) using the ΔNN2 vector, we also observed variation in the expression efficiency. In the second experiment there was an infection efficiency as determined by provirus integration of approximately 60% but very low expression efficiency in CFU-S.

Table 2 Summary of bone marrow infections using ΔNN2 vector.

Experiment	CFU-S		Blood IEF	
	Southern Analysis	IEF	4 wk	12 wk
1	16/22 (72%)	16/22	5/5	2/5
2	8/15 (61%)	1/15	5/5	0/5
3	7/8 (85%)	7/8	4/4	4/4

Expression of Human ADA in Blood of Reconstituted Animals.

Irradiated recipient mice transplanted with 1 x 10^6 viable primary bone marrow cells can survive indefinitely after transplantation and are fully reconstituted in all hematopoietic lineages. Following transplantation with infected marrow, the presence of human ADA in these mice was monitored by serial bleeding at defined intervals. The results of these are illustrated in Figure 6. Human ADA was detected in

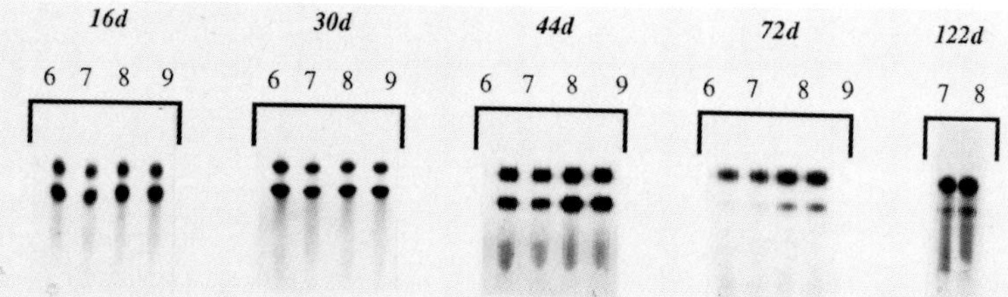

Figure 6. Expression of human ADA in the blood of reconstituted recipients. One x 10^6 infected viable bone marrow cells was injected IV into lethally irradiated recipients. These animals were bled at intervals post transplantation and protein extracts from blood cells were analyzed by IEF. Time after transplantation is indicated above the assays. The four animals (denoted 6,7,8,9) shown were from Experiment 3 in Table 2. The upper band in each lane is the endogenous murine ADA, the lower human ADA.

the blood in 100% of the mice transplanted with the ΔNN2 virus (total 14 animals in three individual experiments shown in Table 2). Over time, the level of human ADA activity decreased. However, 4 of the 14 animals (two each from experiments 1 and 3) have continued to express human ADA, although at diminished levels, up to four months post-transplant. Scanning densitometry of selected IEF gels allows estimates of human ADA activity of 10-100% of the endogenous mouse enzyme. In addition, serial serum assays from the ΔNN2 and ΔXT1 recipient mice using the marker rescue assay indicated a continued presence of replication competent virus in the blood of all the animals tested (see Discussion).

DISCUSSION

Of 16 original vector designs, only 4 (N2Fos2, XT1, ΔXT1 and ΔNN2) fulfilled the criteria to enable their use in bone marrow experiments, namely the ability to be used to generate high titers of ecotropic virus following packaging by helper cell lines and to transduce human ADA activity to target fibroblast cells. Surpringly, human ADA promoter containing constructions failed to meet these criteria. One potential explanation for this may be the relatively large size of the vector (9.4 kb genomic RNA). Should this size be close to, or exceed the packaging size limit for the Moloney genome, then a viral species which has lost sequences might be the preferred substrate for packaging. Vectors containing a shorter ADA promoter are currently under evaluation. Similarly, the orientation of the cFos promoter containing minigene also appeared important - the reverse orientation (relative to the LTR elements) construct were stable, however, the forward orientation was not. This instability might be explained by the presence of cryptic splice donor or acceptor signals, polyadenylation signals, or other interaction between these sequences in the forward orientation cFos-promoter mini-gene and the vector LTR's. In this respect, it is helpful to know the primary sequence of mini-genes to be inserted into retroviral vectors to minimize the likelihood of vector instability mediated by such sequences. As all the vector with 3' LTR deletions were made in the non-gag$^+$ type vector, it is impossible to assess negative effects of these deletions on virus production, (viral titers being too low to proceed with bone marrow experiments). To readdress this question, these same deletions are currently being evaluated with vector containing gag$^+$ sequences.

The 4 vectors selected were used to infect primary mouse

bone marrow cells. Infected marrow was either plated in semisolid culture media for enumeration of $G418^R$ (where appropriate) CFU-C or injected into irradiated recipients. Some animals received graded doses to enable studies of discrete CFUS - others received reconstituting doses to enable long term analysis of human ADA expression.

Two vectors, N2Fos2 and XT1 failed to express human ADA in CFU-C or CFU-S despite efficient infection of both classes of cells, (as judged by $G418^R$ CFU-C and by Southern analysis of CFU-S). In contrast, the two remaining vectors, ΔXT1 and ΔNN2 were both used with success to infect and generate high levels of expression of human ADA in CFU-C. In a first experiment, 100% of CFU-S positive for human *ada* DNA sequences by Southern analysis also exhibited human ADA activity by IEF.

To determine if expression of human ADA in the spleens of animals transplanted with the ΔXT1 vector-infected marrow was obtained from transcription from the LTR or the HSV tK promoter, an RNase A mapping study was performed. This strongly suggests that predominant RNA species originated in the LTR and that the HSV tK mediated transcript contributes little to production of human ADA. Southern analysis perfomed on DNA isolated from individual CFU-S indicated multiple proviral inserts with an average of 2 per cell. Since CFU-S have been shown to be clonally derived (50) this result suggests that the virus titer was not limiting but that there exist limitations as to the receptivity of certain CFU-S progenitors to infection. This can vary from experiment to experiment and suggests, as yet, uncontrolled aspects of the infection protocol which must be optimized in future experiments. In addition, we also observed large fluctuations in the expression efficiency between experiments (Table 2). This suggest that beyond the variability in the infectivity of these cells, there is also biological variation in the regulation of expression of identical vector sequences. An understanding of this variation in a prerequisite to successful application of somatic gene transfer and is the focus of present scrutiny.

At 4 weeks post-reconstitution with ΔNN2 virus infected bone marrow, 14 out of 14 recipient mice had easily detectable levels of human ADA in their blood (Table 2). The ADA enzyme from whole blood is derived almost entirely from red blood cells (RBC's) and thus, does not reflect the level of expression or the longevity of expression in the lymphocyte, granulocyte/monocyte or platelet lineages. Over time, the level of human ADA activity decreased (see Figure 8) suggesting that the cells initially expressing it originated from more mature progenitors which were infected with higher efficiency or in whom no negative regulatory effects or expression had arisen. However, 4 of the 14 animals have continued to express human ADA, although at diminished levels, up to four months post-transplant.

The average life span of a murine RBC is 45 days (51,52) so that in these animals the human ADA detected probably originates from cells whose primitive progenitors were infected with the virus. The level of human ADA observed in these mice, 10-100%, is such that in a human recipient, it would be expected to ameliorate the disease.

It is formally possible that replication competent virus introduced at the time of infection with the ADA-transducing defective virus allows spread of the defective *in vivo*. Indeed, serial assay serum in the ΔNN2 and ΔXT1 recipients by marker rescue assay does indicate continuing viremia with the replication-competent virus in all these animals (data not shown). However, we do not believe that *in vivo* spread contributes much to human ADA expression in the blood for the following reasons: (a) the orderly and consistent reduction in expression seen in all animals at around 10 wks; (b) the stability of expression observed beyond 10 wks in individual animals without apparent fluctuation; (c) the presence of replication-competent virus in animals failing to express human ADA in the blood; and (d) the lack of multiple, less than single copy insertions in individual CFU-S and in the organs of long term recipients (manuscript in preparation)

These experiments demonstrate long term expression of a human disease-related gene product in mouse hematopoietic cells. These results are consistent with previous published reports of expression of neomycin phosphotransferase in these cells (18) and extend the observations on long term stability of expression. Lim *et al.* (40), have recently shown that vectors which utilize the PGK promoter for ADA expression can also express in CFU-S. These results are similar to our experience with the HSVTK promoter.

These results clearly demonstrate that there is no adverse effect imposed upon expression from the Moloney LTR promoter in murine hematopoietic progenitors. This is in contrast to the poor expression from this promoter observed in embryonal carcinoma cells, early mouse embryo cells, and retrovirus transgenic animals (28-37). However, regulation of expression in the differentiated progeny of true pluripotent stem cells may be governed by the same mechanisms which restrict expression in the transgenic animals. This problem may be approached by further analysis of the types of vectors which express introduced genes in embryonal carcinoma cells compared with those that express identical genes in hematopoietic cells.

A number of technical questions remain to be addressed before an attempt at human disease correction can be made. The long term stability and quantity of expression must be carefully studied and the origin of the expressing cells must be unequivocally related to pluripotent stem cell infection. Likewise, experiments involving humans must be performed in a strictly helper virus-free situation. Future experiments will focus on optimization of stem cell infection by enrichment using monoclonal antibodies and cell sorting and by hormonal stimulation. We are optimistic that improvements can be made that will allow exploitation of this set of methods for the therapy of a variety of human genetic diseases.

ACKNOWLEDGEMENTS

GRM is the recipient of an Arthiritis Foundation postdoctoral fellowship; JWB is an Assistant Investigator and CTC is an Investigator in HHMI. The authors wish to thank Jenny Henkel-Tigges and Michelle Rives for technical assistance and Elsa Perez for help in preparation of the manuscript. The authors also wish to thank J. Dick, A. Bernstein, and L. Donehower for useful discussions during the completion of these studies.

REFERENCES

1. Kredich NM, Hersfield MS (1983). Immunodeficiency Diseases Caused by Adenosine Deaminase Deficiency and Purine Nucleoside Phosphorylase Deficiency. In Stanbury JB, Wyngaarden JB, Fredrickson DS, Goldstein JL, Brown MS (eds): The Metabolic Basis of Inherited Disease, McG p 1157.
2. Hershfield MS, Kredich NM (1978). S-Adenosylhomocysteine Hydrolase Is an Adenosine Binding Protein: A target for Adenosine Toxicity. Science 202:757.
3. Hershfield M (1979). Apparent Suicide Inactivation of Human Lymphoblast S-Adenosylhomocysteine Hydrolase by 2'-Deoxyadenosine and Adenine Arabinoside. J Biol Chem 254:22.

4. Honig J, Martiniuk F, DEustachio P, Zamfirescu C, Desnick R, Hirschhorn K, Hirschhorn LR, Hirschhorn R (1984). Confirmation of the regional localization of the genes for human acid alpha-glucosidase (GAA) and adenosine deaminase (ADA) by somatic cell hybridization. Ann Hum Genet 48:49.
5. Mohandas T, Sparkes RS, Suh EJ, Hershfield MS (1984). Regional localization of the human genes for S-adenosyl-homocysteine hydrolase (cen----q131) and adenosine deaminase (q131----qter) on chromosome 20. Hum Genet 66:292.
6. Petersen MB, Tranebjaerg L, Tommerup N, Nygaard P, Edwards H (1987). New assignment of the adenosine deaminase gene locus to chromosome 20q13 X 11 by study of a patient with interstitial deletion 20q. J Med Genet 24:93.
7. Adrian GS, Wiginton DA, Hutton JJ (1984). Structure of adenosine deaminase mRNAs from normal and adenosine deaminase-deficient human cell lines. Mol Cell Biol 4:1712.
8. Wiginton DA, Kaplan DJ, States JC, Akeson AL, Perme CM, Bilyk IJ, Vaughn AJ, Lattier DL, Hutton JJ (1987). Complete sequence and structure of the gene for human adenosine deaminase. Biochemistry 88 25:8234.
9. Markert ML, Hershfield MS, Wiginton DA, States JC, Ward FE, Bigner SH, Buckley RH, Kaufman RE, Hutton JJ (1987) Identification of a deletion in the adenosine deaminase gene in a child with severe combined immunodeficiency. J Immunol 24 138:3203.
10. Akeson AL, Wiginton DA, States JC, Perme CM, Dusing MR, Hutton JJ (1987). Mutations in the human adenosine deaminase gene that affect protein structure and RNA splicing. Proc Natl Acad Sci USA 84:5947.
11. Hirshhorn R, Roegner-Maniscalco V, Kuritsky L, Rosen F (1981). Bone Marrow Transplantation only Partially Restores Purine Metabolites to Normal In Adenosine Deaminase-deficient Patients. J Clin Invest 68:1387.
12. Fischer A, Blanche S, Veber F, LeDeist F, Gerota I, Lopez M, Durandy A, Griscelli C (1986). Correction of immune disorders by HLA matched and mismatched bone marrow transplantation. In Gale RP, Champlin R (eds): "Progress in Bone Marrow Transplantation," New York: Alan R. Liss, p 911.
13. Hershfield MS, Buckley RH, Greenberg ML, Melton AL, Schiff R, Hatem C, Kurtzberg J, Markert ML, Kobayashi RH, Kobayashi AL (1987). Treatment of adenosine

deaminase deficiency with polyethylene glycol-modified adenosine deaminase. N Engl J Med 316:589.

14. Joyner A, Keller G, Phillips RA, Bernstein A (1983). Retrovirus transfer of a bacterial gene into mouse haematopoietic progenitor cells. Nature 305:556.
15. Miller AD, Eckner RJ, Jolly DJ, Friedmann T, Verma IM (1984). Expression of a Retrovirus Encoding Human HPRT in Mice. Science 225:630.
16. Williams DA, Lemischka IR, Nathan DG, Mulligan RC (1984). Introduction of new genetic material into pluripotent haematopoietic stem cells of the mouse. Nature 310:476.
17. Dick JE, Magli MC, Huszar D, Phillips RA, Bernstein A (1985). Introduction of a selectable gene into primitive stem cells capable of long-term reconstitution of the hemopoietic system of W/Wv mice. Cell 42:71.
18. Keller G, Paige C, Gilboa E, Wagner EF (1985). Expression of a foreign gene in myeloid and lymphoid cells derived from multipotent heamatopoietic precusors. Nature 318:149.
19. Eglitis MA, Kantoff P, Gilboa E, Anderson WF (1985). Gene Expression in Mice after High Efficiency Retroviral-Mediated Gene Transfer. Science 230:1395.
20. Lemischka IR, Raulet DH, Mulligan RC (1986). Developmental Potential and Dynamic Behavior of Hematopoietic Stem Cells. Cell 45:917.
21. Williams DA, Orkin SH, Mulligan RC (1986). Retrovirus-mediated transfer of human adenosine deaminase gene sequences into cells in culture and into murine hematopoietic cells in vivo. Proc Natl,Acad Sci USA 83:2566.
22. Hock RA, Miller AD (1986). Retrovirus-mediated transfer and expression of drug resistance genes in human haematopoietic progenitor cells. Nature 320:275.
23. Kwok WW, Schuening F, Stead RB, Miller AD (1986) Retroviral transfer of genes into canine hemopoietic progenitor cells in culture: a model for human gene therapy. Proc Natl Acad Sci USA 55 83:4552.
24. Chang SM, Wager SmithK, Tsao TY, Henkel TiggesJ, Vaishnav S, Caskey CT (1987). Construction of a defective retrovirus containing the human hypoxanthine phosphoribosyltransferase cDNA and its expression in cultured cells and mouse bone marrow. Mol Cell Biol 7:854.
25. Hawley RG, Covarrubias L, Hawley T, Mintz B (1987). Handicapped retroviral vectors efficiently transduce foreign genes into hematopoietic stem cells. Proc Natl

Acad Sci USA 84:2406.

26. Kantoff PW, Gillio AP, McLachlin JR, Bordignon C, Eglitis MA, Kernan NA, Moen RC, Kohn DB, Yu SF, Karson E (1987). Expression of human adenosine deaminase in nonhuman primates after retrovirus-mediated gene transfer. J Exp Med 166:219.
27. Magli M-C, Dick JE, Huszar D, Bernstein A, Phillips RA (1987). Modulation of Gene Expression in Multiple hematopoietic cell lineages following retroviral gene transfer. Proc Natl Acad Sci USA 84:789.
28. Stuhlmann H, Cone R, Mulligan RC, Jaenisch R (1984). Introduction of a selectable gene into different animal tissue by a retrovirus recombinant vector. Proc Natl Acad Sci USA 81:7151.
29. Rubenstein JL, Nicolas JF, Jacob F (1984). Construction of a retrovirus capable of transducing and expressing genes in multipotential embryonic cells. Proc Natl Acad Sci USA 81:7137.
30. Stewart CL, Vanek M, Wagner EF (1985). Expression of foreign genes from retroviral vectors in mouse teratocarcinoma chimaeras. EMBO J 4:3701.
31. Jahner D, Haase K, Mulligan R, Jaenisch R (1985). Insertion of the bacterial gpt gene into the germ line of mice by retroviral infection. Proc Natl Acad Sci USA 82:6927.
32. van derPutten H, Botteri FM, Miller AD, Rosenfeld MG, Fan H, Evans RM, Verma IM (1985). Efficient insertion of genes into the mouse germ line via retroviral vectors. Proc Natl Acad Sci USA 82:6148.
33. Wagner EF, Vanek M, Vennstrom B (1985). Transfer of genes into embryonal carcinoma cells by retrovirus infection: efficient expression from an internal promoter. EMBO J 4:663.
34. Huszar D, Balling R, Kothary R, Magli MC, Hozumi N, Rossant J, Bernstein A (1985). Insertion of a bacterial gene into the mouse germ line using an infectious retrovirus vector. Proc Natl Acad Sci USA 82:8587.
35. Rubenstein JL, Nicolas JF, Jacob F (1986). Introduction of genes into preimplantation mouse embryos by use of a defective recombinant retrovirus. Proc Natl Acad Sci USA 83:366.
36. Soriano P, Cone RD, Mulligan RC, Jaenisch R (1986). Tissue-specific and ectopic expression of genes introduced into transgenic mice by retroviruses. Science 234:1409.
37. Bonnerot C, Rocancourt O, Briand P, Grimber A, Nicolas

J-F (1987). A β-galactosidase hybrid protein targeted to nuclei as a marker for developmental studies. Proc Natl Acad Sci USA 84:6795.
38. Belmont JW, Henkel TiggesJ, Chang SM, Wager SmithK, Kellems RE, Dick JE, Magli MC, Phillips RA, Bernstein A, Caskey CT (1986). Expression of human adenosine deaminase in murine haematopoietic progenitor cells following retroviral transfer. Nature 322:385.
39. Belmont JW, Henkel-Tigges J, Wager-Smith K, Chang SM-W, Caskey CT (1987). Transfer and Expression of Human Adenosine Deaminase Gene in Murine Bone Marrow Cells. In Gale RP, Champlin R (eds): "Progress in Bone Marrow Transplantation," New York: Alan R. Liss, p 963
40. Lim B, Williams DA, Orkin SH (1987). Retrovirus-Mediated Gene Transfer of Human Adenosine Deaminase: Expression of Functional Enzyme in Murine Hematopoietic Stem Cells In Vivo. Mol Cell Biol 7:3459.
41. Miller AD, Buttimore C (1986). Redesign of retrovirus packaging cell lines to avoid recombination leading to helper virus production. Mol Cell Biol 6:2895.
42. Mann R, Mulligan RC, Baltimore D (1983). Construction of a retrovirus packaging mutant and its use to produce helper-free defective retrovirus. Cell 33:153.
43. Miller AD, Trauber DR, Buttimore C (1986). Factors involved in the production of helper virus-free retrovirus vectors. Somat Cell Mol Genet 12:175.
44. Bender MA, Palmer TD, Gelinas RE, Miller AD (1987). Evidence that the packaging signal of Moloney murine leukemia virus extends into the gag region. J Virol 61:1639.
45. Armentuno D, Yu S-F., Kantoff PW, Von Runden T, Anderson WF, Gilboa E (1987). Effect of internal viral squences on the utility of retroviral vectors. J Virol 61:1647.
46. Yee JK, Moores JC, Jolly DJ, Wolff JA, Respess JG, Friedmann T (1987). Gene expression from transcriptionally disabled retroviral vectors. Proc Natl Acad Sci USA 84:5197.
47. Yu SF, von Ruden T, Kantoff PW, Garber C, Seiberg M, Ruther U, Anderson WF, Wagner EF, Gilboa E (1986). Self-inactivating retroviral vectors designed for transfer of whole genes into mammalian cells. Proc Natl Acad Sci USA 83:3194.
48. Kriegler M, Perez CF, Hardy C, Botchan M (1984). Transformation mediated by the SV40 T antigens: separation of the overlapping SV40 early genes with a retroviral vector. Cell 38:483.

49. Cepko CL, Roberts BE, Mulligan RC (1984). Construction and applications of a highly transmissible murine retrovirus shuttle vector. Cell 37:1053.
50. Till JE, McCulloch EA (1961). A direct measure of the radiation sensitivity of normal mouse marrow cells. Radiat Res 14:213.
51. Bannerman RM (1983). Hematology. In Foster HL, Small JD, Fox JG (eds): "The Mouse in Biomedical Research:Volume III, New York: Academic Press, p 293.
52. Vacha J (1983). Red cell life span. In Agar NS, Board PG (eds): "Red Blood Cells of Domestic Mammals," New York: Elsevier Science Publishers B.V., p 67.
53. Deschamps J, Meijlink F, Verma I (1985). Identification of a transcriptional enhancer element upstream from the proto oncogen fos. Science 230:1174.
54. Chirgwin JM, Przybyla AE, MacDonald RJ, Rutter, WJ (1979). Isolation of biologically active ribonucleic acid from sources enriched in ribonuclease. J Bio Chem 18:5294.
55. Gibbs RA and Caskey CT (1987). Identification and localization of mutations at the Lesch-Nyhan locus by ribonuclease A cleavage. Science 236:303.
56. Hogan B, Constantini F, Lacy E. In "Manipulating the Mouse Embryo". New York: Coldspring Harbor Laboratory Press.

Gene Transfer and Gene Therapy, pages 397–408

EXPRESSION OF THE HUMAN GLUCOCEREBROSIDASE GENE BY RETROVIRUS VECTORS

Donald B. Kohn[1], Jan A. Nolta[1]
Chang Mu Hong[2] and John A. Barranger[2]

[1]Division of Research Immunology & Bone Marrow Transplantation, [2]Division of Medical Genetics Childrens Hospital of Los Angeles, University of Southern California School of Medicine, Department of Pediatrics, Los Angeles, California 90027

ABSTRACT

Gaucher disease, caused by glucocerebrosidase deficiency, may be a candidate for early trials of human gene therapy. We have constructed a series of retrovirus vectors, which contain either a normal human glucocerebrosidase cDNA or a minigene with a 5' genomic glucocerebrosidase gene fragment fused to the 3' portion of the cDNA. These genes were under transcriptional control of either heterologous internal promoters or the glucocerebrosidase 5' flanking region. Transfection or infection of the vectors into murine fibroblasts results in expression of glucocerebrosidase activity to levels equal to that of normal murine or human fibroblasts. The rank order of promoter activity is: glucocerebrosidase > SV40 > thymidine kinase. The conferred glucocerebrosidase activity is immunoprecipitable by a monoclonal antibody specific for the human enzyme. Western blots show the expressed protein is of the expected size range (59-66 kd). Southern blotting reveals that cells which express activity in the normal range contain a single intact copy of the vector.

Retrovirus vectors are capable of high efficiency transduction of the human glucocerebrosidase gene and may be useful for clinical gene therapy.

INTRODUCTION

Gaucher disease (GD), the most prevalent sphingolipidosis, results from deficiency of the lysosomal enzyme glucocerebrosidase. In type I Gaucher disease (chronic, non-neuronopathic), abnormal accumulation of the glucocerebroside in tissue macrophages leads to hepatosplenomegaly and bone involvement (1). This form of GD has been shown to be at least partially responsive to allogeneic bone marrow transplantation (BMT) for patients who have an HLA-matched donor (2-4). Following BMT, the tissue macrophages of the recipient become replaced by cells from the donor, and there is normalization of plasma glucocerebroside levels. Clinical improvement has been achieved in some of the transplanted children.

These results suggest that Gaucher disease may also be treated by genetically corrected, autologous bone marrow transplantation (gene therapy). This treatment would be most suitable for patients lacking an HLA-matched sibling donor, for whom a haploidentical transplant would have high risks of graft-versus-host disease or graft failure. Autologous gene transfer/BMT should be unlikely to have such problems.

Two previous reports have shown that the human glucocerebrosidase gene may be transduced into human Gaucher fibroblasts by retrovirus vectors, leading to increased levels of glucocerebrosidase enzyme activity (5,6). Here we present our studies of transfer and expression of the normal human glucocerebrosidase gene by a series of retrovirus vectors, based on the high titer N2 vector (7).

METHODS

Two different human glucocerebrosidase genes were put into the vectors (Figure 1). The first (pCD-GC, herein called GC) is a 1.6 kb cDNA cloned via the Okayama-Berg expression vector (8). The second (G-N) is a composite gene ("minigene") which we have constructed with the following features: 1) the 5' end is a genomic fragment (1.1 kb) which contains two introns (IVS), the 1st exon and part of the 2nd, and also includes an upstream, in-frame ATG (ATG') of unknown importance, 2) the middle portion is from the same cDNA as GC, and 3) the 3' end is from a different glucocerebrosidase cDNA, pGC-1 (9), and is approximately 200 bp longer than that of GC.

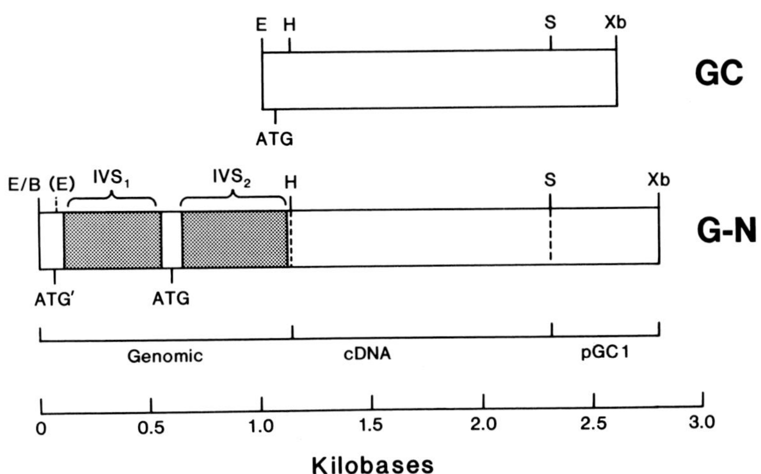

FIGURE 1. Glucocerebrosidase genes.

As an aid to construction, the N2 vector was modified to increase the number of unique restriction sites for gene insertion (Figure 2). The multi-cloning site (mcs) from the plasmid Bluescript (Stratagene) was isolated as a 445 base pair, PvuII-PvuII fragment and ligated into the blunt-ended XhoI site of pN2. A clone containing the mcs in the forward orientation was obtained resulting in adjacent unique restriction sites for: 5'-ApaI-XhoI-BamHI-NotI-BstXI-SstII-3'. This construct (N2-f-mcs) is produced by the amphotropic packaging line PA317 in equivalent titers to the parent N2 vector (data not shown). Promoter fragments were fused to the glucocerebrosidase gene in the Bluescript plasmid and transferred into pN2-f-mcs with cohesive termini. The promoters used were the AccI-HindIII SV40 promoter fragment from pSV2CAT and a 1.5 kb XbaI-BamH1 genomic fragment from the 5' flanking region of the human glucocerebrosidase gene. Synthetic oligonucleotide linkers were put on the GC cCNA which was then ligated into the BglII and XhoI sites of pPXT1, a derivative of N2 with an internal herpes simplex thymidine kinase promoter (10).

The retroviral provirus plasmids were transfected into cells of the ecotropic packaging line Psi2 (11) with calcium-phosphate (12). Transiently-produced virus was obtained two days later and used to infect cells of the amphotropic packaging line PA317 (13) in the presence of polybrene (4 ug/ml). The infected PA317 cells were split the following day and placed under selection with G418 (500 ug/ml). Resistant clones were picked with cloning rings and expanded. The vectors were produced from transfected Psi2 cells and infected PA317 cells at titers of 10^5-10^6/ml (data not shown).

Glucocerebrosidase enzymatic activity was measured in duplicate detergent extracts of cell sonicates using the fluorogenic substrate 4-methylumbelliferyl-beta-D-glucopyranoside, with or without prior immmunoprecipitation of the human enzyme with sepharose-conjugated 8E4 monoclonal antibody (14). Western blots were performed on cell sonicates using the same monoclonal antibody as described (15). Southern blots were performed with ^{32}P-labelled GC cDNA as probe.

Parent vectors

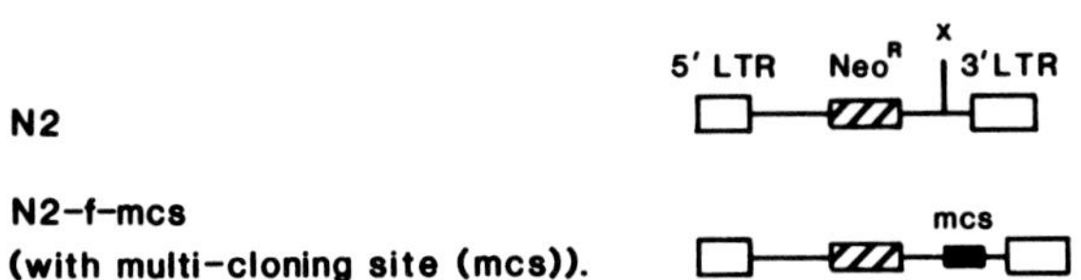

N2-based retrovirus vectors with GC

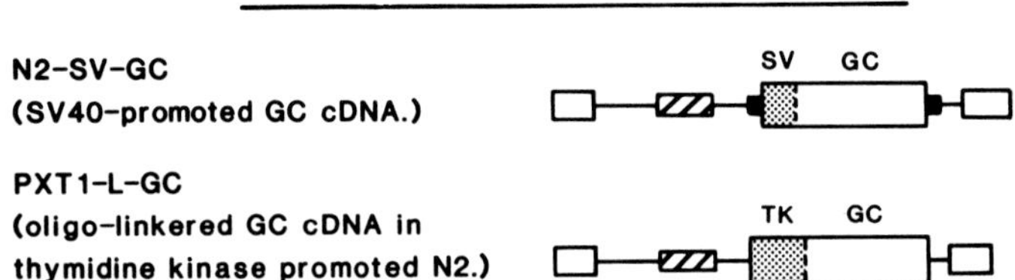

N2-based retrovirus vectors with G-N
(composite gene with 5' genomic end)

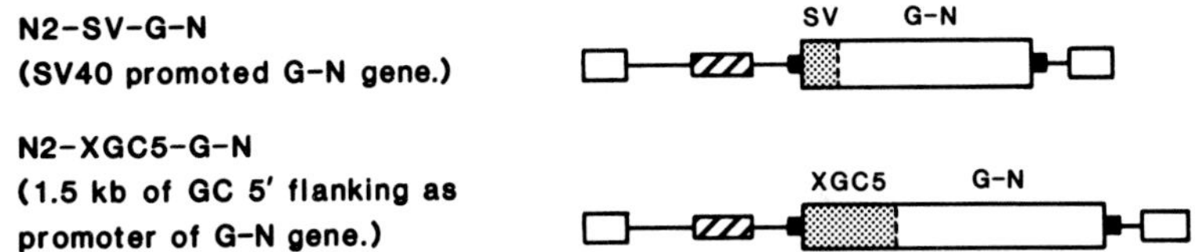

FIGURE 2. Glucocerebrosidase vector constructs

RESULTS

Human glucocerebrosidase enzymatic activity.

We used a monoclonal antibody (8E4) specific for human glucocerebrosidase to distinguish the human protein produced by the vectors from the endogenous glucocerebrosidase activity of the target cells. Figure 3 shows that >90% of the glucocerebrosidase activity from normal human fibroblasts can be immunoprecipitated as has been reported (15). The specificity of the monoclonal antibody is illustrated by the absence of significant immunoprecipitation of the endogenous glucocerebrosidase enzyme from murine 3T3 cells.

PA317 cells infected with the N2-SV-GC vector from Psi2 cells possess total glucocerebrosidase activity over twice that of uninfected cells. Most of the increased activity is immunoprecipitated by the 8E4-sepharose beads, suggesting that it represents human glucocerebrosidase enzyme produced by the vector. Multiple clonal isolates of PA317 cells infected by N2-SV-GC expressed human glucocerebrosidase activity at levels about equal to the endogenous murine activity. These were found by Southern blot to each contain a single copy of vector provirus, integrated at unique sites (data not shown).

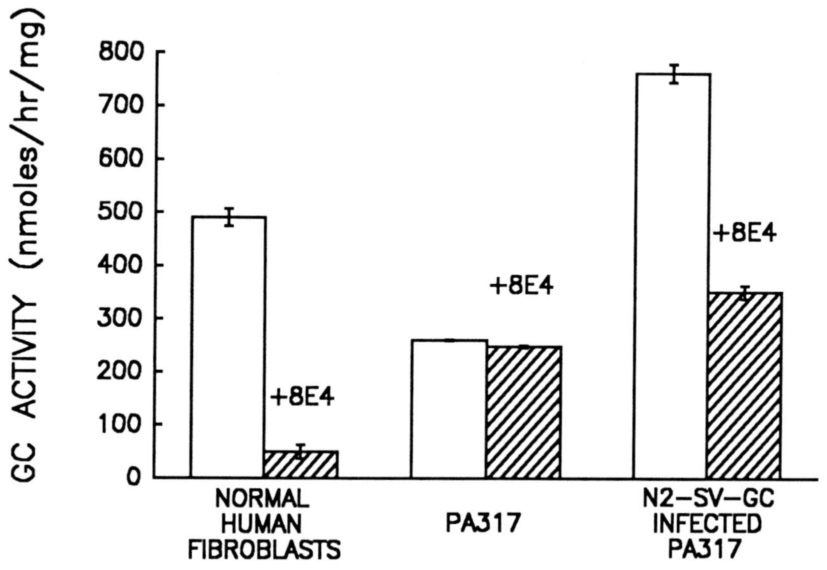

FIGURE 3. Glucocerebrosidase enzymatic activity. Open bars: total activity; hatched bars: activity remaining after immunoprecipitation with 8E4.

Similar assays were performed on cells containing the entire series of glucocerebrosidase vectors. These cells all possessed total glucocerebrosidase activity exceeding the parental murine line, except the Psi2 cells transfected with the thymidine kinase promoted GC gene (PXT-GC) (Table 1). This vector did produce modest levels of increased glucocerebrosidase activity after infection into PA317 cells. The two different glucocerebrosidase gene constructs under control of the SV40 promoter (N2-SV-GC and N2-SV-G-N) produced levels of glucocerebrosidase activity in transfected Psi2 cells that were equivalent to those of normal fibroblasts. The vector with the glucocerebrosidase 5' flanking region as promoter (N2-XGC5-G-N) was quite active and expressed even higher levels of glucocerebrosidase.

TABLE 1
EXPRESSION OF GLUCOCEREBROSIDASE ACTIVITY BY MURINE FIBROBLASTS

Cells	Vector	Glucocerebrosidase Activity[a]	(n)
	Uninfected cells		
3T3	None	319.0	(4)
Psi2	None	328.6	(6)
PA317	None	335.8	(14)
	Transfected cells		
Psi2	N2-TK-GC	267.0	(2)
Psi2	N2-SV-GC	722.4	(2)
Psi2	N2-SV-G-N	762.6	(2)
Psi2	N2-XGC5-G-N	962.6	(2)
	Infected cells		
PA317	N2-TK-GC	511.3	(10)
PA317	N2-SV-GC	844.5	(8)

[a] Nanomoles/hour/mg protein

Western blot for human glucocerebrosidase activity

The monoclonal antibody 8E4 was used to demonstrate the production of human glucocerebrosidase immunoreactive protein by Western blotting. Normal human fibroblasts show the characteristic cross-reacting bands with molecular weight (MW) of 59-63 kd, representing the various biosynthetic forms of glucocerebrosidase (Figure 4, lane 4). For comparison, human glucocerebrosidase produced from the GC cDNA in E. coli is shown in lane 2. This form is non-glycosylated and appears as a sharp band at 55 kd M.W. Murine fibroblasts (3T3 and PA317) do not contain any material which cross-reacts with 8E4 (lanes 5 & 6). Psi2 cells which were transfected with the N2-SV-GC plasmid and then G418 selected have acquired proteins of the 59-66 kd MW range which cross-react with 8E4 (Lane 8). PA317 cells infected with the N2-SV-GC vector from the Psi2 cells show a greater amount of cross-reactive protein (Lane 7). This correlates well with the higher level of enzymatic activity seen in the infected PA317 cells compared to the transfected Psi2 cells.

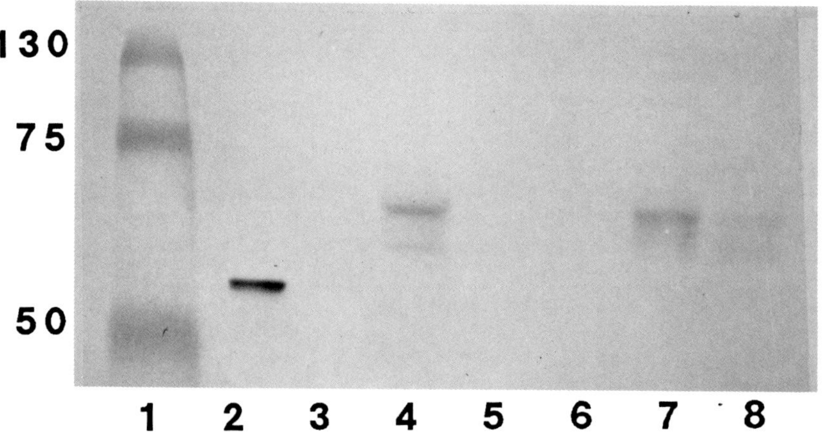

FIGURE 4. Western blots of cell extracts with the 8E4 monoclonal antibody. MW standards in kilodaltons.

DISCUSSION

Our results show that the human glucocerebrosidase gene may be transferred and expressed to normal levels with retroviral vectors containing a variety of internal promoters. The conferred glucocerebrosidase enzyme is immunoprecipitated by a monoclonal antibody specific for the human enzyme and is of the same apparent molecular weight as the endogenous human protein. Cells infected with these vectors and containing a single integrated provirus show levels of activity from the transferred gene equal to that of the endogenous murine activity.

The series of glucocerebrosidase vectors allows comparisons to be made among various promoters. The glucocerebrosidase promoter fragment produced the highest level of enzymatic activity. This is the first demonstration of the function of this region. The use of the endogenous glucocerebrosidase promoter holds the promise of regulated expression of a transferred gene. We are currently defining this region in more detail with a series of CAT constructs. The SV40 promoter was also active, but the thymidine kinase (TK) promoter was somewhat weaker. A similar order of promoter activity has been seen in vectors containing the human adenosine deaminase (ADA) cDNA in human T lymphocytes (16). Although the TK promoter is less strong in cultured cells, it appears to retain activity _in vivo_ in murine hematopoietic cells, embryonal stem cells and transgenic mice, under conditions where a number of other promoters are inactive (17, 18).

Tissue-specific enhancers have been found in the introns of a number of mammalian genes. For example, following retrovirus-mediated transfer of the human beta-globin gene into murine bone marrow, _in vivo_ tissue-specific expression is obtained only when genomic flanking and intron sequences are included (19). To attempt to achieve optimal expression of the human glucocerebrosidase gene, we are assessing whether it has similar enhancers. As the initial step, we have made a pair of vectors which are identical except for the presence in one

(N2-SV-G-N) of a 5' genomic fragment which contains the first two introns. Glucocerebrosidase expression by these two vectors was studied in cells after transfection, rather than retroviral infection, so that the introns would not be lost by splicing during virus production. No increase in the level of conferred glucocerebrosidase enzymatic activity was seen with the minigene construct. This result may reflect the absence of enhancer elements in these introns or their lack of activity in murine fibroblasts. We are currently comparing the activity of the cDNA and minigene in similar vectors in the murine bone marrow transplantation model to determine whether the genomic sequences will enhance expression *in vivo*.

ACKNOWLEDGEMENTS

We wish to thank Drs. J. Tomich and J. Groffen for helpful discussions, M. Cooper for preparation of graphics and Debbie Carroll for typing the manuscript.

REFERENCES

1. Barranger JA, Ginns EI (1988). Glucosylceramide lipidosis: Gaucher disease. In Scriver CR, Beaudet AL, Sly WS, Valle D, Stanbury JB, Wyngaarden JB, Frederickson DS (eds): "The Metabolic Basis of Inherited Disease," New York: McGraw-Hill, in press.

2. Rappeport JM, Barranger JA, Ginns EI (1986). Bone marrow transplantation in Gaucher disease. Birth Defects 22:101-109.

3. Svennerholm L, Mansson JE, Nilsson O, Tibblin E, Eriksson A, Groth CG, Lundgren G, Ringden O (1984). Bone marrow transplantation in the Norrbottnian form of Gaucher's disease. In Barranger JA, Brady RO (eds): "Molecular Basis of Lysosomal Disorders", New York: Academic Press, p 441-459.

4. Hobbs JR, Shaw PJ, Jones KH, Lindsay I, Hancock M (1987). Beneficial effect of pre-transplant splenectomy on displacement bone marrow transplantation for Gaucher's syndrome. Lancet 1:1111-1115.
5. Choudry PV, Barranger JA, Tsuji S, Mayor J, LaMarca ME, Cepko CL, Mulligan RC, Ginns EI (1986). Retrovirus-mediated transfer of the human glucocerebrosidase gene to Gaucher fibroblasts. Mol Biol Med 3(3):293-299.
6. Sorge J, Kuhl W, West C, Beutler E (1987). Complete correction of the enzymatic defect of type I Gaucher disease fibroblasts by retroviral-mediated gene transfer. Proc Natl Acad Sci 84:906-909.
7. Eglitis ME, Kantoff PW, Gilboa E, Anderson WF (1985). Gene expression in mice after high efficiency retroviral-mediated gene transfer. Science 230:1395-1398.
8. Choudry PV, Tsuji S, Martin BM, Guild BC, Mulligan RC, Murray GJ, Barranger JA, Ginns EI (1986). The molecular biology of Gaucher disease and the potential for gene therapy. Cold Spring Harbor Symp Quant Biol 51 pt2:1047-1052.
9. Tsuji S, Choudry PV, Martin BM, Winfeld S, Barranger JA, Ginns EI (1986). Nucleotide sequence of cDNA containing the complete coding sequence for human lysosomal glucocerebrosidase. J Biol Chem 261:50-53.
10. Boulter CA, Wagner EF (1987). A universal retroviral vector for efficienct constituitive expression of exogenous genes. Nucleic Acids Res 15:7194.
11. Mann R, Mulligan R, Baltimore D (1983). Construction of a retrovirus packaging mutant and its use to produce helper-free defective retroviruses. Cell 33:153-159.
12. Chen C, Okayama H (1987). High-efficiency transformation of mammalian cells by plasmid DNA. Mol Cell Biol 7:2745-2752.
13. Miller AD, Buttimore C (1986). Redesign of retrovirus packaging cell lines to avoid recombination leading to helper virus production. Mol Cell Biol 6(8):2895-2902.
14. Aerts J, Donker-Koopman WE, Murray GJ, Barranger JA, Tager JM, Schram AW (1986). A

procedure for the rapid purification in high yield of human glucocerebrosidase using immunoaffinity chromatography with monoclonal antibodies. Anal Biochem 154:655-663.
15. Ginns EI, Tegelaers FPW, Barneveld R, Galjaard H, Reuser AJJ, Tager M, Barranger JA (1983). Determination of Gaucher's disease phenotypes with monoclonal antibody. Clin Chim Acta 131:283-287.
16. Kohn DB, Kantoff P, Zweibel JA, Gilboa E, Anderson WF, Blaese RM (1988). Transfer and expression of the human adenosine deaminase (ADA) gene in ADA-deficient human T lymphocytes by retroviral vectors. In "Gene Transfer in Animals". UCLA Symposia on Molecular and Cellular Biology, New Series, Vol. 87. Eds: Verma I, Mulligan R, and Beaudet A. Alan R. Liss, New York, NY (in press).
17. Keller G, Wagner EF (1986). Efficient expression of foreign genes in mice reconstituted with retrovirus- infected bone marrow cells. Cold Spring Harbor Symp Quant Biol 51 pt 2:412-416.
18. Stewart CL, Schuetze S, Vanek M, Wagner EF (1987). Expression of retroviral vectors in transgenic mice obtained by embryo infection. EMBO J 6:383-388.
19. Dzierzak EA, Papayannopoulou, Mulligan RC (1988). Lineage-specific expression of a human beta-globin gene in murine bone marrow transplant recipients reconstituted with retrovirus-transduced stem cells. Nature 331:35-41.

Gene Transfer and Gene Therapy, pages 409–416

IMPLANTATION OF GENETICALLY MODIFIED CELLS INTO THE RAT BRAIN

AN APPROACH TO RESTORATION OF CNS FUNCTIONS[1]

T. Friedmann*, M. Rosenberg*, J. Wolff*, A. Fagan+, S. Shimohama+, F. Gage+

Departments of Pediatrics* and Neurosciences+, School of Medicine, University of California San Diego, La Jolla, 92093

ABSTRACT Established rat fibroblasts have been infected *in vitro* with MLV-based retroviral vectors expressing the *E.coli* β-galactosidase gene and the human HPRT cDNA. Infected cells have been implanted into the basal ganglia, hippocampus and other regions of the rat brain by stereotaxic injection. Grafted cells containing the single-gene HPRT cDNA vector were found to express the human HPRT enzyme activity *in vivo* at least up to seven weeks after implantation, *although* at moderately reduced levels, while the β-galactosidase activity in cells infected with the two gene β-galactosidase vector was easily detectable for 1-2 weeks post-grafting. The size of the grafts decreased appreciably after the first several weeks following implantation. These results indicate the present combination of donor cells, vectors and implantation techniques can provide continued proviral expression in infected cells for several weeks, long enough to test for some kinds of functional complementation. However, there is evidence for both limited graft survival and some instability of proviral expression *in vivo*.

---

1 This work was supported by Weingart and Gould Family Foundations and by NIH grants HD 20034, HD 00669, GM11013 and NIA 06088.

INTRODUCTION

The introduction of functional foreign genes into recipient animals can be accomplished by genetic modification of appropriate "donor" cells in vitro followed by the replacement of the genetically modified cells into suitable organs of a recipient animal. The development of methods to introduce genes into fertilized mouse eggs or pre-implantation mouse embryos has led to the very powerful transgenic mouse model that has proven so useful for the study of mechanisms of gene regulation and identification of tissue-specific sequences that regulate eukaryotic gene expression(1). Several other useful model systems involving the introduction of foreign genes into animals postpartum have been developed recently, and have been aimed at gene introduction into the cells of accessible organs such as the mammalian bone marrow and skin followed by implantation of genetically altered cells into suitable sites of recipient animals(2,3). These systems have taken advantage of the efficiency of retroviral vectors for the introduction and expression of the foreign genes in mammalian cells in vitro, and have been extremely useful for investigating questions of cell lineage in the bone marrow and in the retina(4,5).

There is a need to develop methods to introduce new genetic functions into organs other than the bone marrow and skin for the purpose of potentially therapeutically useful gene transfer. The mammalian brain is an organ affected seriously by a number of genetic and developmental defects, and methods for the stable and efficient introduction of functional foreign genes may allow a genetic attack on such disorders. It is already clear that the introduction of cells expressing a specific central nervous system function implanted into diseased brains can modify the course of disease, as in the case of adrenal or fetal cell implantation for the treatment of Parkinson's disease(6,7). This approach has been possible in Parkinson's disease because suitable donor cells exist. A similar approach for many other CNS disorders is likely to be difficult because of the lack of availability of cells expressing the appropriate function. In such cases, the design and preparation of donor cells through the introduction of appropriate genetic functions into appropriate "donor" cells would provide a general and versatile approach to complementation of genetic defects. For that purpose, we have begun to study the fate of established fibroblasts and other cells infected in vitro with retroviral vectors and implanted into specific regions of the rat brain. We report here some features of the stability and proviral gene expression in a prototype system, in which we have found that the implanted cells have a short survival time and demonstrate reduction and loss of proviral gene expression.

MATERIALS AND METHODS

Vectors and Donor Cells

The HPRT vector, LPL, contains the full-length human HPRT cDNA expressed from the retroviral LTR transcriptional signals(8). The β-galatosidase vector, BAG, was kindly provided by C. Cepko and contains the E.coli β-galactosidase gene expressed from the retroviral LTR as well as the neomycin resistance gene expressed from the SV-40 early promoter(5). Transmissible virus was produced from these constructs as previously described(9).

HPRT-deficient rat 208F fibroblasts were maintained and infected with the viral vectors by procedures previously described.(10) Infected cells were selected by growth in the presence of HAT medium for the HPRT vector or G-418 in the case of infection with the neomycin-containing BAG vector. Cells were grown overnight in serum-free medium prior to harvesting and implantation.

Cell Implantation

Cells were removed from plates by trypsinization and stored on ice in phosphate-buffered saline prior to injection, as described(11). A total of $1\text{-}4 \times 10^5$ cells were injected with stereotaxic guidance in a total volume of 3 microliters over a period of 3 minutes into the brains of Sprague-Dawley rats.

Assays of Gene Expression

HPRT expression was determined by direct demonstration of HPRT activity in isoelectric focusing gels as previously described(12). β-galactosidase activity was demonstrated by modifications of direct histochemical staining methods in 40 micrometer-thick microtome sections fixed in 3% buffered glutaraldehyde and exposed to the β-galactosidase substrate X-gal (5-bromo, 4-chloro, 3-indolyl β-D-galactoside) in PBS(13). In addition, some samples were stained by immunohistochemical methods, using rabbit antiserum to E. coli β-galactosidase or fibronectin, biotinylated anti-rabbit IgG and complexes of avidin and biotinylated horseradish peroxidase (Vector Labs ABC kit).

RESULTS

Figure 1 illustrates the HPRT activity in grafts of 208F cells infected with the LPL vector, assayed at 3 and 7 weeks post-implantation. Total enzymic activity of the entire graft at 7 weeks was reduced when compared with that present at 3 weeks. Histological examination of the grafts at both 3 and 7 weeks revealed fibronectin-positive 208F fibroblasts surrounded by cells staining with glial fibrillary acidic protein (GFAP), a glial-specific function(14).

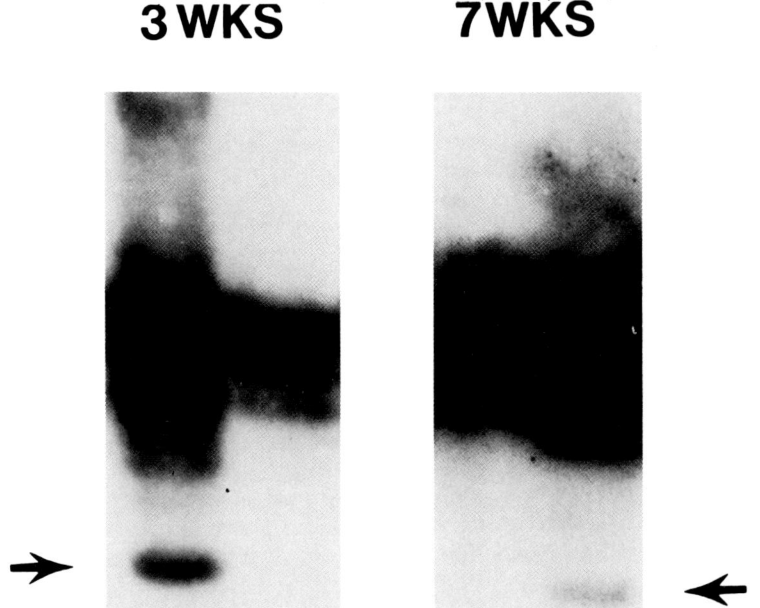

FIGURE 1. Expression of human HPRT in implanted 208F cells after 3 and 7 weeks. Arrows indicate its position of human HPRT enzymic activity in polyacrylamide gels. Plus indicate the implants of 208F cells infected with the HPRT vector, while minus indicates uninfected cells implanted to the contralateral side of the brain.

Results with the BAG vector are presented in Figure 2, in which implanted cells have been examined with anti-fibronectin stains as well as by histochemical and immunohistochemical staining for β-galactosidase enzymic activity and antigenic material. The sizes of the grafts at later times were reproducibly smaller than at 1 week, and the intensity of β-galactosidase enzymic activity was also markedly reduced. The sections showed moderate accumulation of hemosiderin-laden macrophage-like cells but little if any lymphocytic infiltration.

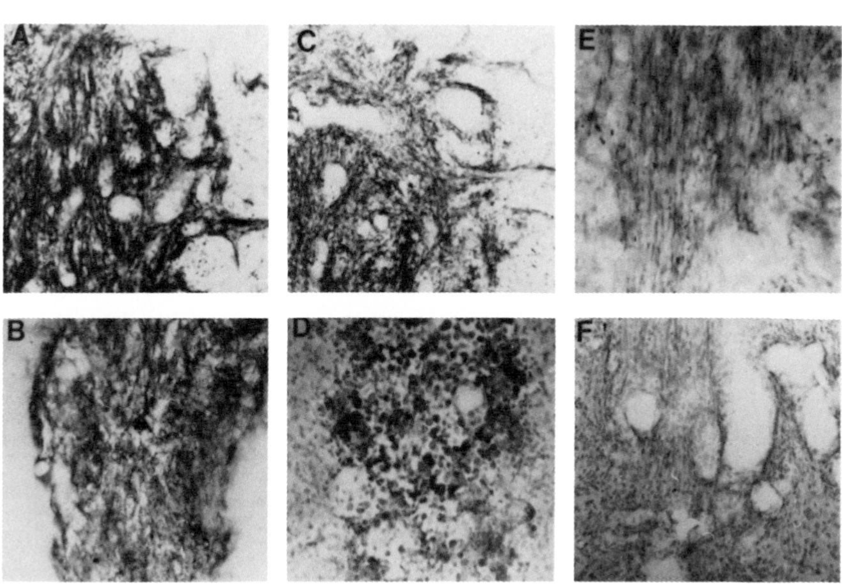

FIGURE 2. Morphology and E.coli β-galactosidase activity in implants at 1 week (A,C,E) and 3 weeks (B,D,F). A and B illustrate fibronectin stain, C and D demonstrate histochemical evidence of the E.coli β-galactosidase activity and E and F illustrate immunocytochemical detection of E.coli β-galactosidase.

DISCUSSION

Using the present combination of donor cells, vectors and grafting methods, we have found that there is continued gene expression and graft survival for several weeks, although we have also found a reproducible and continued decrease in proviral gene expression after the first several weeks. This can result from cell loss or from shutdown of proviral gene expression, or a combination of the two. In the study reported here, there is obviously a component of cell loss, as indicated by the reduction of graft size revealed by cresyl-violet and fibronectin staining. The rapidity of the cell loss and the absence of infiltration with lymphocytes argues against, but obviously does not rule out, an immune basis for the cell loss. Rat 208F cells are originally derived from Fisher rats, and the recipient animals in our experiments have been Sprague-Dawley rats. Even in the event that such a transplant presented histo incompatible cells to the recipient animal, the brain is a somewhat privileged site immunologically. Even mouse tissue implanted to the rat brain is slowly and incompletely rejected(15). A more likely cause of cell loss is reduced viability of cells at the time of grafting resulting from the culture or harvesting conditions. We have determined, however, by trypan-blue dye exclusion and continued in vitro culturing that the cells at the time of implantation are viable and demonstrate growth properties similar to those cells maintained under culture conditions (data not shown). We therefore expect that neither immune nor mechanical factors plays a major role in the loss of cells from the grafts.

The implanted 208F fibroblasts prepared by our current methods seem not to be able to establish themselves in the environment of the rat brain parenchyma. Histologic examination reveals the presence of extensive vascular channels in the grafts at times when the cells are disappearing rapidly. Although the grafted cells are obviously quite crowded, we consider it unlikely that limited availability of nutrients to the cells can explain the cell loss.

The expression of β-galactosidase in surviving grafted cells at later times, as estimated by histochemical staining, has been markedly and consistently reduced after the first several weeks following implantation. Since many of these cells are destined to die or otherwise be removed from the graft site and since many remaining cells have very low or undetectable β-galactosidase enzymic activity, the reduction in enzymic activity seems to result from a combination of cell loss and a shutdown in the expression of the provirus. It is an increasingly common experience in laboratories dealing with the expression of retroviral vectors in vitro to find rapid and marked decrease in proviral expression in vitro from double-gene vectors of the sort represented by the BAG vector, especially in those cases in which the reporter gene is driven from an internal promoter, but certainly also when the reporter gene is expressed from the retroviral LTR promoter in the absence of continued selection for the second gene. We are therefore examining the effects of vector construction on stability of expression both in vitro and after implantation in vivo in the rat brain.

The implantation of genetically modified cells into the mammalian brain is likely to be an effective method of complementing some kinds of genetic or developmental defects, especially those involving absence of diffusible gene products or metabolites whose deficiency causes a disease phenotype. For this approach to become possible, techniques must be established for the stable and long-term expression of the newly introduced gene function. These studies have suggested to us that there are important cellular and genetic mechanisms that must be operating during the loss of proviral gene expression by current methods, and these mechanisms and procedures for circumventing them are under study.

REFERENCES

1. Brinster R, Chen H, Trumbauer M, Senear A, Warren R, Palmiter R (1981). Somatic expression of herpes thymidine kinase in mice following injection of a fusion gene into eggs. Cell 27:223.

2. Miller A, Eckner R, Jolly D, Friedmann T, Verma I (1984). Expression of a retrovirus encoding human HPRT in mice. Science 225:630.

3. Morgan J, Barandon Y, Green H, Mulligan R (1987). Expression of an exogenous growth hormone gene by transplantable human epidermal cells. Science 237:1476.

4. Dzierzak E, Papayannopoulou T, Mulligan R (1988). Lineage- specific expression of a human β-globin gene in murine bone marrow transplant recipients reconstituted with retrovirus- transduced stem cells. Nature 331:35.

5. Turner D, Cepko C (1987). A common progenitor for neurons and glia persists in rat retina late in development. Nature 328:131.

6. Marazo I, Drucker-Colin R, Diaz V, Martines-Mata J, Torres C, Becerril J (1987). Open microsurgical autograft of adrenal medulla to the right caudate nucleus in two patients with intractable Parkinson's disease. N Engl J Med 316:831.

7. Madrazo I, Leon V, Torres C, del Carmen Aguirlera M, Varela G, Alvarez F, Fraga A, Drucker-Colin R, Ostrosky F, Skurovich M, Franco R (1988). Transplantation of fetal substantia nigra and adrenal medulla to the caudate nucleus in two patients with Parkinson's disease. N Engl J Med **318**:51.

8. Miller A, Jolly D, Friedmann T, Verma I (1983). A transmissible retrovirus expressing human HPRT: Gene transfer into cells obtained from humans deficient in HPRT. Proc Nat Acad Sci USA **80**:4709.

9. Yee J-K, Moores, J, Jolly D, Wolff J, Respess J, Friedmann T (1987). Gene expression from transcriptionally disabled retroviral vectors. Proc Nat Acad Sci USA **84**:5197.

10. Quade K (1979). Transformation of mammalian cells by avian myelocytomatosis virus and avian erythroblastosis virus. Virology **98**:461.

11. Gage F, Wolff J, Rosenberg M, Li X, Yee J-K, Shults C, Friedmann T (1988). Grafting genetically modified cells to the brain: possiblities for the future. Neuroscience 23: 795.

12. Jolly D, Yee J-K, Friedmann T (1987). High efficiency gene transfer into cells. In Green R, Widder K. (eds): Drug and Enzyme Targeting. Methods in Enzymology, vol 149, part B. San Diego:Academic Press, p10.

13. Dannenberg A, Suga M (1981). Histochemical stains for macrophages in cell smears and tissue sections. In Adams D, Edelson P, Koren M. (eds). Methods for studying mononuclear phagocytes. New York:Academic Press, p375.

14. Lewis S, Balcarek J, Krek V, Shelanski M, Cowan N (1984). Sequence of a cDNA clone encoding mouse glial fibrillary acidic protein: structural conservation of intermediate filaments. Proc Nat Acad Sci USA **81**:2743.

15. Bjorklund A, Stenevi U, Dunnett S, Gage F (1982). Cross-species neural grafting in a rat model of Parkinson's disease. Nature **298**:652.

Gene Transfer and Gene Therapy, pages 417–430

TRANSFER OF CHROMOSOMAL SEQUENCES ONTO PLASMIDS BY GENE CONVERSION[1]

Brent Seaton and Suresh Subramani

Dept. of Biology and Center for Molecular Genetics, University of California, San Diego, La Jolla, CA 92093.

ABSTRACT

Gene rescue may be defined as the transfer of chromosomal sequences onto extrachromosomal viral or plasmid vectors. We have observed interactions between two homologous, yet individually nonfunctional, gene sequences which result in the formation of an intact gene (the Neo gene) encoding resistance to the drug G418 in mammalian cells. The recipient cells are derivatives of the COS African green monkey cell line which contain a single integrated copy of the 3′ two-thirds of the Neo gene. The introduced plasmid molecules contain the SV40 origin and early promoter upstream of the 5′ two-thirds of the Neo gene, allowing for two regions of homology (420 and 135 bp) separated by a 1217 bp gap. We present evidence for a targeting event involving gene conversion, in which DNA sequences in the genome are copied onto the incoming plasmid to produce a functional Neo gene. The resulting plasmid integrates into the cell genome after the gene conversion event and the Neo sequence is amplified.

INTRODUCTION

Before the current gene transfer techniques can be exploited fully for the correction of genetic defects or for the study of gene regulation it is essential to develop

[1]Supported by grant GM31253 and an RCDA (CA01062) to SS. BS was supported by an NIH training grant GM07240.

methods for the efficient targeting of genes to specific predetermined chromosomal locations and for the recovery of chromosomal sequences onto plasmids by homologous recombination. Though gene targeting has been achieved in mammalian cells, it is still an inefficient process whose mechanism is not understood clearly. Two types of events, however, have been well documented. The first involves the transfer of exogenous DNA sequences into the chromosome by reciprocal or nonreciprocal recombination (1-6). In the second category are events involving the transfer of chromosomal sequences onto extrachromosomal viral or plasmid molecules (7-9). Because the latter event could occur either by reciprocal or nonreciprocal mechanisms, we asked whether one or both mechanisms could explain such an event.

Previously, we reported the rescue of a 1018 bp segment of the simian virus 40 (SV40) T-antigen gene from the chromosome of monkey COS cells onto two different extrachromosomally-replicating SV40 DNA molecules lacking this 1018 bp sequence (9). Because the wild-type SV40 produced by the recombination event lysed the COS cells, the reciprocal or nonreciprocal nature of the recombination event could not be ascertained. In this paper, we have used an analogous system in which the rescue of chromosomal information by double-reciprocal recombination or gene conversion could be distinguished. The system is based on recombination between a single chromosomal DNA segment containing the 3′ two-thirds of the neomycin resistance gene (Neo) and a replicating plasmid carrying an overlapping, nonfunctional 5′ portion of the same gene. We present evidence for a gene conversion event in which DNA sequences in the genome are copied onto the incoming plasmid to produce a functional Neo gene. The resulting plasmid integrates into the cell genome after the gene conversion event and the Neo coding sequence is amplified.

MATERIALS AND METHODS

<u>The recombination substrates</u>. The recipient cell line, COS/Neo2, was produced by the insertion of the plasmid pNeo2tkgptβG-1 into the genome of a COS cell line (Fig. 1A). This plasmid contains the 3′ two-thirds of the Neo gene (nucleotides 2179 to 2684) followed by the SV40 small t intron and polyadenylation signal. The plasmid also contains the bacterial xanthine guanine phosphoribosyl transferase (gpt) marker gene under the control of the

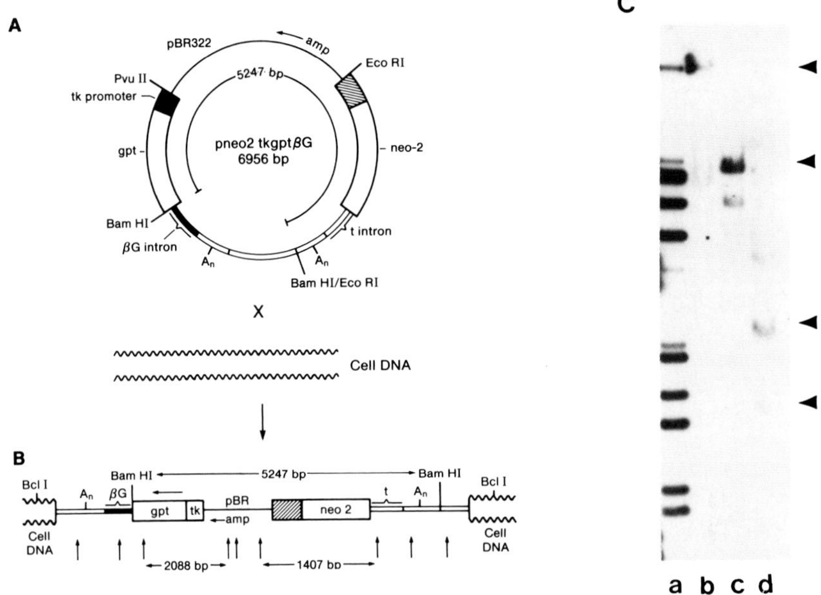

Fig. 1 - (A) Structure of the plasmid pNeo2tkgptβG-1. (B) Structure of the integrated plasmid and flanking cellular DNA in the COS/Neo2 line. Vertical arrows below the figure denote DraI sites. Fragments larger than 1 kb are shown. (C) Southern blot of genomic COS/Neo2 DNA digested with BclI (lane b), BamHI (lane c) and DraI (lane d) confirm the structure of the integrated DNA. Lane a has λ HindIII-EcoRI markers. The blot was probed with labeled pNeo2tkgptβG-1 DNA. Arrows show the presence of predicted diagnostic fragments.

Herpes simplex virus thymidine kinase (tk) promoter. COS-1 cells were transfected with the above plasmid and gpt-positive clones were isolated. Southern analysis of the genomic DNA from these clones identified one clone (COS/Neo2) which contains a single copy of the plasmid stably integrated into the genome via its SV40 sequences (Fig. 1B and C). COS/Neo2 cells are easily propagated in Dulbecco's modified Eagle's medium (DME) containing 5% fetal calf serum (FCS).

The plasmid, pSV2Neo1X, contains the 5′ two-thirds of the Neo gene under the control of the SV40 early promoter. The plasmid also contains a functional SV40 origin of DNA replication which allows the plasmid to replicate in the COS/Neo2 cell line. The truncated 5′ fragment of the Neo gene in the plasmid contains a 420 bp overlapping homology with the Neo fragment found in the COS/Neo2 cell line.

Transfection protocol. Growth conditions for the cell line prior to and following transfection with pSV2Neo1X were modified to increase the probability of a targeting event. Cells were grown initially in DME containing 5% FCS until a confluency of 20-30% was reached, at which time the medium was changed to DME + 0.1% FCS. The cells were grown in this "starvation" medium for 24 hours, after which they were transfected with pSV2Neo1X by the calcium phosphate method (10). Ten ug of supercoiled or XhoI-digested linear pSV2Neo1X were added to each plate of cells. After addition of the DNA, cells were incubated in DME containing 0.1% FCS and 0.2 mM sodium butyrate for four hours. The cells were shocked for 30 seconds with 15% glycerol, and incubated in DME containing 0.1% FCS and 0.2 mM sodium butyrate for 36 hours. The cells were then incubated in DME + 5% FCS for 48 hours before trypsinizing the cells and replating in selective medium containing 400 ug/ml G418 (11).

Southern blots, DNA probes and DNA recovery from recombinant lines. Low molecular weight DNAs, representing extrachromosomal plasmids, were extracted from the $G418^r$ lines by the method of Hirt (12). Genomic DNA was isolated using minor modifications of the procedures of Wigler et al. (13). DNA samples (10 ug) for Southern analyses were electrophoresed on 0.8% agarose gels and depurinated before transfer to nitrocellulose paper. Filters were hybridized, washed and exposed to X-ray film as previously described (14,15). Neo probe DNA corresponds to the 1339 bp fragment from nucleotide position 1193 to 2532 in the Neo gene (16).

Plasmid rescue. Plasmids containing the recombination product were rescued from the recombinant cell lines by digesting genomic DNAs with excess DraI. Ten ug of genomic DNA were digested with 40 units fo DraI for 36 hours. The digested DNA was diluted to a concentration of 1 ug/ml and recircularized by the addition of DNA ligase. Ligated DNA was ethanol precipitated and used to transform *E. coli* strain DH5 to neomycin-resistance.

RESULTS

Experimental strategy. When pSV2Neo1X is linearized by digestion with XhoI, the ends of the plasmid contain sequences which could interact with homologous sequences in the COS/Neo2 genome. The first region is the 420 bp overlap within the Neo gene, and the second is the 135 bp region containing the SV40 polyadenylation signal sequence. Interactions between these homologous sequences can produce an intact Neo gene either by a gene conversion event or a double-reciprocal recombination event (Fig. 2).

Initially, we transfected ten plates of COS/Neo2 grown in DME + 5% FCS with 10 ug/plate of either supercoiled or XhoI-digested pSV2Neo1X. No $G418^r$ colonies were observed in several such experiments. We modified the protocol to include "serum starvation" of the cells as described in Materials and Methods. This protocol, which increases the frequency of extrachromosomal recombination 5-10 fold (Seaton and Subramani, unpublished data), yielded 40 $G418^r$ colonies from 10 plates of COS/Neo2 transfected with linear pSV2Neo1X (Table 1). No $G418^r$ colonies were observed on plates transfected with supercoiled pSV2Neo1X or no DNA. Parallel transfections of COS/Neo2 with supercoiled and linear pSV2Neo yielded 52,000 and 10,400 $G418^r$ colonies, respectively (Table 1). Thus, of those cells which are able to take up and express supercoiled or linear pSV2Neo, approximately one in 1300 cells or one in 260 cells, respectively, show a targeting event (Table 1).

TABLE 1
PREDICTED DIAGNOSTIC RESTRICTION FRAGMENTS HYBRIDIZING TO THE NEO PROBE.

Enzyme	pSV2Neo1X	COS/Neo2	Double Reciprocal Recombination		Gene Conversion	
			Plasmid	Genome[a]	Plasmid	Genome
DraI	3362	1407	3066	989	3066	1407
KpnI	--	>4615	>5727	>3398	>5727	>4615[b]

[a]Chromosome of COS/Neo2 cells.
[b]Same fragment as in COS/Neo2 cells.

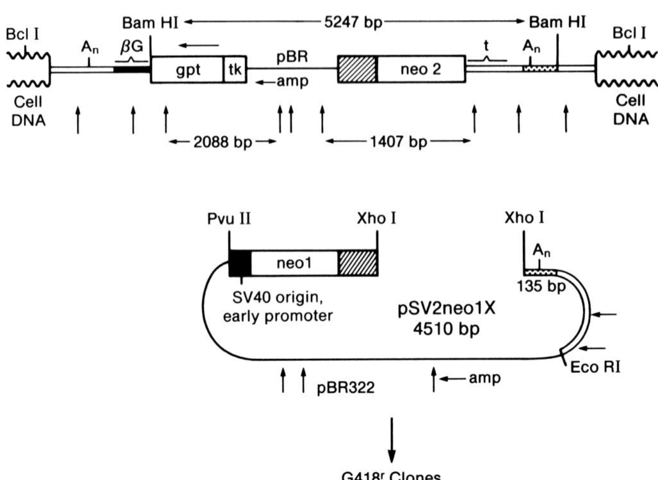

Fig. 2 - Experimental strategy used to detect recombination between pSV2Neo1X and the Neo sequences in COS/Neo2.

The recombination product is generated by gene conversion and is integrated into the chromosome. Because pSV2Neo1X contains a functional SV40 ori, the recombination product was expected to be an extrachromosomal plasmid containing an intact Neo gene. Hirt DNAs were isolated from nine of the recombinant cell lines, electrophoresed on 0.8% agarose gels and transferred to nitrocellulose paper. When the filter was probed with Neo DNA sequences, the Hirt DNAs did not hybridize but the control pSV2Neo did (data not shown), suggesting that the recombination product was not present extrachromosomally.

Restriction digests of genomic DNA from COS/Neo2 and the G418^r recombinant lines were used to differentiate between the two possible recombination mechanisms which could generate an intact Neo gene (see predictions in Table 2). In digests with DraI, the recombination product containing the functional Neo gene should yield a 3066 bp fragment. If the mechanism is double-reciprocal

TABLE 2
FREQUENCY OF RECOMBINATION USING pSV2NEO1X IN COS/NEO2 CELLS.

DNA	State of DNA	G418^r col. 10 plates	Recombination Frequency	
			Relative to SC pSV2Neo	Relative to LIN pSV2Neo
Mock	--	0	--	--
pSV2Neo1X	SC[a]	0	0	0
pSV2Neo1X	LIN[b]	40	1/1300	1/260
pSV2Neo	SC	5.2×10^4	--	--
pSV2Neo	LIN	1.0×10^4	--	--

[a]Transfection with supercoiled DNA.
[b]Transfection with linear DNA.

recombination, the 1407 bp fragment containing the parental Neo2 sequence should decrease to 989 bp, whereas if the recombination event is due to gene conversion, the parental Neo2 sequence should remain the same length. If the pSV2Neo1X plasmid integrates into the genome at the 420 bp of Neo overlap (single crossover), thereby producing a functional Neo gene, one would expect the diagnostic 3066 bp Neo fragment and the loss of the parental 1407 bp fragment corresponding to the Neo2 segment. The results (Fig. 3A and B) suggest that the intact Neo gene results from a nonreciprocal gene conversion event in that the parental 1407 bp Neo2 fragment remains in each target line and no 989 bp band is observed. The predominant recombination product (3066 bp) is highly amplified in each of the lines. Densitometric measurement of the bands on the Southern blots suggest amplification ranging from 10-100 fold.

Similar conclusions could be drawn from KpnI digests of genomic DNAs from the parental and recombinant cell lines. The integrated DNA in the COS/Neo2 line has a single KpnI site in the gpt gene. In a KpnI digest of COS/Neo2 DNA, a single band (greater than 4615 bp) should hybridize with the Neo probe (Fig. 3C, lane 1). If single or double-reciprocal recombination had occured, this band should increase in size by 4510 bp or decrease in size by 1217 bp, respectively. In contrast, if gene conversion had occured the parental band should remain in the recombinant cell lines. The retention of the parental COS/Neo2 band in all the recombinants (Fig. 3C) indicates that gene conversion had occured.

Further evidence for the presence of functional Neo genes in the recombinant cell lines was obtained from experiments in which DraI fragments containing the pBR322 origin of DNA replication and an intact Neo gene were rescued from the genomes of the recombinant cell lines by transformation of bacteria to neomycin resistance. Restriction digests of DNA from these rescued plasmids suggest that the targeting products do have an intact Neo gene, and that this gene is downstream from the SV40 early promoter and origin of replication as expected (Fig. 4).

Do cells exhibiting these gene conversion events catalyze other types of recombination events at a higher than normal frequency? The ability of the G418^r and the COS/Neo2 cell lines to support recombination was addressed using extrachromosomal recombination assays (17). Plasmids with two truncated but overlapping segments of the firefly luciferase marker gene (18) were used to determine the frequency of extrachromosomal intramolecular recombination in the cell lines. Results of these assays (data not shown) suggest that frequencies of extrachromosomal intramolecular recombination exhibited by the G418^r cell lines are similar to those observed in the parental COS/Neo2 cell line.

<u>The recombination products are chromosomal though they contain a functional origin and the recombinant cell lines support replication of plasmids containing the SV40 ori</u>. Why do the recombination products remain in the cell genome when they should be able to replicate extrachromosomally? To address this question, we asked whether the recombinant cell lines had lost their ability to support replication of plasmids containing the SV40 ori (eg. by mutation of the T antigen gene) and whether the recombination products still retained a functional ori. Wild type SV40 virus was able to replicate in each recombinant cell line, suggesting that the G418^r lines retain the machinery to support replication from an SV40 origin of replication (data not shown).

The recombinant cell lines and the COS/Neo2 cells were also transfected with the plasmids rescued by transformation into bacteria. Sensitivity of the Hirt DNA, containing plasmids passaged through these cells, to MboI suggests that the plasmids were able to replicate (Fig. 5).

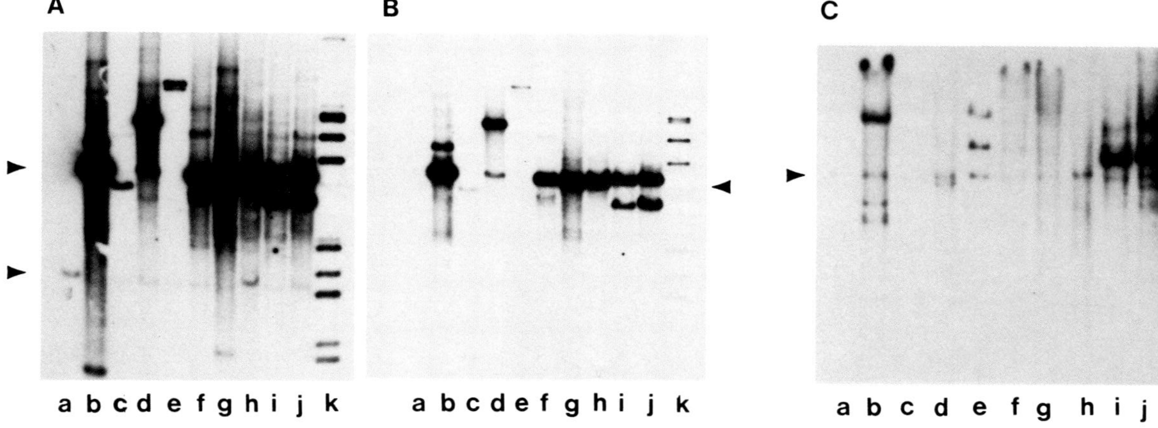

Fig. 3 - Southern blots of genomic DNAs from the parental COS/Neo2 line and its G418^r progeny. (A) 3 day exposure of DraI digests probed with Neo sequences. Lanes a-j correspond to DNAs from COS/Neo2 and G418^r lines 1,2,3,4,7,8,11,10 and 12, respectively. Lane k has λ HindIII-EcoRI markers. Arrows indicate fragments diagnostic for gene conversion (see Table 2). (B) one day exposure of the blot in (A). (C) KpnI digests of genomic DNAs. Lanes a-j correspond to DNAs from COS/Neo2 and G418^r lines 1,2,3,4,7,8,10,11 and 12, respectively. The retention of the 6300 bp band (arrow) in the recombinant lines indicates that gene conversion has occured.

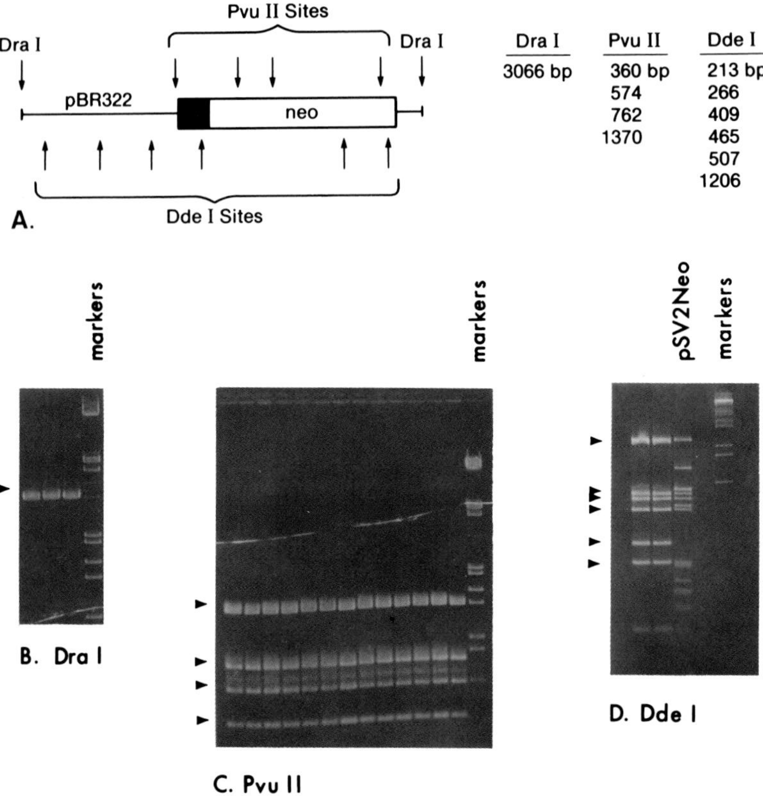

Fig. 4 - (A) Predicted fragments that should be generated in restriction digests of rescued plasmids containing the intact Neo gene and SV40 ori. DraI (B), PvuII (C) and DdeI (D) digests of the representative plasmids rescued from several recombinant lines. Arrows show predicted fragments. Marker DNA is a HindIII-EcoRI digest of lambda DNA.

DISCUSSION

In this study, we have shown that the rescue of a 1217 bp sequence from the genome of COS/Neo2 cells to the plasmid pSV2Neo1X is accomplished by nonreciprocal gene conversion

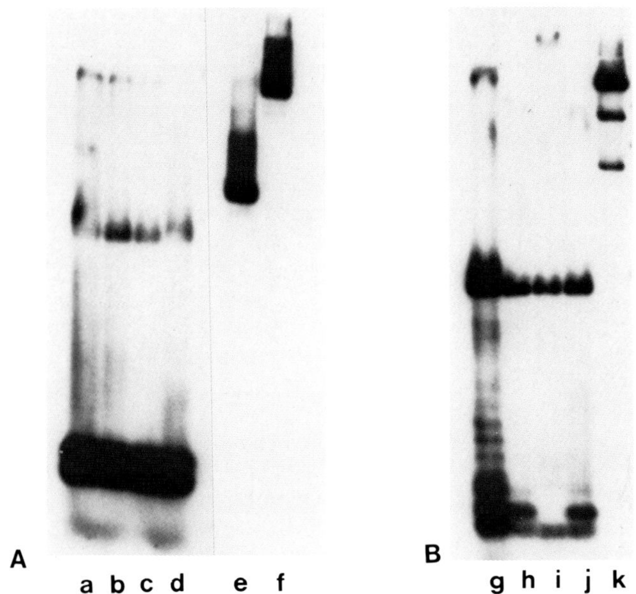

Fig. 5 - Rescued plasmids replicate in recombinant cell lines and in COS/Neo2 cells. Southern blots of MboI digested plasmid DNAs passaged through these cells were probed with Neo sequences. (A) Replication of plasmids (12-4) and (8-1) rescued from recombinant lines 12 (lane a) and 8 (lane b), respectively, in recombinant line 4. Lanes c and d are markers showing replication of control plasmids pSVDNeo (corresponds to the DraI fragment of pSV2Neo and should be identical to the rescued plasmids) and pSV2Neo, respectively, in the same cells. MboI digests of pSV2Neo1X (lane e) and pSV2Neo (lane f) not passed through these cells. (B) Replication of rescued plasmids and controls in COS/Neo2 cells. MboI digests of plasmid 12-4 (lane g), 8-1 (lane h), pSV2Neo1X (lane i) and pSV2Neo (lane j). Lane k has MboI digest of pSV2Neo1X not passed through these cells.

and not by double-reciprocal recombination. The conditions under which this event occurs correspond closely with previous data concerning extrachromosomal gene conversion (19). No rescue events were detected when supercoiled plasmid molecules were used as the exogenous substrate.

Only when the plasmids were introduced in a linear form did gene rescue occur. This is not surprising since the introduction of double strand gaps has been shown to increase the frequency of gene conversion events (20). Furthermore, the serum starvation of the cells, prior to and immediately after the introduction of the second substrate into the cells, stimulated the frequency with which the recombination event occured significantly.

Another interesting observation from this study is the paradox concerning the inability of the recombination products to replicate extrachromosomally even though they contain a functional origin of replication and the recombinant cell lines are capable of supporting the replication of plasmids containing the SV40 ori. Instead the recombination product integrated into the host genome. The increased copy number of the DraI restriction fragments containing the complete Neo gene (Fig. 3A and B) may result from replication of this DNA either in the chromosome or transiently outside the chromosome following the recombination event. It is possible that integration of the recombination product supresses the level of DNA replication from the SV40 ori in a manner analogous to the suppression exerted by the bovine papilloma virus origin on replication from the SV40 ori (21).

REFERENCES

1. Doetschman T, Gregg RG, Maeda N,Hooper ML, Melton DW, Thompson S, Smithies O (1987). Targeted correction of a mutant HPRT gene in mouse embryonic stem cells. Nature (London) 330: 576.
2. Lin F-L, Sperle K, Sternberg N (1985). Recombination between DNA introduced into cells and homologous chromosomal sequences. Proc Natl Acad Sci USA 82: 1391.
3. Smith AJH, Berg P (1984). Homologous recombination between defective neo genes in mouse 3T6 cells. CSH Symp Quant Biol 49: 171.
4. Smithies O, Gregg RG, Boggs SS, Koralewski MA, Kucherlapati RS (1985). Insertion of DNA sequences into the human chromosomal β-globin locus by homologous recombination. Nature (London) 317: 1230.
5. Thomas KR, Capecchi MR (1986). Introduction of homologous DNA sequences into mammalian cells induces

mutations in the cognate gene. Nature (London) 324: 34.

6. Thomas KR, Folger KR, Capecchi MR (1985). High frequency targeting of genes to specific sites in the mammalian genome. Cell 44: 419.
7. Jasin M, DeVilliers J, Weber F, Schaffner W (1985). High frequency of homologous recombination in mammalian cells between endogenous and introduced SV40 genomes. Cell 43: 695.
8. Shaul Y, Laub O, Walker MD, Rutter WJ (1985). Homologous recombination between a defective virus and a chromosomal sequence in mammalian cells. Proc Natl Acad Sci USA 82: 3781.
9. Subramani S (1986). Rescue of chromosomal T-antigen sequences onto extrachromosomally replicating, defective simian virus 40 DNA by homologous recombination. Mol Cell Biol 6: 1320.
10. Parker BA, Stark GR (1979). Regulation of simian virus 40 transcription: sensitive analysis of the RNA species present early in infection by virus or virus DNA. J Virol 31: 360.
11. Southern PJ, Berg P (1982). Transformation of cells to antibiotic resistance with a bacterial gene under control of the SV40 early region promoter. J Mol Appl Genet 1: 327.
12. Hirt B (1967). Selective extraction of polyoma DNA from infected mouse cell cultures. J Mol Biol 26: 365.
13. Wigler M, Silverstein S, Lee LS, Pellicer A, Cheng Y, Axel R (1977). Transfer of purified herpes thymidine kinase gene to cultered mouse cells. Cell 11: 223.
14. Southern EM (1975). Detection of specific sequences among DNA fragments separated by gel electrophoresis. J Mol Biol 98: 503.
15. Wahl GM, Stern M, Stark GR (1979). Efficient transfer of large DNA fragments from agarose gels to diazobenzyloxymethyl paper and rapid hybridization by using dextran sulfate. Proc Natl Acad Sci USA 76: 3683.
16. Beck E, Ludwig G, Auerswald EA, Reiss B, Schaller H (1982). Nucleotide sequence and exact localization of the neomycin phosphotransferase gene from transposon Tn5. Gene 19: 329.
17. Rubnitz J, Subramani S (1985). Rapid assay for extrachromosomal homologous recombination in monkey cells. Mol Cell Biol 5: 529.

18. de Wet JR, Wood KV, DeLuca M, Helinski DR, Subramani S (1987). Firefly luciferase gene: structure and expression in mammalian cells. Mol Cell Biol 7: 725.
19. Rubnitz J, Subramani S (1986). Extrachromosomal and chromosomal gene conversion in mammalian cells. Mol Cell Biol 6: 1608.
20. Rubnitz J, Subramani S (1987). Correction of deletions in mammalian cells by gene conversion. Som Cell Mol Genet 13: 183.
21. Roberts JM, Weintraub H (1986). Negative control of DNA replication in composite SV40-bovine papilloma virus plasmids. Cell 46:741.

Index